Education in a Competitive and Globalizing World

nova
science publishers
www.novapublishers.com

Education in a Competitive and Globalizing World

Students' Attitudes Towards Learning and Education
Walter S. Rolph (Editor)
2022. ISBN: 978-1-68507-618-4 (Softcover)
2022. ISBN: 978-1-68507-636-8 (eBook)

Charted and Uncharted Territories in Translation and Interpreting Research Methods
Alireza Akbari (Author)
2022. ISBN: 978-1-68507-473-9 (eBook)

Teacher Identity Development Within a Community of Practice
Ronél Ferreira (Editor)
2022. ISBN: 978-1-68507-487-6 (Hardcover)
2022. ISBN: 978-1-68507-511-8 (eBook)

School Leadership in the 21st Century: Challenges and Coping Strategies
Adam E. Nir, PhD (Editor)
2021. ISBN: 978-1-68507-368-8 (Hardcover)
2021. ISBN: 978-1-68507-423-4 (eBook)

Parents and Teachers: Perspectives, Interactions and Relationships
Francesco Arcidiacono (Editor)
2021. ISBN: 978-1-68507-358-9 (Hardcover)
2021. ISBN: 978-1-68507-418-0 (eBook)

Handbook of Homework: Theoretical Principles and Practical Applications
Susana Rodríguez Martinez, PhD (Editor)
Antonio Valle Arias, PhD (Editor)
Isabel Piñeiro Aguin, PhD (Editor)
Bibiana Regueiro Fernandez, PhD (Editor)
2021. ISBN: 978-1-68507-380-0 (Softcover)
2021. ISBN: 978-1-68507-395-4 (eBook)

More information about this series can be found at
https://novapublishers.com/product-category/series/education-in-a-competitive-and-globalizing-world/

Charles E. Baukal, Jr.
Editor

Continuing Engineering Education Handbook

nova
science publishers
www.novapublishers.com

DOI: https://doi.org/10.52305/ZRNH8663

Library of Congress Cataloging-in-Publication Data

ISBN: 978-1-68507-793-8

Published by Nova Science Publishers, Inc. † New York

Contents

Preface

The book is divided into four sections. Section one includes two chapters on history. The opening chapter of the book introduces the subject of continuing engineering education (CEE). There are general sections on continuing education and engineering education and then the subject of CEE is presented. The chapter concludes with some sources for further information including relevant organizations, journals, and workshops. The second chapter in the section is on the 50-year history of engineering distance education at Iowa State University. A timeline with notable events is presented along with the key people involved in the program.

Section two concerns instructional design and contains five chapters. The first is on the learning preferences of working engineers. It presents the results of a study on the learning strategy and verbal-visual preference of engineers working in the oil and gas industry. The results show that the engineers surveyed have statistically significantly different preferences compared to the general population. They prefer problem solving and are highly visually oriented. The second chapter shows how learning styles have been incorporated into a course on process heaters. The learning styles include print, aural, interactive, and visual. The use of multiple learning styles helps training content to match the preferences of most participants. The third chapter in the section provides guidelines for selecting multimedia for CEE. The primary takeaway is that engineers are highly visual so training content should be designed accordingly. Examples include drawings, photos, videos, animations, and virtual reality simulation. The fourth chapter provides recommendations for effective instructional practices for CEE. It discusses hybrid, converged, and hyflex delivery modes. The 60-year curriculum and the growing impacts of alternative credentials and virtual/ augmented/extended reality are presented. It concludes with some work and learn models in effective instructional practices. The fifth chapter in the section presents the use of virtual reality (VR) and augmented reality (AR) in CEE. Brief histories of VR and AR are given. The human factors of information overload, verbal and non-verbal communications, sight, sound/auditory, gated response, and touch are discussed. Cognitive psychology subjects of sensory memory, iconic memory, echoic memory, split attention effect, automated behaviors, kinesthetics, chunking, and serial position effects are briefly described. VR/AR integration and extensibility show the power of contextualizing information for users.

Section three focuses on CEE delivery and administration. The first chapter in the section discusses the results of a survey of working engineers to determine their CEE practices and preferences. It includes results such as how many courses engineers take each year, what types of courses they take (technical; management; environment, health, and safety; legal; and other), what fraction of the course are mandatory including what fraction is required by the participant's organization and by outside organizations, the fraction of courses taken that are taken for credit, and how many courses are taken after normal working hours. It also presents data on what those same working engineers preferred regarding their CEE. The second chapter provides detailed information on how to conduct effective virtual (distance) training courses including the instructional design and delivery platforms. The third chapter in the section describes combustion CEE training designed for the oil and gas industry. While it is for a specific industry, the general principles apply universally. It also includes some case studies of classes conducted at clients' sites. The fourth chapter discusses the many types of potential CEE providers categorized by internal, external, and hybrid providers. Providers include the engineer's own organization, universities/community colleges, training organizations, vendors, professional societies, trade organizations, government, and consultants. It is recommended that organizations choose the combination of providers that makes the most sense for their particular situation which usually will include a combination of providers. The fifth and last chapter in the section describes how to build a sustainable administrative organization for delivering CEE. It provides cultural implications, organizational design models, the four-phase natural evolution of an organization delivering online CEE, a matrix organizational design model, the new professional working adult learner, and changing demographics.

The fourth and last section concerns the future of CEE and includes three chapters. The first chapter describes a sustainable future through the 60-year curriculum model. It discusses the changing postsecondary landscape, the relevance of the model, adapting to the curriculum, the role of CEE in pioneering the 60-year curriculum, and mapping higher education's future in the context of the 60-year curriculum. The second chapter in the section discusses sustainability education and research in action (SERINA), developed by the International Association on Continuing Engineering Education (IACEE). The executive committee used the United Nations 17 Sustainable Development Goals (SDGs) as a benchmark tool to evaluate SERINA. The purpose of SERINA is to provide practical tools for engineers to help meet the SDGs. The final chapter of the section and of the book suggest some future research areas for CEE. They are broadly categorized under learning preferences, curriculum, VR and AR, and multimedia.

Chapter 1

Continuing Engineering Education

Charles E. Baukal, Jr.*

Koch Engineered Solutions Institute℠, John Zink Hamworthy Combustion,
Tulsa, Oklahoma, USA

Abstract

This chapter is designed to give a brief introduction to the subject of continuing engineering education. It first considers the continuing education of professionals in a broad sense, not specifically engineers. Engineering education, primarily for undergraduates, is then discussed, particularly regarding some of the weaknesses that have been identified in the literature and the many calls for improvement. Then, the continuing engineering education of working engineers is considered specifically. The chapter concludes with some sources for further information including organizations like the American Society for Engineering Education, journals like the *International Journal of Continuing Engineering Education and Life-Long Learning*, and workshops such as Frontiers in Engineering Education.

Keywords: continuing engineering education

Introduction

Engineering is often described as problem solving (Shepherd et al. 2009, p. 3). The National Academy of Engineering report (2004, p. 43) *The Engineer of 2020* describes engineering as "problem recognition, formulation, and solution." A key ABET (2021) student outcome is "an ability to acquire and apply new knowledge as needed, using appropriate learning strategies." In a report published by UNESCO (2010, p. 24), engineering is described as "the field or discipline, practice, profession and art that relates to the development, acquisition and application of technical, scientific and mathematical knowledge about the understanding, design, development, invention, innovation and use of materials, machines, structures, systems and processes for specific purposes."

Jonassen (2014, p. 103) writes, "Learning to solve workplace problems is an essential learning outcome for any engineering graduate." Those problems are much more complex and multi-disciplinary in industry than those studied in undergraduate education. There are no answers in the back of the book for real-world problems. In some cases, there may be multiple possible solutions that will vary depending on who is solving the problem. Those problems often require advanced training beyond what was learned in undergraduate studies. As the problems become more complex, the need for continuing engineering education (CEE) grows.

---

* Corresponding Author's Email: charles.baukal@kes.global.

Engineers who fail to deliberately and systematically maintain and upgrade their engineering skills, abilities, and knowledge risk becoming obsolete. Similarly, organizations that fail to provide their engineers with CEE opportunities including supporting their engineers with sufficient time and financial resources also risk falling beyond their competitors, particularly related to innovative new processes and technologies.

This chapter describes continuing education in general, regardless of profession. It also describes engineering education more broadly than just CEE and then discusses CEE in particular. It concludes with some useful CEE resources including organizations, journals, and workshops.

Continuing Education

Various terms are used to describe the ongoing education of an individual over the course of their lifetime. Some of these include: continuing education, continuous professional development, lifelong learning, adult education, and workforce development. Some might argue there is a distinction, for example, between lifelong education and lifelong learning where the former is more formal and the latter more informal (Hager, 2021). This book is concerned more with formal lifelong education. Some might also include the word *training*, where Rothwell and Whiteford (2021) argue training is between education which is more general and the world of work. They also note that *training* and *development* are not synonymous, where the former is short-term and the latter is long-term. This section will broadly survey the subject of continuing education. The more specific topic of continuing engineering education (CEE) is discussed in a later section.

Professions such as law, medicine, and engineering are often characterized by "rigorous professional knowledge" (Schön, 1987, p. 3) that changes, sometimes rapidly, with time. Del Gaizo and Laudermith (2021, p. 367) write:

> Lifelong learning for professionals is not a new concept. Many professional organizations have required for many years that their members take continuing education (CE) classes in order to maintain their license or practice. Professionals themselves have also sought CE for reasons of their own. In fact, without any mandate from overseeing bodies or legislation, many professionals would continue to seek opportunities to learn or refresh their skills.

Research continues on how one becomes an expert in a given discipline (Ericsson, 2009) where a person becomes a subject matter expert (SME). Lifelong learning is required to maintain proficiency and in many cases licensure (Baukal, 2012; Bennett & LeGrand, 1990). Learning may be informal or formal where informal is usually done without an instructor while formal usually includes an instructor. Examples of informal learning, sometimes referred to as *workplace learning* (Malloch, Cairns, Evans, and O'Connor, 2011), include reading trade magazines and journals and searching the Internet. Formal learning includes taking classes that may be either for-credit or not-for-credit. The "classes" also include one-on-one on-the-job mentoring by more experienced colleagues.

The Institute of Continuing Professional Development (ICPD, 2011) defined continuing professional development (CPD) as "the systematic maintenance and improvement of knowledge, skills and competence, and the enhancement of learning, undertaken by an individual throughout his or her working life." The on-going education of professionals sets them apart from other jobs, leading to elevated status and prestige. This dynamic process is called *professionalization* where professionals continuously upgrade their knowledge and skills (Houle, 1980). Continuing education includes "all those processes that contribute to the advancement of an individual's knowledge, skill, understanding, competence, and general professional and personal development" (Padfield and Schaufelberger, 1998, p. 8). Continuing education, CPD, and professional development are all related terms that generally concern learning outside the normal workplace (Dirkx, 2011).

The Organisation for Economic Co-operation and Development (OECD; 2021) issued a report concerning the importance of lifelong learning starting in childhood. Some drivers include increased life expectancy, rapid technological changes, globalization, migration, environmental changes, and digitalization. The COVID-19 pandemic is an example of how rapidly things can change requiring adaptability such as how to work from home. Some of the key recommendations from the report include:

- putting learners at the center of learning,
- the importance of developing foundational skills such as a habit of learning, and
- developing coordinated policies to support high quality and inclusive learning.

A report by the European Commission specifically identifies the importance of digital lifelong learning (Beblavý et al. 2019).

Training can be defined as "a planned effort by a company to facilitate employees' learning of job-related competencies" where the competencies include "knowledge, skills, or behaviors that are critical for successful job performance" (Noe, 2005, p. 3). The continuing education and training of engineers is of specific interest here.

Eduard Lindeman (1961) wrote a classic book on adult education. He argued all of life is concerned with learning which is called *adult learning* and believed the curriculum should be built around the learner, not the institution. Allen Tough (1979) wrote another adult education classic. One of his major contributions to adult learning theory is that adults are self-directed in their learning as compared to children who learn at the direction of their teachers. Adults want to learn things that are useful to them personally and therefore must be sufficiently motivated in their adult education classes. Peter Jarvis (2010), Ross-Gordon et al. (2017), and Michael Galbraith (2004) provide helpful books on the theory and practice, the theory, and the practice of adult education, respectively.

Malcolm Knowles wrote what many consider to be another classic book on adult learning. It continues after his death into the ninth edition (Knowles, Holton, Swanson, and Robinson, 2020). Knowles is known as the father of the theory known as *andragogy* which postulates that adults learn differently than children. Some use pedagogy to describe any teaching, regardless of the age of the learner, but Knowles argued adults are distinctly different learners than children. His andragogical model is based on the following assumptions:

1. Adults need to know why they need to learn something before undertaking to learn it.
2. Adults have a self-concept of being responsible for their own decisions and their own lives.

3. Adults come into an educational activity with both a greater volume and different quality from that of youths.
4. Adults become ready to learn those things they need to know and be able to do in order to cope effectively with their real-life situation.
5. Adults are life-centered (or task-centered or problem-centered) in their orientation to learning.
6. Adults are responsive to some external motivators (better jobs, promotion, higher salaries, etc.), but the most potent motivators are internal pressures (the desire for increased job satisfaction, self-esteem, quality of life, etc.).

Based on these assumptions, Knowles argued that adults should be taught differently than children.

Patricia Cross (1981) suggested three categories for the perceived barriers to continuing education: situational (e.g, costs too much), institutional (e.g., courses not scheduled at a convenient time), and dispositional (e.g., don't enjoy studying). Some of these can be accommodated such as an employer paying for training courses and institutions offering asynchronous online training where students can attend at their convenience. However, other barriers, particularly dispositional, are more intrinsic to the learner and may require a change in mindset to improve the motivation for learning.

Ralph Brockett (2015) has written a practical guide for those teaching adults and lists the following as the qualities of a good teacher: trust, empathy, authenticity, confidence, humility, enthusiasm, and respect. His four keys to effective teaching are: know the content, know the adult learner, know teaching, and know yourself. He lists twelve things teachers should know about adult learners:

1. Most adults are actively involved in learning and undertake at least one learning project every year.
2. Adults have the ability to learn successfully throughout their lives.
3. Adult learners bring a wealth of experience to the teaching-learning setting, and this experience can be a valuable resource.
4. It is important to recognize that decisions you make about how you will teach are based on whether you are trying to change attitudes and values, skill and performance or knowledge and factual information.
5. Most adult learning is self-directed.
6. The need for adult learning is often triggered by some kind of developmental transition or crisis.
7. Adults choose to learn for many different reasons, and you need to know what these are.
8. It is important to understand something about motivation and how you can use motivation strategies to help excite the learners about what you are teaching and, in some case, to help break down learner resistance.
9. Teachers of adults need to understand the many kinds of barriers that can limit whether adults choose a learning activity and stay with the activity.
10. Understanding learning style is important, but it is often misunderstood and misused.
11. Most successful adult learning takes place in a collaborative or cooperative setting, where sharing and synergy are crucial.

12. The ultimate purpose of adult education is to help learners think for themselves.

He offers some techniques to consider for teaching adults: lecture, a panel of experts, demonstration, simulation, discussion, and role playing. He also notes the importance of creating a positive learning environment and motivating learners.

Merriam and Baumgartner (2020) have written a classic text in its fourth edition on adult education, intended primarily for adult educators. They note one of the bigger drivers for adult learning is rapid technological changes and the move toward a more digitally-driven society. That driver is particularly important for engineers given their intimate role in developing new technologies.

Bonvillian and Sarma (2021) have written a book on workforce education with a particular emphasis on the United States. They discuss the many problems such as the inequities in who is trained and insufficient employee training by most employers. They strongly believe the current workforce education system is broken, but that it can be fixed. An important objective of the book is to recommend solutions to the problems identified. One of the industries given particular attention is manufacturing, which is of interest to engineers who play a key role in that segment. There are many stakeholders in the education of the workforce including the employees, employers, government at all levels, and higher education among others. One of the recommended solutions is the increased use of educational technologies such as virtual and augment realities (see Chapter 7), educational games, simulations, and collaborative tools. Another recommended solution is micro-credentials where workers can get very specific training and certification on a particular subject to help them with their current job or for a future role they aspire to.

Engineering Education

This section will give a comprehensive but not exhaustive review of significant engineering education publications that are primarily books and reports. The references are presented in chronological order and focus primarily on undergraduate eduction.

Lawrence Grayson (1993) has written a book on the history of engineering education in the U.S. and Canada. A consistent theme is the interplay between broad social change and the resulting stresses that helped shape engineering education. The two World Wars and the space race had significant impacts. The book starts in 1893 which is the year the Society for the Promotion of Engineering Education (SPEE) which later became known as the American Society for Engineering Education (ASEE) was started but does consider earlier times as well including the Revolutionary War. The book is generously illustrated with 335 photos of the many engineering projects that have been completed over the years and many of the institutions involved in engineering education. It also includes brief biographies of some of the key people in the history of engineering education. The keystone recommendation was that the National Science Foundation should expand and reinvigorate its efforts to stimulate and disseminate innovation in engineering education.

Fox and Hackerman (2003) edited a book on excellence in teaching undergraduate science, mathematics, engineering, and technology. They present rigorous methods for evaluating engineering education because the perception of many faculty at research institutions is that research is valued above teaching in part because there is no rigorous way to evaluate teaching and learning. A rigorous and fair method of evaluation is needed to properly reward faculty and to drive curriculum changes. Most faculty who teach undergraduate STEM disciplines receive little formal training in teaching and assessment techniques. Overall recommendations of the study included:

- Teaching effectiveness should be judged by the extent of student learning, not on the specific styles or methods used as many may be effective
- Scholarly activities related to improving teaching and learning should be recognized and rewarded
- Summative assessments of teaching should rely not only on student evaluations but also on peer evaluations and teaching portfolios
- Faculty should be rewarded for improving student learning
- Faculty should accept the responsibility to improve their teaching skills

The National Academy of Engineering Committee on Engineering Education (2004) issued a report looking at the future of engineering and recommends changes to engineering education to meet the forecast changes. The task force attempted to answer the question, "What will or should engineering be like in 2020" (p. 2). A companion report by the National Academy of Engineering Committee on Engineering Education (2005) gave specific recommendations on how the undergraduate engineering curriculum should be re-engineered (pp. 2-3):

- B.S. degree should be considered as a pre-engineering or "engineering in training" degree
- M.S. degree should be accredited and recognized as the engineering professional degree
- Students should be introduced to the essence of engineering early in the undergraduate program
- Engineering education research should be recognized and endorsed by colleges and universities
- Institutions should teach students how to be lifelong learners
- Interdisciplinary learning should be encouraged
- Four-year institutions should work to develop articulation agreements with two-year community colleges engineering programs
- Institutions should encourage U.S. students to obtain M.S. and Ph.D. degrees
- Efforts should be encouraged to improve the public's understanding of engineering
- The National Science Foundation should collect or assist the collection of data on program approaches and student outcomes to help freshmen better understand available engineering baccalaureate programs.

Reinventing the engineering curriculum requires that industry and academy work together.

Gretar Tryggvason and Diran Apelian (2006) wrote in a journal opinion piece that engineering education needs to be re-engineered to meet the challenges of the 21st century. They noted that a country's prosperity is due in part to the education of its engineers. They believe the entrepreneurial engineer can find information quickly, understands engineering basics, has effective communication skills and understands global and current issues, and has the imagination and managerial skills to identify and meet needs with new solutions.

The National Science Board (2007) issued a report detailing findings and recommendations for the National Science Foundation to support innovations in engineering education programs. It is the result of two public workshops which had the goal of "moving forward the national conversation on engineering issues by calling attention to how engineering education must change in light of changing workforce demographics and needs" (p. 1). Three key challenges for engineering education were identified: responding to the changing needs for engineers, changing the perception of engineering, and retaining top students.

A report by the Royal Academy of Engineering (2007) is the result of a working group commissioned to address educating engineers in the 21st century. It notes the growing concern in the UK that the education system responsible for producing new generations of engineers is failing to keep pace with the rapid changes in society and technology and the increasing need for engineers. Nine key messages from the study are:

1. There will be an increasing shortage of high caliber engineering graduates entering industry
2. More funding is needed to ensure effective engineering education courses
3. Ensure the UK MEng qualification is not undermined by the development of the proposed European Qualifications Framework under the Bologna agreement
4. Engineering courses need to adapt to changing industry requirements
5. Teaching prestige has been compromised by a disproportionate emphasis on research
6. More effective interaction between industry and university engineering departments is needed
7. Core engineering courses need to continue to be emphasized and not diluted by peripheral subject matter
8. Accreditation should inform the development of course content to ensure graduates meet industry needs
9. More needs to be done to ensure engineering is perceived as an exciting, stimulating, and well-paid career.

Specific recommendations are given to government, engineering institutions, and industry to improve engineering education.

James Duderstadt (2008), former president of the University of Michigan, wrote a report as part of the Millennium Project designed to be a roadmap for the future of engineering practice, research, and education. Of the context he writes, "The changing workforce and technology needs of a global knowledge economy are dramatically changing the nature of engineering practice, demanding far broader skills than simply the mastery of scientific and technological disciplines. The purpose of the study was to pull together the principal findings and recommendations of the various reports concerning the profession of engineering, the technology and innovation needs of the nation, and the role played by human and intellectual capital, into an analysis of the changing nature of engineering practice, research, and education. The following are the key conclusions:

1. The transformation of knowledge into products, processes, and services is critical to competitiveness, long-term productivity growth, and the generation of wealth

2. American engineers must be able to add significantly more value than their counterparts abroad through their great intellectual span, their capacity to innovate, their entrepreneurial zeal, and their ability to address the grand challenges facing the world
3. It is essential to elevate the status of the engineering professions, providing it with the prestige and influence to play the role it must in an increasingly technology-driven world while creating sufficiently flexible and satisfying career paths to attract a diverse population of outstanding students
4. Engineering education should encompass a broader educational experience with more liberal arts in the undergraduate education which is common in other professions such as law and medicine

Patricia Galloway, past and first woman president of the American Society for Civil Engineers, has written a book (2008) calling for reform in engineering education. Her premise is that engineering education has not kept pace with the numerous changes that have occurred in technology and society. She believes engineers need to be better communicators and better prepared to work in a global economy. She also discusses the importance of ethics, professionalism, diversity, leadership, the engineer's role in public policy. She also proposes a Master of Professional Engineering Management.

Spurlin, Rajala, and Lavelle (2008) have edited a book on effective assessment of engineering education programs designed primarily for engineering faculty and department chairs as part of the ABET accreditation process. The book is divided into four sections: assessment basics, barriers and challenges, learning along the continuum of the educational experience, and the future. The book is tailored to the ABET requirements, is generously referenced, includes examples, and also includes rubrics. There is also a chapter on using assessment in graduate education programs. A key barrier to assessment is overcoming the resistance to change. An important element requiring assessment is the capstone project. A useful glossary of terms is included.

Zawojewski, Diefes-Dux, and Bowman (2008) have edited a book on modeling in engineering education. The objective is to bring real-world modeling into the classroom. A process is described that includes the following model-eliciting activities (MEAs): construction, reality, self-assessment, documentation, shareability and reusability, and effective prototype. Chapters are given on applying the process to a first-year and upper-level engineering courses. Chapters on learning from student surveys, interviews, discussion boards, designing in small group mathematical modeling, by observation, from teaching assistants, and from a faculty self-study are also presented. The final chapter concerns future research which focuses on three problems in engineering education research: course quality, diversity, and emerging technology.

Shepherd, Macatangay, Colby, and Sullivan (2009) have written a book sponsored by the Carnegie Foundation for the Advancement of Teaching which contains twenty-two chapters divided into six parts: preparing the new-century engineer, a foundation to build on, a place to explore, a way to create, affecting the world, and bringing professional practice forward. The primary goal of the book is to "understand, through field research, how the educational practices of the schools from future engineers" (p. xix). The primary finding was "that in the midst of a profound, worldwide transformation in the engineering profession, undergraduate engineering education in the United States is holding on to an approach to problem solving and knowledge acquisition that is consistent with practice the profession has left behind" (p. xxi). While there are pockets of innovation, much more needs to be done to reform the traditional curriculum. Five attributes are listed for the new-century engineer: strong analytical skills, practical ingenuity, creativity, communicator, and master of principles of business and management (p. 9). The authors found a widespread lack of enthusiasm for reforming the engineering education curriculum and call for a reformed curriculum that makes a much stronger connection between what is taught in the classroom and what is practiced in the profession.

Ryan Davidson (2010) reported on a day and a half workshop sponsored by the National Science Foundation to consider how engineering curricula could be enhanced to better prepare future engineers. General themes of the workshop included: focusing more on inductive teaching and learning which is more learner-centered, integrating relevant topics across STEM fields, and making more extensive use of learning technologies. Suggestions for promoting curricular innovation included: expanding academic communication networks, increasing faculty incentives, and promoting interactions among all engineering education stakeholders.

Manjit Sidhu (2010) has written a book on computer-aided learning applied to engineering education. It includes chapters on multimedia and learning styles, user interface design, hardware and software for multimedia design, evaluating interactive multimedia packages, and various aspects of the TAPS (technology-assisted problem solving) packages. There is a chapter specifically on the use of TAPS in a mechanical engineering course.

UNESCO (2010) issued an extensive (392 pages) report with the premise that "mobilizing the engineering community to become more effective in delivering real products and services of benefit to society, especially in the developing world, is a vitally important international responsibility" (p. 5). The purpose of the report is "to contribute to greater international understanding of the issues, challenges and opportunities facing engineering, with a particular focus on contributions of our discipline to sustainable development" (p. 5). While it considers all aspects of engineering, the discussion on transforming "engineering education, curricula and teaching methods to emphasize relevance and a problem-solving approach to engineering" (p. 6) is particularly relevant. Chapter seven focuses on transforming engineering education and training with specific discussions of problem-based learning, sustainability, curriculum renewal, environmental engineering education, and engineering education research.

Ruth Graham (2012) authored a report sponsored by the Royal Academy of Engineering that reiterates the numerous calls to reform undergraduate engineering education but notes the challenge in making that happen. The focus of the report is on how successful change can be initiated, implemented, and sustained. The report was based on interviews with 70 experts from 15 countries with experience in successful engineering education curriculum changes. Four common features were identified from these changes:

- Successful systemic changes often resulted from circumstances, such as significant threats in the marketplace to a school or program, where change was unavoidable
- Changes needed to be driven across the entire curriculum and by the entire faculty; those only involving a handful of courses and faculty were not typically sustainable
- Support from the department and department head were critical factors in successful change

- Problems were often encountered after successful implementation due to a slow drift back to the traditional curriculum

Singer, Nielsen, and Schweingruber (2012) reported on two workshops in 2008 that looked at promising practices, particularly research-based instructional strategies, in undergraduate engineering (along with science, technology, and mathematics). One aspect of the study was to look at instructional practices shown to increase the performance of specific student groups, especially low socioeconomic status, minority, and female students. A more student-centered instructional approach is advocated where there is more interaction, student participation is encouraged, and more student collaboration activities. Unfortunately, national surveys show that current engineering faculty are the least likely to use student-centered or collaborative instruction and rely primarily on lectures.

Crawley, Malmqvist, Östlund, Brodeur, and Edstrom (2014) have written a book advocating for reforms in engineering education using the CDIO (conceive-design-implement-operate) approach. They, like many others, argue there is need to reform engineering education and propose that the CDIO approach is the best way to accomplish this. They list the following professional aspects as the context for engineering education:

- Focus on the needs of customers,
- Delivery of products, processes, and systems
- Incorporation of new inventions and technologies
- Focus on the solution, not disciplines
- Working with others
- Effective communication
- Working within resources

Examples are given of the implementation of CDIO at multiple institutions under a variety of situations ranging from introductory courses to capstone projects.

Paulo Davim (2014) has edited a book on curriculum, pedagogy, and didactic aspects of engineering education. There are chapters on the influences of personality traits on academic performance through imaginative capability, developing a personalized and adapted curriculum through an ambient intelligence environment, evaluation supporting stakeholder-centered design and continuous quality improvement, software engineering education, central occupation requirements for engineering jobs, and energy engineering as an emerging discipline.

Goldberg and Sommerville (2014) tell the stories of the Olin College of Engineering (Needham, MA) and the iFoundry incubator at the University of Illinois which both have developed innovative engineering education curricula that have the potential to significantly improve student learning and effectiveness. Olin was initially founded specifically to develop a radically different engineering education curriculum. iFoundry was started with a grant from the National Science Foundation's Engineering Research Center program. Olin and iFoundry have partnered on innovating the engineering curriculum. The authors argue engineering education has missed three important revolutions: entrepreneurial, quality, and information technology. They advocate for a change from a carrot and stick approach to teaching and learning to one where students are intrinsically motivated to learn where the professor is more of a coach than an expert. While change is certainly needed, it will not be easy.

Johri and Barbara Olds (2014) have edited an extensive book on engineering education research with numerous international authors, thirty-six chapters, and six parts: engineering thinking and knowing, engineering learning mechanisms and approaches, pathways into diversity and inclusiveness, engineering education and institutional practices, research methods and assessment, and cross-cutting issues and perspectives. A very wide range of aspects of engineering education research is considered including, for example, learning theories, problem solving, problem- and project-based learning, curriculum design, engineering design, retention, social justice and inclusion, community engagement, research methods, assessment, communication, global issues, ethics, and interdisciplinarity. Numerous references are provided.

Flumerfelt, Kahlen, Alves, and Siriban-Manalang (2015) have written a book on lean engineering education designed to drive content and competency mastery specifically for mechanical engineering students. It calls for plans to improve the future of mechanical engineering education to prepare graduates for job opportunities in 2030. It identifies the key stakeholders as students, faculty, employers, and society who all need to have input into improving mechanical engineering education. Competencies that will be needed include systems, sustainability, and ethics. A suggested lean process for improving ME education is the Plan, Do, Check, Act process suggested by the Toyota Education Model. Many examples are presented for content- and competency-based lean engineering education processes.

Linda Kober (2015) has written a book sponsored by the National Research Council on research-based instructional methods for undergraduate science and engineering students. Its primary audience is those teaching those students with the goal of more student-centered instruction to promote more learning. Learning goals should be set for each course to drive instruction. Instructors can collaborate with fellow instructors to form learning communities. Learning workshops are recommended before implementing a new instructional method. Problem solving is a key skill for engineers which can be taught. A wide range of instructional techniques and strategies are presented including example implementations including making lectures more effective. Collaborative peer learning, supplemental tutorials, and authentic experiences are examples of research-based methods that have been proven to enhance learning. Assessments should be used to improve instructional methods. While not all methods will be effective in every course, instructors are encouraged to try new techniques that have the potential to significantly increase student learning. Strong institutional support is needed to make many of the changes recommended to improve engineering education.

Valenzuela-Valdés and Fernández (2015) have edited a book on project-based learning (PBL) including an introduction to and foundations of PBL. One chapter gives an example of PBL applied to telematics engineering and another on information and communications technology (ICT). All of the authors were from Spanish institutions except for one from Switzerland.

Vanajakumari, Johnston, Lawrence, and Menon (2015) describe a teaching methodology they developed for continuing education courses in the engineering technology department. They incorporated andragogical principles by engaging the students and encouraging them to use their experiences in the classes. They note two substantial challenges for CEE instructors which are the diversity of the educational backgrounds of the participants and the relatively short time they have to teach complicated subjects. They believe that instructional effectiveness is maximized by tailoring course materials to the needs of individual learners.

Annetta and Minogue (2016) have edited a book on connecting science and engineering education practices. While science and engineering are distinct disciplines, they are both important in engineering education. There are chapters on educational games, museum design experiences, principles for supporting design activity, enhancing integrative STEM literacy through engineering design, an integrated STEM instructional model (iSTEM), a STEAM-based curricula for robotics, educational games for learning biology, a language of design within science and engineering, teaching with design thinking, elementary school engineering for fictional clients in children's literature, teaching engineering design in elementary science methods classes, infusing engineering concepts into high school science, and using mathematics in design-based lessons about a biological process. One chapter discusses the following grand design challenges in engineering education:

1. Explaining technology
2. Explaining what engineers do
3. Developing new curriculum materials
4. Teaching the design process
5. Developing assessments
6. Teaching the teachers
7. Balancing technical and academic subjects
8. Engaging technology and career and technology education (CTE) teachers
9. Teaching the teacher educators

Frerich, Meisen, Richert, Petermann, Jeschke, Wilkesmann, and Takkaya (2016) have edited a nearly one thousand page book reporting on the results of a project entitled Excellent Teaching and Learning in Engineering Sciences sponsored by a consortium of RWTH Aachen University, Ruhr-Universität Bochum, and TU Dortmund University which is where the authors came from. There are many chapters divided into the following sections: virtual learning environments, mobility and internationalization, student lifestyle, and professional competency. Some of the chapters are very specific such as the self-optimized assembly planning for a ROS-based robot cell and bitmap-based online analytical processing of time interval data. Some are more general such as a chapter on massive open online courses in engineering education, incorporating the virtual theatre into engineering education, virtual and remote labs, and educating engineers for a globalized world. Most of the chapters are written in English but some are in German.

John Heywood (2016) has written a book about assessing learning in engineering education. An objective of the book is to show how proper assessment of courses and programs can lead to needed improvements. He provides nuts and bolts for how to do assessments including examples and case studies. The book includes discussions of topics such as emotional intelligence, peer and self-assessments, and portfolios. It also includes generous references and is well-documented. The impact of accreditation on assessment is also considered.

Baukal (2017) has written a book based on a survey of working engineers from a single company to determine their learning strategy, verbal-visual, and multimedia preferences (see Chapters 4 and 6). The working engineers surveyed were found to be statistically significantly different than the general population. They preferred problem solving and much more visual content than the average person.

Guerra, Ulseth, and Kolmos (2017) have edited a book on problem-based learning (PBL) in engineering education from an international perspective. The book provides specific examples of how PBL has been used at various institutions. The book shows the diversity of ways PBL may be used in an engineering curriculum.

Heywood (2018) has written a book on empowering professional teaching in engineering as part of the annual Synthesis Lectures on Engineering started in 2006. The book covers a wide range of topics such as teaching research, objectives and outcomes, problem solving, critical thinking, decision making, intellectual development, and concept learning.

Mutis, Fruchter, and Menassa (2018) have edited a book on transforming engineering education, with specific emphasis on civil engineering. It includes chapters on foundational courses such as teaching computers to civil engineering students, building information modeling (BIM) curriculum, and augmented and virtual reality applications.

Plato Kapranos (2019) has edited a book on the interdisciplinary future of engineering education. The first chapter makes the case for new pedagogies. The second chapter describes the past and future of engineering education. The third chapter considers the pedagogical and cost advantages of a multidisciplinary approach to delivering practical teaching. Chapter four argues for a more liberal engineering program with almost no lectures and more student-centered learning including problem- and project-based methods. Chapter five concerns interdisciplinary project weeks. Chapter six discusses improving engineering education in the United Kingdom. Chapter seven is on teaching core knowledge efficiently and employability competencies in chemical engineering education. Chapter eight concerns the importance of teaching personal and professional skills. Chapter nine argues for more liberal studies in engineering education. Chapter ten suggests methods of teaching to enhance innovation. Chapter eleven discusses the value in enterprising experiences outside the classroom. Chapter twelve shows how group creativity can be enhanced and managed through off-task breaks.

Kellam, Coley, and Boklage (2019) have edited a book on transformative teaching in engineering as part of the annual Synthesis Lectures on Engineering started in 2006. Excellent engineering educators tell their stories of how they developed their exemplary pedagogies. These educators come from a variety of institutions and teach a variety of courses in multiple disciplines. There are chapters discussing vulnerability and empowerment in teaching, going from the armed forces to the classroom, engaging students through service learning and innovation, engaging students in hands-on learning, and creating a community of collaborators to achieve curriculum change.

Kelly and Green (2019) have edited a book on sociocultural research in science and engineering education with international authorship. The book (p. 1) explores "how science and engineering concepts, processes, and practices are socially constructed through coordinated interactions among students, teachers, curricula, texts, and technologies." The chapters related to engineering education include multimodal analysis of decision making, learning through improvement from failure in elementary engineering design projects, and research methods for the advancement of possibility knowledge and practice in engineering education.

Mohsen, Ismail, Parsaei, and Karwowski (2019) have edited a book with ten chapters with international authorship related to global advances in engineering education. There are general chapters on the history of engineering education dating back to ancient Egypt and Greece, strategies and challenges of engineering education, globalization issues related to engineering education, and augmented reality. There are also some specific chapters on safety engineering and RLaaS-Frame for remote laboratories.

Roberta Powell (2019) has edited a book covering a wide range of topics related to curriculum and pedagogy in engineering education with an international authorship. Some of the 26 chapters are fairly specific such as a chapter on inductive teaching of a remote lab. Others are more general such as going from STEM to STEAM to enhance engineering and technology education. It even includes a chapter on medical engineering education.

Rahman and Ilic (2019) have edited a book on blended learning in engineering education with international authorship. Blended learning includes prescribed learning both inside and outside the classroom, where lectures are typically recorded and watched outside of the classroom while classroom time is dedicated to hands-on activities and working problems. It also includes other innovative teaching methods such as hybrid learning, online learning, experiential learning, gamification, and learning spaces. A sampling of chapters includes PBL applied to statistical hydrology and applied fluid flow, along with gamification to increase student engagement.

Conclusion

Continuing education is critical for working engineers because of the breadth of processes and equipment they design and use and because of rapid changes in technology. For example, plant engineers take courses to learn how to operate different types of equipment specific to their operations (Baukal & Crawford-Fanning, 2013; Valencia, Link, Baukal, & McGuire, 2008). This training takes a variety of forms (see Chapter 11) including traditional classroom, on-the-job, and computer-based training (CBT), sometimes referred to a virtual learning (see Chapter 9). Classroom training is led by an instructor or facilitator. In on-the-job training, a more-experienced employee trains a less-experienced colleague on specific procedures and operations. This is usually less structured but more personalized than classroom training. CBT involves course materials delivered by computer, often over the Internet (Baukal, 2010). CBT instruction is increasingly important in delivering continuing training for working engineers. Willets (1983) noted that courses taken earlier in an engineer's career tend to be more specialized to meet an identified need. Most CEE courses focus on technical and management subjects (Bolle and Eugene, 1990).

There are two general types of online training: synchronous and asynchronous (Moore & Kearsley, 2005). In *synchronous* training, an instructor and one or more students are online at the same time, interacting with each other in real time. In *asynchronous* training, the participants are not usually online at the same time and the courses may or may not be instructor-led. Online courses without an instructor are generally referred to as *standalone* or *self-directed* (Horton & Horton, 2003) and are typically used to provide learners with information but not usually to teach new skills (Colbrunn & Van Tiem, 2002). Many companies use asynchronous self-directed courses to provide their employees with periodic legal, safety, and environmental training. These courses are generally short in length and can be completed at the convenience of the participants.

Instructional and graphic design are major challenges for self-directed courses, which are often poorly designed with too much text, not enough graphics, and very little interaction between learners and course materials or each other. These courses are frequently developed by subject matter experts who know the content, but may not know instructional design principles. Poor design and weak learner appeal are problematic because employees have no option to avoid or opt out. They must take these mandatory courses, frequently with the attitude of completing them as quickly as possible.

Engineering can be defined as "the field or discipline, practice, profession and art that relates to the development, acquisition and application of technical, scientific and mathematical knowledge about the understanding, design, development, invention, innovation and use of materials, machines, structures, systems and processes for specific purposes" (Marjoram & Zhong, 2010). The National Science Foundation (1977, p. 1) defined continuing engineering education and its characteristics as:

- the substantive content deals with engineering knowledge,
- taken after initial employment,
- addresses updating and diversification rather than advanced education,
- is structured, goal oriented, and intentional, and
- has a duration adequate for a significant increment of training.

Saline (1983, p. 123) defined CEE as "professional technical or management education taken in relatively small doses throughout a career."

Many institutions have identified the importance of CEE. A number of U.K. government agencies have recommended CEE (Galloway, 1998). The European Society for Engineering Education sponsored a report which argued action is needed to improve the continuing education of engineers (Padfield & Schaugelberger, 1998). A report by the U.S. National Academy of Engineering (2005) recommended that engineering institutions teach students how to be lifelong learners. The American Society of Mechanical Engineers' 2028 Vision included lifelong learning as one of the four areas for focused improvement (ASME, 2008). The U.S. National Academy of Engineering (2008) identified 14 grand challenges for engineering, many of which will require CEE as some of the subjects (e.g., prevent nuclear terror) are not typically covered in university education. "To successfully integrate process and knowledge, engineers must not only stay informed about new and emerging technologies but also be aware of knowledge and skills from other domains" (Sheppard, Macatangay, Colby, and Sullivan, 2009, p. 6). A comprehensive report sponsored by UNESCO (2010) on engineering also emphasized the need for CEE (Jones, 2010, p. 329):

> With the huge explosion of knowledge, engineers have been forced to become more specialized in their professional skills, which in turn forces judicious choices to be made in the topics taught during an initial education. The only way by which the tussle between breadth and depth can be reconciled is the recognition that a first degree provides just an initial education and that Continuing Engineering Education (CEE) or Continuing Professional Development (CPD) is essential.

A report sponsored by the U.S. National Academy of Engineering argued that lifelong learning for engineers is imperative for sustaining American competitiveness in the 21st century (Dutta, Patil, and Porter, 2012). The most recent accreditation requirements for engineering programs include an outcome that students are expected to achieve at the completion of their degree, “a recognition of the need for, and an ability to engage in life-long learning” (Engineering Accreditation Commission, 2012, p. 3).

The demand for CEE continues to increase. There are many reasons for this including rapid technological advances, globalization, increasing environmental emphasis, and sociopolitical changes (Baukal and Ausburn, 2018). Technical obsolescence (Ovesen, 1980) is a continuing problem due to the ever-increasing pace of changes in technology. Failure to continually learn after receiving an engineering degree leads to obsolescence because of the rapid changes in technology (Barton and Bommer, 1992). For example, the fields of nanotechnology and biotechnology are changing so rapidly that continuous training is required to maintain proficiency in those fields. As technology continuously changes engineering, CEE improves job performance and usually leads to increased compensation (Morris, 1978). It is recommended for quality assurance in a profession and in some cases it may be mandated to maintain licensure (Jeris, 2010). Licensure is particularly important in civil engineering because of the many public works projects done for governments. Some argue that CEE is critical to enhancing the innovation (Keating, Stanford, Dunlap, Aherne, and Mendelson, 2001) that keeps a professional field viable and its workforce effective and competitive.

Engineers participate in many types of continuing education, including both informal and formal training (Cervero, Miller, and Dimmock, 1986). Informal training is generally unplanned and initiated by the individual, although it could include sanctioned learning such as mentoring, coaching, and special assignments. For example, reading journals and magazines and attending conferences helps keep engineers current with changes in technology. Formal training is “planned learning activities that are intended to help individuals acquire specific areas of knowledge, awareness, and skills” and “mostly involves institutionally sponsored and endorsed programs, which would include almost all training and development programs that organizations offer” (Jacobs and Park, 2009, p. 140). Engineers may also take formal classes with instructors that may be for credit and possibly even leading to an advanced degree. Formal professional development not leading to a degree is the specific focus for the study reported here.

The challenge is how to deliver CEE when and where it is needed, at a reasonable cost (Baukal, 2012). Continuing education must be relevant and should be designed to solve industrial problems (Weimar, 1992). Continuing professional education (CPE) must be related to professional practice to help professionals make decisions about situations they encounter (Cervero, 1992).

Employer-sponsored CPE dwarfs that offered by any other type of provider and possibly accounts for more than all other providers combined (Cervero, 2001). While much has been written, for example, on CPE offered by universities, relatively little has been written about employer-sponsored CPE (Baukal, 2012) despite its prominence in workforce education and training. While some larger organizations like Microsoft and Motorola provide many and varied CEE courses for their engineers, may smaller organizations do not have the resources to provide CEE training course. Ultimately, it is up to the individual engineer to make sure CEE is a priority and to take the initiative to get the appropriate CEE training throughout their careers (Naimpally, Ramachandran, and Smith, 2012). Fortunately, more and more courses are available over the web which makes it easier and more cost effective for engineers to get the training they need.

CEE courses include all types of technical (e.g., equipment, processes, software), management (e.g., leadership, employment law, and enhancing team performance), and soft skills (e.g., communication and presentation skills) courses. CEE courses combining engineering and management are the most popular types of courses (Beruvides and Ng, 2009). Cole, Moss, Gohs, Lacefield, Barfield, and Blythe (1984) broadly defined four categories of continuing education courses for engineers:

1. remediating and upgrading basic technical knowledge and skills,
2. extending and broadening previously-learned scientific and technical skills,
3. learning new concepts and skills to keep up with technology advancements, and
4. learning new knowledge and skills outside engineering.

CEE is one of the most efficient ways to transfer technology and improve productivity (Markkula, 1985). Besides facilitating maintenance of proficiency, CEE allows engineers to learn about new technologies, whether in their own organization or in others. This might be in the form of traditional classroom instruction, by reading trade journals, or by attending conferences. Productivity can be improved by applying new technologies or taking advantage of existing technologies, processes, or procedures that were not previously used.

There are some specific challenges unique to CEE. Continuing professional development is particularly important for engineers, compared to other professions, because of the rapid changes in technology (Evetts, 1998). The half-life of knowledge is the time it takes for half of one's knowledge to become obsolete. In some engineering fields, the half-life has become so short, due to rapid changes in technology, that some information is becoming obsolete even before completing an undergraduate education (Wulf and Fisher, 2002). However, it should be noted that the basics taught in many college engineering courses, such as the First Law of Thermodynamics, do not change with time.

Another unique challenge is globalization. New technology is being developed all over the world. It is no longer sufficient to keep up with advancements in one's own country. Today's engineers must be familiar with and be able to apply new technologies developed on a global scale. It is not practical for engineers to travel all over the world to learn about the latest advancements, which is one reason why distance CEE is so important.

Learning new technologies can often be more challenging because of the need to use multiple senses to fully understand them. In some cases, this may mean hands-on instruction where students actually manipulate the new technology. For example, learning how to program a controller to automatically adjust a flow rate may require students to actually use the controller. In other cases, instruction may need advanced visualization and animation to aid in understanding the technology (see Chapter 7). This challenge is not usually faced by, for example, accountants or attorneys learning the latest developments in their fields. Bork (2004) believed computer-based distance education is the best way to provide continuing education on a global basis. Industry has been a leader in using media-based continuing engineering education (Biedenbach, 1978).

References

ABET, Criteria for Accrediting Engineering Programs, 2021-2022, https://www.abet.org/accreditation/accreditation-criteria/criteria-for-accrediting-engineering-programs-2021-2022/, accessed December 22, 2021.

American Society for Mechanical Engineers (ASME). (2008). *2028 vision for mechanical engineering*. Retrieved from http://www.asmeconferences.org/asmeglobalsummit/FinalGlobalSummitReport.pdf.

Annetta, Leonard and James Minogue (eds.), Connecting Science and Engineering Education Practices in Meaningful Ways, Heidelberg: Springer, 2016.

Barton, B., & Bommer, M. (1992). Myths about engineers maintaining professional technical competence. *International Journal of Continuing Engineering Education and Lifelong Learning*, *2*(1), 22-30.

Baukal, C. (2010). Continuing engineering education through distance learning. *European Journal of Engineering Education*, *35*(2), 225-233.

Baukal, C. (2012). Strategy for delivering continuing engineering education. *International Journal of Training and Development*, *16*(4), 284-299.

Baukal, C. (2017). *Learner Preferences for Continuing Engineering Education*. Mauritius: Lambert.

Baukal, C., and Ausburn, L. (2018). Effective Strategies for Training Working Engineers. In Abend, M. (ed.), Effective Teaching and Learning: Perspectives, Strategies and Implementation, NY: Nova, 115-137.

Baukal, C. E., & Crawford-Fanning, M. N. (2013). Combustion training. In C. E. Baukal (Ed.), *The John Zink Hamworthy combustion handbook*. Boca Raton, FL: CRC Press, 1: 513-550.

Beblavý, Miroslav, Sara Baiocco, Zachary Kilhoffer, Mehtap Akgüç, and Manon Jacquot (2019). Readiness for Digital Lifelong Learning: Changing How Europeans Upgrade Their Skills. Brussels: Centre for European Policy Studies. https://espas.secure.europarl.europa.eu/orbis/sites/default/files/generated/document/en/Index-of-Readiness-for-Digital-Lifelong-Learning.pdf. Accessed 25 July 2021.

Bennett, N. L., & LeGrand, B. F. (1990). *Developing continuing professional education programs*. Urbana, IL: Illinois University.

Beruvides, M. G., & Ng, E-H. (2009). The continuing education of engineers and engineering managers. *International Journal of Continuing Engineering Education and Lifelong Learning*, *19*(1), 4-18.

Biedenbach, J. M. (1978). Media-based continuing engineering education. *Proceedings of the IEEE*, *66*(8), 961-968.

Bolle, L. and Eugene, C. (1990). Engineering continuing education: a point of view. European Journal of Engineering Education *15*(2), 165-167.

Bonvillian, William and Sanjay Sarma (2021). *Workforce Education: A New Roadmap*. Cambridge, MA: MIT Press.

Bork, A. (2004). A new lifelong learning system for the world. *International Journal of Continuing Engineering Education*, *14*(1/2), 46-57.

Brockett, Ralph (2015). *Teaching Adults: A Practical Guide for New Teachers*. San Francisco: Jossey-Bass.

Cervero, R. M. (1992). Professional practice, learning, and continuing education: An integrated perspective. *International Journal of Lifelong Education*, *11*(2), 91-101.

Cervero, R. M. (2001). Continuing professional education in transition, 1981-2000. *International Journal of Lifelong Education*, *20*(1/2), 16-30.

Cervero, R. M., Miller, J. D., & Dimmock, K. H. (1986). The formal & informal learning activities of practicing engineers. *Engineering Education*, *77*(2), 112-114.

Colbrunn, S. R., & Van Tiem, D. M. (2002). From binders to browsers: Converting classroom training to the web. In A. Rossett (Ed.), *The ASTD E-learning handbook* (pp. 85-95). Alexandria, VA: ASTD Press.

Cole, H. P., Moss, J., Gohs, F. X., Lacefield, W. E., Barfield, B. J., & Blythe, D. K. (1984). *Measuring Learning in Continuing Education for Engineers and Scientists*. Phoenix: Oryx Press.

Crawley, Edward, Johan Malmqvist, Sören Östlund, Doris Brodeur, and Kristina Edstrom, Rethinking Engineering Education: The CDIO Approach, 2nd edition, Heidelberg, 2014.

Cross, K. Patricia (1981). *Adults as Learners: Increasing Participation and Facilitating Learning*. San Francisco: Jossey-Bass.

Davim, J. Paulo (ed.), *Engineering Education: Curriculum, Pedagogy and Didactic Aspects*, Oxford: Chandos Publishing, 2014.

Davison, Ryan, Engineering Curricula: Understanding the Design Space and Exploiting the Opportunities: Summary of a Workshop, Washington, DC: National Academies Press, 2010.

Del Gaizo, E. and Laudermith, A. (2021). Professionals as lifelong learning. In M. London (ed.). *The Oxford Handbook of Lifelong Learning*. NY: Oxford University Press, 195-208.

Dirkx, (2011). Work-related learning in the United States: Past practices, paradigm shifts, and policies of partnerships. In M. Malloch, L. Cairns, K. Evans, & B. N. O'Connor (Eds.), *The Sage handbook of workplace learning* (pp. 293-306). Los Angeles: Sage.

Duderstadt, J. J. (2008). *Engineering for a changing world: A roadmap to the future of engineering practice, research, and education*. Ann Arbor, Michigan: The Millennium Project, The University of Michigan. Retrieved from: http://milproj.dc.umich.edu/.

Dutta, D., Patil, L., & Porter, J. B. (2012). *Lifelong learning imperative in engineering: Sustaining American competitiveness in the 21st century*. Washington, DC: National Academies Press.

Engineering Accreditation Commission. (2012). Criteria for accrediting engineering programs: Effective for reviews during the 2013-2014 accreditation cycle. Baltimore, MD: Accrediting Board for Engineering and Technology.

Ericsson, K. Anders (2009). *Development of Professional Expertise: Toward Measurement of Expert Performance and Design of Optimal Learning Environments*. Cambridge, UK: Cambridge University Press.

Evetts, J. (1998). Continuing professional development for engineers: UK and European dynamics. *European Journal of Engineering Education*, *23*(4), 443-452.

Felder, R. M., R. Brent, and M. Prince, Engineering Instructional Development: Programs, Best Practices, and Recommendations, *Journal of Engineering Education 100*(1), 89-122, 2011.

Flumerfelt, Shannon, Franz-Josef Kahlen, Anabela Alves, and Anna Bell Siriban-Manalang, *Lean Engineering Education: Driving Content and Competency Mastery*, New York: ASME Press, 2015.

Fox, Marye and Norman Hackerman (eds.), Committee on Recognizing, Evaluating, Rewarding, and Developing Excellence in Teaching of Undergraduate Science, Mathematics, Engineering, and Technology, Washington, DC: National Academies Press, 2003.

Frerich, S., T. Meisen, A. Richert, M. Petermann, S. Jeschke, U. Wilkesmann, and A. Takkaya (eds.), *Engineering Education 4.0: Excellent Teaching and Learning in Engineering Sciences*, Cham, Switzerland: Springer, 2016.

Galbraith, Michael, ed. (2004). *Adult Learning Methods: A Guide for Effective Instruction*, 3rd ed. Malabar, FL: Krieger Publishing.

Galloway, Patricia, *The 21st-Century Engineer: A Proposal for Engineering Education Reform*, Reston, VA: ASCE Press, 2008.

Goldberg, David and Mark Sommerville, *A Whole New Engineer: The Coming Revolution in Engineering Education*, Douglas, MI: Threejoy, 2014.

Graham, Ruth, *Achieving Excellence in Engineering Education: The Ingredients of Successful Change*, London: Royal Academy of Engineering, 2012.

Grayson, Lawrence, *The Making of an Engineer: An Illustrated History of Engineering Education in the United States and Canada*, New York: John Wiley & Sons, 1993.

Guerra, Aida, Ronald Ulseth, and Anette Kolmos (eds.), *PBL in Engineering Education: International Perspectives on Curriculum Change*, Rotterdam: Sense Publishers, 2017.

Hager, Paul, Concepts and Definitions of Lifelong Learning, in Manuel London (ed.), *The Oxford Handbook of Lifelong Learning*, Second Edition, Oxford University Press, Oxford, UK, 2021.

Heywood, John, *The Assessment of Learning in Engineering Education: Practice and Policy*, New York: IEEE Press, 2016.

Heywood, John, *Empowering Professional Teaching in Engineering: Sustaining the Scholarship of Teaching*, Morgan & Claypool Publishers, 2018.

Horton, W., & Horton, K. (2003). *E-Learning tools and technologies*. Indianapolis, IN: Wiley.

Houle, C. O. (1980). *Continuing learning in the professions*. San Francisco: Jossey-Bass.

Institute for Continuing Professional Development (2009). www.cpdinstitute.org [Accessed on 16 September 2009].

Jacobs, R. L., & Park, Y. (2009). A proposed framework of workplace learning: implications for theory development and research in human resource development. *Human Resource Development*, *8*(2), 133-150.

Jarvis, Peter (2010). Adult Education and Lifelong Learning: Theory and Practice, 4th ed. London: Routledge.

Jeris, L. H. (2010). Continuing professional education. In C. E. Kasworm, A. D. Rose, & J. M. Ross-Gordon (Eds.), *Handbook of adult and continuing education* (pp. 275-283). Los Angeles: Sage.

Johri, Aditya and Barbara Olds, *Cambridge Handbook of Engineering Education Research*, New York: Cambridge University Press, 2014.

Jonassen, D. H., "Engineers as Problem Solvers," Chapter 6 in *Cambridge Handbook of Engineering Education Research*, edited by A. Johri and B. M. Olds, Cambridge University Press, New York, 2014.

Kapranos, Plato (ed.)., The Interdisciplinary Future of Engineering Education: Breaking Through Boundaries in Teaching and Learning, London: Routledge, 2019.

Keating, D. A., Stanford, T. G., Dunlap, D. D., Aherne, M. J. & Mendelson, M. I. (2001, June). Enhancing U.S. Technology Development Through Lifelong Education of Engineers and Technologists as Creative Professionals. *Proceedings of the 2001 ASEE Annual Convention*. Albuquerque, NM.

Kellam, Nadia, Brooke Coley, and Audrey Boklage, *Transformative Teaching: A Collection of Stories of Engineering Faculty's Pedagogical Journeys*, Morgan & Claypool Publishers, 2019.

Kelly, Gregory and Judith Green (eds.), *Theory and Methods for Sociocultural Research in Science and Engineering Education*, New York: Routledge, 2019.

Knowles, Malcolm, Elwood Holton, Richard Swanson, and Petra Robinson (2020). *The Adult Learner: The Definitive Classic in Adult Education and Human Resource Development*. London: Routledge.

Kober, Linda, Reaching Students: What Research Says About Effective Instruction in Undergraduate Science and Engineering, Washington, DC: National Academies Press, 2015.

Lindeman, Eduard (1961). *The Meaning of Adult Education*. Canada: Harvest House.

Malloch, M., Cairns, L., Evans, K., & O'Connor, B. N. (2011). *The Sage handbook of workplace learning*. Los Angeles: Sage.

Markkula, M. (1985). Continuous studies as a basis for professional development. *European Journal of Engineering Education*, *10*(3/4), 221-227.

Marjoram, T., & Zhong, Y. (2010). 1.1 What engineering is, what engineers do. In *UNESCO, Engineering: Issues, Challenges and Opportunities for Development* (Paris: United Nations

Educational, Scientific and Cultural Organization) (http://unesdoc.unesco.org/images/0018/001897/189753e.pdf), 24-26.

Meriam, Sharan and Lisa Baumgartner (2020). *Learning in Adulthood: A Comprehensive Guide*, 4th edition. San Francisco: Jossey-Bass.

Mohsen, J. P., Mohamed Ismail, Hamid Parsaei, and Waldemar Karwowski (eds.), *Global Advances in Engineering Education*, Boca Raton, FL: CRC Press, 2019.

Moore, M., & Kearsley, G. (2005). *Distance education: A systems view*. Belmont, CA: Thomson Wadsworth.

Morris, A. J. (1978). *The return on investment in continuing education of engineers*. Palo Alto, CA: Genesys Systems, Inc.

Mutis, Ivan, Renate Fruchter, and Carol Menassa (eds.), *Transforming Engineering Education: Innovative, Computer-Mediated Learning Technologies*, Reston, VA: ASCE Press, 2018.

Naimpally, Ashok, Hema Ramachandran, and Caroline Smith (2012). *Lifelong Learning for Engineers and Scientists in the Information Age*. London: Elsevier.

National Academy of Engineering, The Engineer of 2020: Visions of Engineering in the New Century, Washington, DC: National Academies Press, 2004.

National Academy of Engineering, Educating the Engineer of 2020: Adapting Engineering Education to the New Century, Washington, DC: National Academies Press, 2005.

National Academy of Engineering (2008). Grand challenges for engineering. Washington, DC: National Academy of Science.

National Science Board, Moving Forward to Improve Engineering Education, Washington, DC: National Science Foundation report NSB-07-122, November 19, 2007.

National Science Foundation (1977). Continuing education in science and engineering. Washington, DC: National Science Foundation.

Noe, R. A. (2005). *Employee training and development* (3rd ed.). New York: McGraw-Hill.

Organisation for Economic Co-operation and Development (2021). OECD Skills Outlook 2021: Learning for Life. Paris: OECD Publishing. https://doi.org/10.1787/0ae365b4-en. Accessed 25 July 2021.

Ovesen, N. K. (1980). *Advances in the continuing education of engineers*. Paris: United Nations Educational, Scientific and Cultural Organization (UNESCO).

Padfield, C., & Schaufelberger, W. (1998). *Lifelong learning in engineering education: A call to action*. SEFI Document no. 20. Brussels: European Society of Engineering Education (SEFI).

Powell, Roberta (ed.), *Engineering Education: Curriculum and Pedagogy*, Forest Hills, NY: Willford Press, 2019.

Rahman, Ataur and Vojislav Ilic (eds.), *Blended Learning in Engineering Education: Recent Developments in Curriculum, Assessment and Practice*, Boca Raton, FL: CRC Press, 2019.

Ross-Gordon, Jovita, Amy Rose, and Carol Kasworm (2017). *Foundations of Adult and Continuing Education*. San Francisco: Jossey-Bass.

Rothwell, William and Anita Whiteford, Corporate Employee Training and Development Strategies, in Manuel London (ed.), *The Oxford Handbook of Lifelong Learning*, Second Edition, Oxford University Press, Oxford, UK, 2021.

Royal Academy of Engineering, Educating Engineers for the 21st Century, London: Royal Academy of Engineering, June 2007.

Saline, L. E. (1983). Continuing engineering education: One element of lifelong learning for engineers. *IEEE Transactions on Education, E-26*(4), 122-126.

Schön, D. A. (1987). *Educating the reflective practitioner*. San Francisco: Jossey-Bass.

Shepherd, Sheri, Kelly Macatangay, Anne Colby, and William Sullivan, *Educating Engineers: Designing for the Future of the Field*, San Francisco: Jossey-Bass, 2009.

Sidhu, Manjit, *Technology-Assisted Problem Solving for Engineering Education: Interactive Multimedia Applications*, Hershey: Engineering Science Reference, 2010.
Singer, Susan, Natalie Nielsen, and Heidi Schweingruber (eds.), *Discipline-Based Education Research: Understanding and Improving Learning in Undergraduate Science and Engineering*, Washington, DC: National Academies Press, 2012.
Spurlin, Joni, Sarah Rajala, and Jerome Lavelle (eds.), *Designing Better Engineering Education Through Assessment: A Practical Resource for Faculty and Department Chairs on Using Assessment and ABET Criteria to Improve Student Learning*, Sterling, VA: Stylus, 2008.
Tough, Allen (1979). The *Adult's Learning Projects: A Fresh Approach to Theory and Practice in Adult Learning*. 2nd ed. Austin, TX: Learning Concepts.
Tryggvason, Gretar and Diran Apelian, "Re-Engineering Engineering Education for the Challenges of the 21st Century, *Journal of the Minerals, Metals and Materials Society 58*(10), 14-17, 2006.
United Nations Educational, Scientific and Cultural Organization (UNESCO; 2010), *Engineering: Issues, Challenges and Opportunities for Development*, Paris: UNESCO Publishing.
Valencia, R., Link, D., Baukal, C., & McGuire, J. (2008). Consider classroom training for plant operators. *Hydrocarbon Processing*, *87*(11), 55-59.
Valenzuela-Valdés, Juan & Pedro Pado Fernández (eds.), *Project Based Learning on Engineering: Foundations, Applications and Challenges,* New York: Nova Publishers, 2015.
Vanajakumari, M., Johnston, K., Lawrence, F., and Menon, R. (2015). An effective Teaching Methodology for Continuing Education Programs. *Journal of Engineering Technology*, *32*(1), 18-29.
Weimar, B. (1992). Assumptions about university-industry relationship in continuing professional education: A re-assessment, *European Journal of Education*, *27*(4), 379-396.
Willets, J. (1983). Continuing Engineering Education. *European Journal of Engineering Education* 7(4), 331-336.
Wulf, W. A., & Fisher, G. M. C. (2002). A makeover for engineering education. *Issues in Science & Technology*, *18*(3), 35-39.
Zawojewski, Judith, Heidi Diefes-Dux, and Keith Bowman (eds.), *Models and Modeling in Engineering Education: Designing Experiences for All Students,* Rotterdam: Sense Publishers, 2008.

Sources for Further Information

This section is intended to be representative but not exhaustive. Links are provided for further information.

Organizations

This section includes some organizations whose primary mission is the education of undergraduate engineering students, which includes developing a sense of the need for lifelong learning. Undergraduate education is fixed largely on theory while CEE focuses primarily on application of the theory.

ABET[1] is a worldwide accreditor of engineering programs whose purpose is "to assure confidence in university programs in STEM (science, technology, engineering, and mathematics) disciplines." At the time of this writing it had accredited 4,144 programs at 812

[1] https://www.abet.org.

institutions in 32 countries. ABET consists of four accreditation commissions: Applied and Natural Science, Computing, Engineering, and Engineering Technology. The Engineering Accreditation Commission (EAC) is of specific interest here. At the time of this writing, the following are the criteria used by the EAC for Baccalaureate programs:[2]

1. Students
2. Program Educational Objectives
3. Student Outcomes
4. Continuous Improvement
5. Curriculum
6. Faculty
7. Facilities
8. Institutional Support

ABET partners with relevant societies in the accreditation process. For example for mechanical engineering, ABET partners with the ASME which supplies the program evaluators during accreditation visits.

The American Society for Engineering Education or ASEE[3] is an international society dedicated to engineering education. Its mission is to advance "innovation, excellence, and access at all levels of education for the engineering profession." It provides training, sponsors conferences, publishes journals and papers, and recommends public policy. The Continuing Professional Development Division or CPDD[4] is one of the largest within ASEE. The mission of the CPD Division is "to foster the development, delivery, and improvement of continuing education in all areas which pertain to allied branches of science, technology, engineering and math; to provide advocacy for and leadership in lifelong learning; and to create opportunities for exchange of information and expertise among providers and consumers of educational programs."[5] The vison for the CPDD is to be "the preeminent organization for individuals and organizations involved in providing continuing professional development for practicing engineers and technical professionals worldwide."[6] Members include industry representatives, training managers, college faculty, societies and private sponsors, and professionals in continuing and distance education.

The International Federation of Engineering Education Societies or IFEES[7] was founded in 2006. Note there are many individual engineering societies for a variety of countries and regions which are not specifically discussed here. The purpose of IFEES is to encourage collaboration among the many engineering education societies. IFEES annually sponsors the World Engineering Education Forum held at different locations around the world.

The International Association for Continuing Engineering Education (IACEE) "is an international, non-profit and non-governmental organization. The aim of IACEE is to support and enhance lifelong learning education and training worldwide, with special emphasis to

[2] https://www.abet.org/accreditation/accreditation-criteria/criteria-for-accrediting-engineering-programs-2020-2021/
[3] https://www.asee.org/.
[4] http://mechanical.asee.org/.
[5] https://www.asee.org/membership-and-communities/community/Divisions/Continuing-Professional-Development.
[6] cpd.asee.org/about.
[7] http://www.ifees.net/.

address the needs of developing countries."[8] It includes individual members, professional organizations, industrial organizations, and academic institutions. It conducts an annual world conference that rotates locations.

The Institute of Electrical and Electronics Engineers or IEEE has an education branch called IEEE Education Society[9] which is "an international organization that promotes, advances, and disseminates state-of-the-art scientific information and resources related to the Society's field of interest and provides professional development opportunities for academic and industry professionals." It publishes the journal *IEEE Transactions on Education.*

U.S. National Academy of Engineering or NAE[10] was founded in 1964 and provides engineering leadership to the US. Its mission is "to advance the well-being of the nation by promoting a vibrant engineering profession and by marshalling the expertise and insights of eminent engineers to provide independent advice to the federal government on matters involving engineering and technology." It has more than 2,000 peer-elected members. They have published many reports relevant to engineering education.

The United Nations Educational, Scientific and Cultural Organization or UNESCO[11] was formed in 1945 and looks to "build peace through international cooperation in education, the sciences and culture." UNESCO has published some reports on engineering education.[12] In partnership with UNESCO is the World Federation of Engineering Organizations or WEFO[13] whose primary function is to represent engineering worldwide and to encourage humanitarian efforts. It also has an educational component.

The Royal Academy of Engineering or RAE[14] is the UK national academy for engineering and technology. Engineers are elected to become fellows and serve to advance and promote engineering excellence to benefit society. The RAE has issued a wide range of reports related to engineering including those related to education at all levels of schooling. They conduct education research and provide policy guidance to the UK government.

The Center for the Integration of Research, Teaching, and Learning or CIRTL[15] "seeks to enhance excellence in STEM undergraduate education through development of a national faculty committed to implementing and advancing evidence-based teaching practices for diverse learners. It was founded in 2003 and consists of a network of nearly 40 research universities in the U.S. and Canada. Some of its key values include learner-centered education, equity and inclusion, diversity and representation, collaboration, intellectual generosity, inclusive excellence and innovation, and reflective decision-making. CIRTL offers a range of CEE courses for training engineering faculty.

Journals

This section is not intended to be exhaustive but to present some of the more common journals dedicated to engineering education. There are some others dedicated to specific countries (example, Argentina, India, Japan, Korea, and Taiwan) or regions (e.g., Latin America and the

[8] Iacee.org.
[9] http://ieee-edusociety.org/.
[10] https://www.nae.edu/.
[11] https://en.unesco.org/.
[12] http://www.unesco.org/new/en/natural-sciences/science-technology/engineering/engineering-education/.
[13] https://www.wfeo.org/.
[14] https://www.raeng.org.uk/.
[15] https://www.cirtl.net/.

Caribbean). The journals are considered in alphabetical order. Most of them concern predominantly undergraduate education.

Advances in Engineering Education (AEE)[16] is a peer-reviewed, online journal published two or three times a year by the American Society for Engineering Education. Its mission is "to disseminate significant, proven innovations in engineering education practice, including those that are enhanced through the creative use of multimedia."

Australasian Journal of Engineering Education (AJEE)[17] is published by Taylor and Francis under the auspices of the Australasian Association for Engineering Education (AAEE). It supports research designed to improve engineering education by informing practice, policy, and research.

Chemical Engineering Education (CEE)[18] is a quarterly, international, peer-reviewed journal dedicated specifically to chemical engineering education which originated in 1962. The readership is predominantly chemical engineering faculty. It is published by the Chemical Engineering Division of the American Society for Engineering Education.

European Journal of Engineering Education (EJEE)[19] is published by Taylor & Francis and sponsored by the European Society for Engineering Education (SEFI).[20] It is "a forum for scholarly dialogue to further engineering education" that "embraces multiple perspectives to examine relevant conditions, developments, approaches, methods and experiences relevant for the education of engineers." The focus is on higher engineering education.

IEEE Transactions on Education[21] is a publication of the Institute for Electrical and Electronic Engineers (IEEE) which publishes "significant and original scholarly contributions to education in electrical and electronics engineering, computer engineering, computer science, and other fields within the scope of interest of IEEE."

The *International Journal of Continuing Engineering Education and Life-Long Learning* (IJCEELL)[22] is published by Inderscience on a quarterly basis. It is a, open-access premier journal for continuing engineering education, lifelong learning, and professional development for engineers and technologists.

The *International Journal of Electrical Engineering & Education* (IJEEE)[23] is published by Sage. It originated in 1948 as the *International Journal of Electrical Engineering Education.* It provides a "showcase for international developments in the undergraduate teaching of electrical engineering and electronics, from power systems to nanotechnology.

International Journal of Engineering Education (IJEE)[24] is an independent, peer-reviewed journal "serving as an international archival forum of scholarly research related to engineering education." It is published six times a year on a wide range of topics related to engineering education.

International Journal of Mechanical Engineering Education (IJMEE)[25] is an open-source, peer-reviewed journal published by Sage. It is aimed at "teachers and trainers of mechanical

---

[16] https://advances.asee.org/.
[17] https://www.tandfonline.com/loi/teen20.
[18] http://ww2.che.ufl.edu/cee/.
[19] https://www.tandfonline.com/loi/ceee20
[20] https://www.sefi.be/
[21] https://site.ieee.org/review-criteria-toe/.
[22] https://www.inderscience.com/jhome.php?jcode=ijceell.
[23] https://journals.sagepub.com/home/ije.
[24] https://www.ijee.ie/.
[25] https://journals.sagepub.com/home/ijj.

engineering students in higher education and focuses on the discussion of the principles and practices of training professional, technical and mechanical engineers and those in related fields." It includes topics such new experimental methods and laboratory techniques, book reviews, and highlights of recent articles in the field.

Journal of Engineering Education (JEE)[26] is a premier, quarterly, peer-reviewed, international journal published by the American Society for Engineering Education. Its mission is to "cultivate, disseminate, and archive scholarly research in engineering education. It includes quantitative, qualitative, and mixed methods studies.

Studies in Engineering Education (SEE)[27] is a new journal that was launched in June 2019. Its mission is "to expand the conversations among engineering education researchers." It is an international, open-access, peer-reviewed journal.

Workshops

National Effective Teaching Institute or NETI (https://www.asee.org/education-careers/continuing-education/courses-and-workshops/neti) are semi-annual three-day workshops established in 1991 to familiarize engineering faculty members with student-centered teaching strategies. Based on a survey of participants, NETI has achieved its goal of improving the instructional methods of engineering faculty (Felder and Brent, 2010). While the focus of the workshops is how to more effectively teach undergraduates, the courses themselves are CEE for faculty.

Frontiers in Engineering Education or FIE[28] runs an annual conference with published proceedings and has a vision "to advance and re-define engineering and computing education to ensure that all students receive the best possible preparation for their future." FIE is sponsored by the ASEE Educational Research and Methods Division and the IEEE Computer and Education Societies.

ExCEEd[29] is a six day workshop offered multiple times each year sponsored by the American Society of Civil Engineering (ASCE) designed to improve engineering education faculty skills and abilities. There are also two 1.5 day refresher workshops: one is a refresher of the full six day workshop and the other is a condensed version of the full workshop. Those interested in these workshops must apply and be approved to attend.

---

[26] https://www.asee.org/papers-and-publications/publications/jee.
[27] https://www.seejournal.org/.
[28] http://fie-conference.org/.
[29] https://www.asce.org/exceed/.

Chapter 2

Fifty Years in Engineering Distance Education: Past Trends and Future Directions

Paul G. Jewell, Ann D. Thompson, Rema Nilakanta*, Hiroyuki Iino and Darshana Juvale
Engineering LAS-Online Learning (ELO), Iowa State University, IA, USA

Abstract

In this chapter, the history of the distance education work in the College of Engineering at Iowa State University is described and analyzed. The university's and college's early commitment to delivering high quality distance education is explained as is the development of new approaches to delivering distance education over the 50-year history of the program. Data from key leaders in the program combined with records and documents suggest three major themes: 1) the importance of creating key collaborations with industry partners to both serve and create the audience for distance education initiatives; 2) a continuous commitment to high quality programs taught by full time university faculty members with the same quality and rigor as the on-campus programs; and 3) the importance of identifying and sharing key strong program areas from the College of Engineering. Program leaders from the rich 50-year history of distance education in the College of Engineering agree on the importance of strategically building upon the long and successful history of high-quality distance education in the College of Engineering and maintaining Iowa State University's leadership and commitment in this important area.

Keywords: engineering education, distance education, online learning

Introduction

In the past decade, colleges of engineering around the country have created numerous structures designed to support high quality distance education experiences for students. These structures have met with varying degrees of success, with some creating the basis for large, effective distance education programs and others struggling to survive. The College of Engineering at Iowa State University celebrated 50 years of distance education in 2019. In this chapter, we provide a summary of the major milestones in the development of distance education in the College of Engineering and describe the major themes that surround the development of the structure. We conclude the chapter using data from the history of distance education in the College of Engineering to provide recommendations for future directions.

---

* Corresponding Author's Email: rema@iastate.edu.

Methods

In order to define these milestones themes and recommendations, we used historical methodology to determine significant events in the development of the current program. Through the study of past events and personalities, historical methodology helped shed light on current happenings and future trends.

Data sources included people involved in the development of the program and archived records from the program. In order to tell the story of the development of the current Engineering-LAS Online Learning (ELO), the distance education support center in the College of Engineering at Iowa State University, the research team met regularly to engage in collaborative reflection and developed a timeline of important events (see Appendix A) that marked the development of distance education in the College of Engineering. We also interviewed key faculty, staff, and graduate students who were involved and/or are still involved with the development of distance education work in the College of Engineering.

The research team that included current ELO staff members reflected and explored jointly the following questions:

1. What was the early history of ELO when it was known as Professional Programs in Engineering (PPE) and later as Engineering Distance Education (EDE)?
2. What were the important milestones in terms of the development of ELO?
3. What are the significant themes in the history of ELO?
4. In the context of history, what are the most promising future directions for distance education in the College of Engineering?

In addition to our meetings and document searches, we used an individual interview procedure to collect data on perceptions of key program leaders. We chose to interview Edwin Jones, Ted Okishi, Jim Melsa, Tom Brumm, Ron Cox, Loren Zachary, Sriram Sundararajan, and faculty leader Doug Jacobson. Questions asked at the interviews included:

- Please describe your role in distance education at Iowa State University's (ISU's) College of Engineering.
- At the time, what was your vision for the future of distance education in Engineering?
- During the time you were a leader, you made some bold moves. What were those? How did you feel about it at that time?

Following each research group meeting and interview, the research team met to define and discuss major themes and events revealed through each session.

Results

Combining our data from program participants, program leaders, and program documents, we were able to construct a timeline (see Appendix A) that summarizes the major developments in the history of distance education initiatives in the College of Engineering.

The first question posed to the research group was: "When did distance education work begin at Iowa State University?" We were somewhat surprised when one participant suggested that the first major initiative began in 1907 with the establishment of the Seed Corn Gospel Train (Pounds, 2007). As the name suggests, the Seed Corn Gospel Train was designed to transport ISU faculty around the state to share current information on successfully raising corn. Although this early effort was not from the College of Engineering, this early commitment to extending educational resources was a precursor to the long term commitment of Iowa State University to its Land Grant charge to: "... follow the principle that all people in the state are, in reality, students of the college. Therefore, we must go to them and help them where they are, under their own conditions, with their own problems" (Pounds, 2007).

Over the next several decades, our data describe a consistent record of effective and innovative distance education efforts that emerged. A cadre of talented staff along with inspiring leadership led to new opportunities. Of special note was a collaborative program with Collins Radio[1] in the 1950s that involved faculty travel to Cedar Rapids. Ed Jones remembers these efforts: "These classes had been taught by people that actually rode the trains over to Cedar Rapids. People like Grover Brown and Al Read and some of the others. The trains ran at 120 miles an hour so it didn't take long to get there" (phone interview, May 5, 2017).

Approximately 10 Master of Science degrees in Engineering were earned each year in this program. According to Jones, the average of 10 degrees awarded annually started in the 1950s and continued through the 1960s. In 1969, videotape delivery started and ran concurrently with face-to-face for a couple of years. The 1970s were characterized exclusively by videotape delivery as a way to save money during the recession. Jones often quipped: "We delivered engineering education by "Planes, Trains, Automobiles… and then Videotape." Course lectures on VHS tapes were delivered through UPS. Engineers at Rockwell Collins would gather on Fridays as a cohort to watch these lectures and work on their advanced degrees. Faculty members visited Cedar Rapids twice a semester in order to establish some personal contact.

Beginning in 1969, the College of Engineering affirmed the importance of its distance education work through the creation of administrative structures that guided and implemented distance education in the college. Our research group work identified four major administrative structures/organizations in the ensuing history of the distance education program in the College of Engineering:

- Professional Programs in Engineering, 1969 – 1995
- Engineering Distance Education (EDE), 1995 - 2011
- Engineering Online Learning (EOL), 2011 - 2012
- Engineering-LAS Online Learning (ELO), 2012 – Present.

Notable Events

Examining the 50-year history of distance education in the College of Engineering, four notable events emerge.

---

[1] Collins Radio became Rockwell Collins and is now called Collins Aerospace after it merged with UTC Aerospace Systems.

1. College of Engineering's leadership and national reputation was evidenced by the invitation to become a charter member of the National Technological University (NTU), which was founded in 1984 as a partnership between academia, business, and government. Engineering graduate level courses were delivered from NTU participating universities via satellite in both live and taped mode 24 hours, six days per week to 190 industrial locations in the U.S. and Canada. It was considered an honor to have courses selected for NTU offerings. ISU had significant contributions in electrical, computer, and systems engineering; ISU courses were heavily enrolled. The consortium grew to 54 participating universities by 2003 and NTU started a transition to a "for profit" business model. As additional distance education opportunities developed, ISU dropped out of NTU in 2004.
2. The development of the Iowa Communications Network (ICN) in 1994 provided the infrastructure for bringing live classrooms to distant locations. ICN is a fiber-optic network that linked some 800 sites with central switching from the National Guard Armory in Camp Dodge, Iowa. The ICN was used to deliver the undergraduate BSEE program to cohorts of engineering students at Rockwell-Collins in Cedar Rapids. The department was able to show that the learning outcomes of these students was substantially equivalent to that of on-campus students, and they were included in the positive accreditation action of the overall program. Thus, the BSEE program was one of the first in the country to be accredited with distance education students.
3. The sale of WOI-TV in 1994 produced a funding source for distance education efforts across the University. The College of Engineering funds were used to build delivery infrastructure (studio classrooms), off campus labs and ICN classrooms (Kirkwood) for the BSEE program. The funds also supported early efforts at Internet delivery, expanded partnering with community colleges for ICN delivery, adaptations of active learning strategies for distance education, development of additional off-campus undergrad programs (B. S. Mechanical Engineering).
4. A confluence of internet and web-based technologies in the late 1990s and the leadership of Jim Melsa, Dean of College of Engineering 1995-2004 led to a dramatic increase in the number of distance education courses offered around the state. Melsa was enthusiastic about increasing distance education opportunities and created metrics to measure success. He wanted 50 strategic partnerships and 1,000 active distance learning enrollments. Both were accomplished within 5 years.

Development of EDE and ELO

Around 2000, some of the first computer-based approaches emerged in what had become Engineering Distance Education (EDE). EDE introduced the learning management system WebCT to distance education faculty as another method of course delivery along with videotapes. Soon after integrating WebCT, EDE also introduced streaming media to deliver captured lectures over the Internet. Streaming media delivery provided working students the flexibility to view lectures anytime and place.

In addition, demonstrating the commitment to online course design, faculty development, and new distance education pedagogy, EDE hired in 2002, the first College of Education graduate student. From 2000 to 2010, EDE continued to grow student enrollment numbers with an average of 11% annual enrollment increase. It supported more programs and a variety of courses to meet distance students' demand. The number of courses offered through EDE doubled over the period. However, from 2010 – 2018 enrollment growth became relatively static.

In 2010, the College of Engineering appointed Dr. Tom Brumm to lead EDE. Early in his tenure as Professor-in-Charge, Dr. Brumm removed the word "distance" from Engineering Distance Education believing all enrolled students should have access to the online learning environment regardless of location. The new name became EOL (Engineering Online Learning).

In 2012 EOL merged with the College of Liberal Arts and Sciences (LAS) DE unit and became ELO (Engineering-LAS Online Learning). This new administrative structure brought with it increased access to instructional designers and asynchronous models of distance education.

In 2012 the university adopted the Resource Management Model to develop its annual operating budgets and the financial incentives for engineering faculty teaching through ELO disappeared. Both our focus group data and our interview data from leaders suggest that this change in faculty funding for teaching at a distance created a challenge for the development of distance education efforts. Leaders tended to agree that the faculty incentive model was in need for a change and that there needed to be a new, carefully created incentive program for faculty who teach and provide leadership for the distance education program.

Themes

In addition to providing a chronology of major events in the development of the history of distance education in the College of Engineering, data from records, our focus groups and interviews revealed three major themes that surround this chronology. These themes were defined and refined by our research group as we continued to add additional interviews of program leaders.

1. A major theme that emerged from both our focus group work and the interviews with program leaders was the importance of creating key collaborations with industry partners to both serve and create the audience for distance education initiatives. During his time as Dean of the College of Engineering, Dr. James Melsa emphasized the importance of university collaborations with business and industry. Dr. Melsa described his vision for these partnerships: "The idea was to build strategic partnerships with particularly the companies in the state of Iowa, and because we wanted to get places that our students could do coop and intern experiences, we wanted to get their knowledge about how the practice of engineering really happened perhaps to our faculty and our students. We wanted to offer to them, as part of this partnership, excellent educational experiences so that their employees could get advanced degrees, could expand their professional repertoire, and of course we hoped at some point they might donate money back to us" (phone interview, June 6, 2017).
2. A second major theme involved a continuous commitment to high quality programs taught by full time university faculty members with the same quality and rigor as the on-campus programs. From the beginning of distance education work, leaders in the

College of Engineering demonstrated strong support for the creation of distance education programs that paralleled the quality of on campus offerings. This use of full-time faculty as distance education instructors also included faculty development and support programs and structures for participating faculty. Sriram Sundararajan, Associate Dean at the College of Engineering at the time of writing, summarized this commitment, "I guess I would think distance learning is when you're engaging people that are not on your campus with a good learning experience that would be equal to if they were on" (interview, December 1, 2017). Dr. Doug Jacobson, University Professor of ECpE has national recognition for his distance learning programs in Information Security. Since 1995, Jacobson has worked to create educational pathways for a varied workforce and developed curriculum using a truly interdisciplinary approach through his Information Assurance Center. He reinforces this theme for equal rigor and quality: "And again, we work very hard and pride ourselves on the fact that ours are still hands on…the students off campus do the same labs, do the same experiments that our on-campus students do" (interview, September 29, 2017). Several interviewees agreed that continuing to offer high quality, innovative distance education programs in the College of Engineering should include reasonable, strategic incentives for faculty who are contributing to the growth and development of distance education in the College of Engineering.

3. The third major theme that emerged was the importance of identifying and sharing key strong program areas from the College of Engineering. Program leaders emphasized the importance of identifying program niches of excellence and leveraging these programs in distance education efforts. Tom Brumm, ELO's Professor-In-Charge from 2010-2015 stressed "I think you'd be more successful if you said, 'Here's what we're really good at, and we're going to kick butt" (interview, November 10, 2017). Dr. Jacobson extended this point to suggest a strategy for future growth: "I think there is a handful of programs that we are uniquely positioned to offer to the world. We would first identify what those are. Where our market segment would be. And I think that we build an infrastructure beyond just the infrastructure that exists here. That would accommodate a rapid growth. Then I think you need to go out and market and you need to push these things" (interview, September 29, 2017). The areas identified for excellence in the 2018 College of Engineering Strategic Plan ("advanced materials and manufacturing, energy systems, resilient infrastructures, engineered medicine, engineering education, and secure cyberspace and autonomy") provide suggestions for potential distance education focus.

Conclusion and Recommendations

Beginning with the Seed Corn Gospel Train in 1907, land grant Iowa State University has demonstrated a strong commitment to innovative and effective distance education initiatives. For more than 50 years, the College of Engineering has worked to identify areas of need, share high quality programs that address these needs, and ensure quality in distance education offerings. Throughout its history in distance education, the College of Engineering has worked with and supported faculty members to meet the needs of students at a distance. The long history of successful work in distance education provides a strong foundation for future directions. Dr. Ted Okiishi reinforced this point: "There are few things in higher ed where you can kind of look at it and see. We improved. We grew. We're in a position of national leadership now. Why is that? Because a lot of colleges of engineering around the country don't have that sort of a history with this. It tends to be more, 'We'll try this, and then we won't do it for a while, and then we'll try this'" (interview, May 9, 2017).

Taken together, data on the history of distance education in the College of Engineering provide a strong platform for future directions. These data provide clear and focused recommendations for the next steps to continue Iowa State University College of Engineering's history of innovation and excellence in the area of distance education. These recommendations are especially important in the light of the current static rate of growth in distance education in the College of Engineering at ISU.

1. Use areas of excellence within the College and create (or expand) distance education programs from these areas.
2. Work closely with business and industry partners to create and deliver programs meeting the needs of these partners.
3. Provide support for full time faculty members to help them develop effective approaches to teaching at a distance and deliver high quality, rigorous courses.

In conclusion, the data gathered for this paper reveal an interesting and relevant story about a 50 year history of innovative and effective distance education. These data reveal a history characterized by dedicated, innovative leaders and staff, commitment to serving the people of Iowa, commitment to providing rigorous, courses and programs and faculty development and support. Lessons learned from this rich history provide clear direction for the future.

References

Horton, R., Brocklin, L. V. & Jones, E. C. (1997). Challenges of delivering undergraduate engineering programs to place-bound members of the technical work force. *Proceedings Frontiers in Education 1997 27th Annual Conference. Teaching and Learning in an Era of Change*, 2, 588-592. doi: 10.1109/FIE.1997.635868.

Meier, R. D. (1997). Good morning Cedar Rapids. Do you have audio? Reflections on teaching a university distance education course. *Proceedings Frontiers in Education 1997 27th Annual Conference. Teaching and Learning in an Era of Change,* 3, 1469-1472. doi: 10.1109/FIE.1997.632712.

Pounds, D. (2007, June 7). *The Seed Corn Evangelist.* Retrieved June 1, 2018 from https://archive.inside.iastate.edu/2007/0607/holden.shtml.

Appendix A: ELO Timeline

Appendix B: ELO Leaders

(Dates indicate years of service at ELO)

Dr. Charles Lee Townsend, 1966-1985, served as the first director of ELO. He provided the vision, direction, and motivation to extend educational opportunities to working engineers back in 1966 and when video tape technology was introduced in 1968/69, he ushered in video based engineering programs. Townsend was on the faculty at ISU for 37 years and retired as an emeritus professor in 1992. His university activities included considerable committee work at the department, college and university level. He received the Award for Creativity, National University Extension Association, 1969. Charles was recognized as an Outstanding Professor in Electrical Engineering, College of Engineering, in 1986. Dr. Townsend passed away in 2017.

Edwin C. Jones, Jr., 1985-2000, University Professor Emeritus, EDE Professor in Charge, took over the distance education program in 1985 when it was still known as the Professional Programs in Engineering. Jones was also one of the professors teaching courses at Rockwell Collins back in 1966. He expanded the first "studio classrooms" in Engineering Annex created by Charles Townsend. Jones was passionate about creating learning environments that mixed the views of traditional students with those of students working in industry. "Many a times, I have had a student from industry interject our discussion with 'You know, at work yesterday, I saw one of the engineers do …. Why?' This usually leads to a discussion of an important issue that occurs in industry but may be rarely discussed in standard lectures" (phone interview, May 5, 2017).

George Burnet, 1986-1995, Anson Marston Distinguished Professor Emeritus, College of Engineering, George Burnet served in senior administrative positions in the College of Engineering including associate dean and interim dean in the mid-1990s when NTU and ICN distinguished ISU's Engineering Distance Education as a premier program in the country. A former president for the American Society of Engineering Education (ASEE), Burnet was a great champion for distance learning and used his considerable influence and authority to position EDE for success.

Ted Okiishi, 1995-2000, Associate Dean for Research and Outreach, College of Engineering, helped get the Bachelor of Science in Electrical Engineering (BSEE) program off the ground. He was an expert in turbine (jet engine) technology. Ted knew how to work with industry. It was not only technology transfer that interested him, he wanted to prepare students to be innovators when they left ISU for that industry job. This led to work with John Deere in the 1980s. "They wanted a more synchronous delivery like F2F, ICN, videoconferencing. Iowa State delivered live classes to them in the mid 1990s using the ICN to connect to Deere's videoconferencing network" (interview, May 9, 2017).

James L. Melsa, 1995-2004, Dean, College of Engineering, returned to academia from a successful career in business to make a difference. He restructured the engineering college empowering faculty and administrators with resources to achieve and stretch goals. He raised money to build the Engineering Teaching and Research Complex (ETRC) and promoted Virtual Reality Applications Center (VRAC) and Information Assurance as "Best in Class" national programs. Melsa believed partnering with business and industry would help us prepare better future engineers. A key value proposition for industry partners was fostering a lifelong education path through engineering distance education. Jim Melsa's leadership mantra was: *Hire good people, monitor best practices, create lasting partnerships, stay connected.*

Ronald A. Cox, 2003-2005, Interim Director, EDE, provided leadership during installment of a new Dean of Engineering and then again as Associate Dean for Economic Development. Ron Cox was and continues to serve at the time of writing, as Director for The Center for Industrial Research and Service (CIRAS). A history buff, Dr. Cox reminds colleagues of our rich history in engineering extension starting in 1904 with Anson Marston, Iowa State's first engineering dean, creating the Engineering Extension Service (EES) which led to correspondence study and the first distance education programs. Ron Cox continues to work to provide "a simple system" for companies to engage with Iowa State University.

Loren W. Zachary, 2005-2010, Assistant Dean, College of Engineering – EDE Professor-in-Charge, had a passion for the mission and the people of EDE. He focused on finding talented people and creating healthy team dynamics. Dr. Zachary also created policy for distance education that provided incentive for faculty, departments, and the college to continue investing effort into distance education. He was keen on understanding our market, potential for impact, and how to fill the need.

Thomas J. Brumm, 2010-2015, Associate Professor, College of Engineering – ELO Professor-in-Charge, Dr. Tom Brumm is a leader in assessment and competency-based learning at Iowa State and incorporates service-learning activities into engineering and technology curricula. He has extensive experience teaching online and managing online course delivery programs. During his tenure as Professor-in-Charge, the unit merged distance learning operations with the College of Liberal Arts and Sciences (LAS) and changed its name from EOL (Engineering Online Learning) to ELO (Engineering-LAS Online Learning).

Sriram Sundararajan, Associate Dean for Academic Affairs, College of Engineering, ELO Professor-in-Charge, 2015 – Present. Dr. Sundararajan has developed graduate courses on tribology, scanning probe microscopy and broader impacts of engineering. He has also been engaged in efforts to transform the undergraduate mechanical engineering curriculum and introduce engineering concepts to the K-12 community. Dr. Sundararajan is a Fellow of the American Society of Mechanical Engineers (ASME) and serves as an ABET program evaluator. He is an executive committee member of the Mechanical Engineering Division of the American Society of Engineering Education (ASEE).

Chapter 3

Working Engineers' Learning Preferences

Charles E. Baukal, Jr.[1,*] and Lynna J. Ausburn[2]

[1]Koch Engineered Solutions Institute[SM], John Zink Hamworthy Combustion, Tulsa, Oklahoma, USA

[2]Professor Emerita, Oklahoma State University, Stillwater, Oklahoma, USA

Abstract

This chapter presents two major components. First, it establishes a new environment for continuing education that features distance learning resulting from COVID-19, new visual and realistic technologies such as virtual reality, and learner preferences and characteristics. Then the chapter reports an example study in this new line of research using a quantitative descriptive research design. The learner characteristics of preferred learning strategy and verbal-visual preferences of 291 working engineers were studied in relation to the demographics of gender, age, ethnicity, native country, and native language. Learning strategy and verbal-visual preferences were measured with the ATLAS and the Verbal-Visual Learning Style Rating instruments, respectively. The overall learning strategy preference profile for the engineers surveyed was significantly different from the established norms for the general population and was characterized by a preponderance of Problem Solvers. This suggests that learning content for engineers should include problem solving. The overall verbal-visual preference profile was strongly visual and mixed, with little verbal preference. This profile is also statistically different from the general population and suggests learning content for engineers should include highly visual instructional materials. Then the research study (Baukal and Ausburn, 2015; Baukal, 2017) reports on the relationships found between learner characteristics and demographics. Finally, it presents recommendations for instructional practice and future research.

Keywords: continuing engineering education, preferences, working engineers

Introduction and Background

New Learning Environments and Requirements: COVID-19 and Continuing Education

It has long been recognized that effective continuing professional development of all skilled professionals is critical to ensure proper and current skills, procedures, and knowledge are learned to facilitate safe, efficient, and environmentally-friendly operations with minimal unplanned downtime (Baukal, 2012). This requirement is extremely important for continuing

[*] Corresponding Author's Email: charles.baukal@kes.global.

professional development of engineers as they cope with rapidly-changing technological advances.

With the arrival of COVID-19, rapid and dramatic changes arose for engineers in learning environments and techniques to accompany the ever-present advancements in professional content. Decentralized and individual, even isolated, teaching/learning models and technologies suddenly took on new importance, and new technologies such as the highly visual and realistic virtual reality (VR) family of technologies (see Chapter 7) suddenly came into much sharper focus as tools for improving learning at-distance and separated from instructors and fellow learners. In the post-COVID environment of individual, and frequently solitary learning, it soon became apparent that issues of learner isolation, personal engagement and immersion, and clarity or fidelity of content presentation would assume increased importance in the efficacy of continuing education programs. Thus, COVID-imposed restrictions contributed directly to increased importance of new and more detailed research on effective use of already-popular digital visual technologies, particularly VR, to improve continuing education and learning at-a-distance. As this new research on advanced multimedia has developed, it has shifted from early simplistic questioning about its technical and equity- of-access issues; it has shifted from *whether* multimedia and VR work to complex multivariate examinations of *when, why, how, and for whom they work*, bringing multimedia research squarely under the umbrella of individual learner characteristics, differences and preferences (Martens et al. 2019).

VR Characteristics and Theory Base

VR technologies encompass a range of experiences from multi-sensory experiences in enclosed physical environments with complex headgear and sensory suits, to screen-based environments operated on desktop computers, to inexpensive stereoscopic viewers paired with smartphone apps. These technologies share the general definitional characteristics of digital 3D graphic spaces that users can "walk through" in real time, explore at will, and experience with a sense of immersion or actually being somewhere rather than just "seeing it," generally referred to in the research literature as *presence.* (Baukal et al., 2019; International Society for Presence Research, 2019; Dionne et al., 2021). The critical features are real-time user interaction with, and control of, both content and operation. Ausburn et al. (2009) reported that VR and the virtual worlds (VWs) it creates have been demonstrated to present benefits that include not only successful learning performance, but also affective appeal and motivation, transfer to real-world situations, and safe experience in places and situations that are dangerous or even impossible to access in the physical world. Those authors pointed out that, importantly, VR environments can successfully remove locational walls and allow learning and experience to become uncoupled from traditional institutional settings across a wide variety of subject matters and types of learning. An engineering-based example is using VR for training purposes to simulate a fired heater explosion that would be impossible to demonstrate on an operating heater (Baukal et al., 2019). The defining visual and realism features of VR suggest its value for continuing education, particularly at-distance.

Visual VR technologies and VWs also share support from a rich variety of theoretical stances. These theories include Computers as Social Actors (CASA) theory, anthropomorphism, ethopoeia (putting self in the place of another), homophily (tendency to relate to others like us), appearance and personality, knowledge creation theory, social cognitive theory, wayfinding and navigating, individual differences, culture theory, and presence or "being there" theory (Ausburn et al., 2009; Ausburn and Ausburn, 2014; Dionne et al., 2021).

Dale's Cone of Experience and a Proposed Multimedia Cone of Abstraction

Strong theoretical support for visual media such as VR from the perspective of multimedia-based learning stems from Dale's (1969) classic Cone of Experience. Dale's Cone categorizes learning experiences by their relative level of abstractness, ranging from direct real-world objects and experiences (realia) at the supporting base of the cone to the complete abstraction of written words at its apex. Dale's Cone of Experience, shown in Figure 1, proposes that learning benefits from high levels of concreteness or reality in learning experiences, and that concreteness is particularly beneficial when learners are unfamiliar with relevant information content.

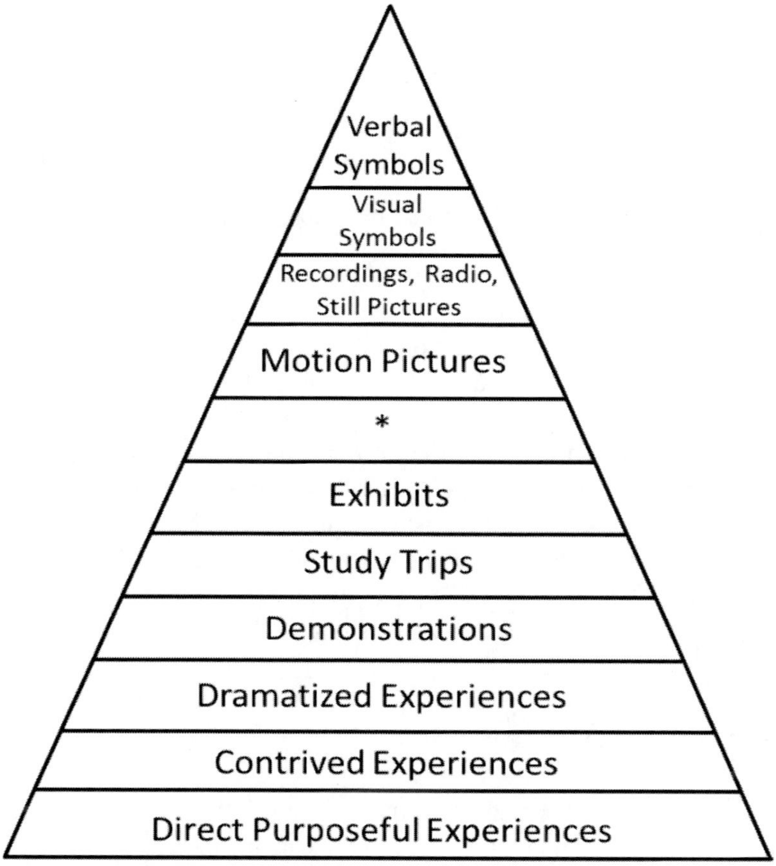

Figure 1. Dale's Cone of Experience (Baukal et al., 2013).

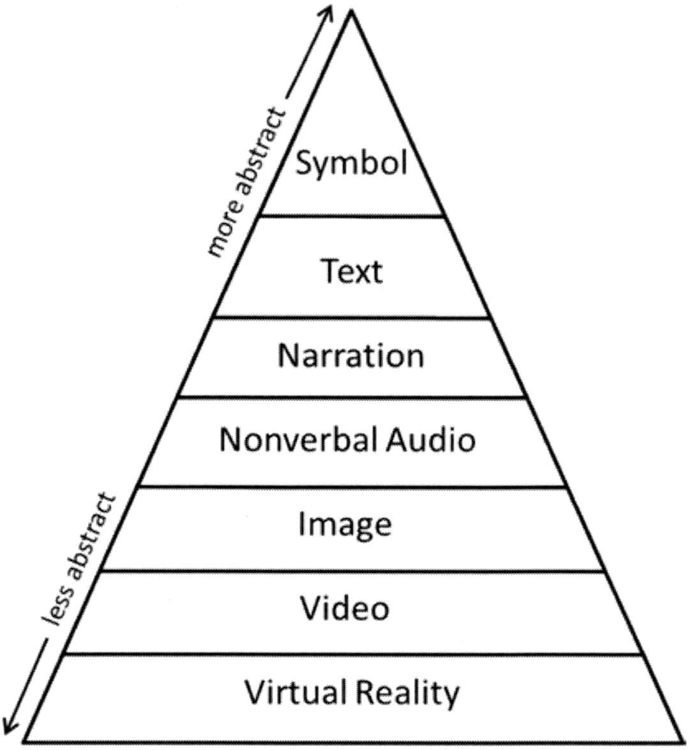

Figure 2. Multimedia Cone of Abstraction (Baukal et al., 2013).

One of the primary characteristics of VR is concreteness or high level of reality. This suggests that the proposition of the learning benefit of concreteness presented in Dale's Cone has relevance in theorizing the efficacy of VR. However, at the time Dale's Cone was developed, current digital media, including VR, were not yet available. Thus, application of the Cone's theory necessitated an update of its components. In a recent re-conceptualization of Dale's Cone, the focus is on new technologies in a Multimedia Cone of Abstraction. In this theory update, shown in Figure 2, the highly visual technology of VR sits at the supporting base of the cone as the most concrete and realistic of media environments (Baukal, Ausburn, & Ausburn 2013), suggesting it may be beneficial for and preferred by some learners and/or for some environments, content and tasks. This may point to a preference for visual learning in occupational groups such as working engineers.

Need and Relevance of the Present Study

One of the challenges of teaching CEE content is the relatively short instruction time to teach a complicated subject (Vandajakumari et al., 2015). This means effective techniques must be employed. One way to address effective professional development design is through attention to *learner preferences*. Research has shown that engineering students have different learner preferences compared to the general population (Baukal et al., 2013). However, in an extensive review of available literature, the authors found no studies reporting specifically on the learner preferences of working engineers. Thus, it appeared that while much research discussed effectively educating engineering university students, very little was available in the literature on the continuing education of working engineers. This is not surprising as engineering disciplines have received relatively little attention from learning sciences researchers (Johri and Olds, 2011). This situation defined the need for the present study of engineers' learning styles and visual/verbal preferences.

Need for this study is supported by the recognized importance of continuing engineering education (CEE). This importance has been well documented and reported by several relevant institutions. A number of U.K. government agencies have recommended CEE (Galloway, 1998). A report by the U.S. National Academy of Engineering (2005) recommended that engineering institutions teach students how to be lifelong learners. An important finding of that study was the importance of addressing *how* students learn in addition to *what* they learn and a call for more research into engineering education. This introduces analysis of learner preferences as an important factor in determining how to better serve individuals with different preferences and how to determine instructional approaches that excite and motivate them. Such an analysis is the focus of the study reported here.

Learning Preferences

Learning Strategy Preference

Learning or cognitive styles, psychological types, and learning strategies are similar concepts that address individual preferences in how people like to, or excel at, perceiving and processing information. *Learning/cognitive style*, discussed at length by Ausburn and Ausburn (1978) and also referred to as *psychological type* by some investigators (McCaulley et al, 1983), refers to how individuals preferentially perceive (e.g., sensory vs. intuitive and verbally vs. visually), organize (e.g., inductive vs. deductive), process (e.g., actively vs. reflectively), and progressively understand (e.g., sequentially vs. globally) (Felder and Silverman, 1988) information. These styles may reflect capability barriers or simple preference; are relatively stable; and concern cognitive, affective and psychological behaviors about how learners perceive, interact with, and respond to a learning environment.

Numerous previous studies have considered learning styles for engineering students. One example is a study of a small sample of engineering students at the University of Texas (Stice, 1987). In that study, Kolb's Learning Style Inventory or LSI (Kolb and Kolb, 2005) comprising four learning styles of convergent, divergent, assimilation, and accommodation (Kolb, 1984) was used to determine the students' learning styles. The overwhelming majority was almost equally split between convergers (learning style characterized by problem solving, decision-making, and practical application of ideas) and assimilators (learning style characterized by inductive reasoning and the ability to create theoretical models). Another example study was done at the University of Cincinnati under a grant from the U.S. National Science Foundation (Rutz and Westheider, 2006). Again, most engineering students were found to be assimilators or convergers. This was comparable to other studies that found the learning styles distribution of engineering students was statistically significantly different from those of the general population. Another example study (Cagiltay, 2008) using Kolb's LSI to determine the learning styles of engineering students at Atilim University in Turkey found that assimilators were predominant. In another study (Hargrave et al., 2008) that also used Kolb's LSI, engineering students at Morgan State University were predominantly assimilators. Larkin-Hein and Budny (2001) gave specific instructional design recommendations for each type of learning style for engineering students. However, Holvikivi (2007) argued that despite its popularity, the use of learning styles testing in engineering education is poorly understood. Another problem with learning styles is that they have been defined and tested in a variety of ways which makes it difficult to compare studies and generalize results (Ausburn and Brown, 2006). An additional problem with learning styles as they are frequently defined and measured by instruments such as the LSI is a lack of practicality and feasibility in designing and implementing differentiated instruction to accommodate various styles. The styles are so conceptually complex that many instructors are reluctant to use them to differentiate their instructional strategies.

A potentially beneficial alternative to the standard definitions and assessments for learning styles is known as *learning strategies*. *Learning strategy preferences*, like traditional learning styles, are important characteristics that vary among learners. Conti and Fellenz (1991, p. 1) defined learning strategies as "techniques or skills that an individual elects to use in order to accomplish a learning task." Learning styles are believed to be stable and deeply ingrained processes for processing information (Ausburn and Ausburn, 1978; Kramer, 2002). In contrast, learning strategies are believed to be less rigid and are more related to personal preferences and choices made by learners during learning tasks (Smith, 1982; Fellenz and Conti, 1993; Conti and Kolody, 1995). Additionally, learning strategies as identified by Conti and his associates can be easily assessed and lend themselves to simple and practical instructional adjustments and alternatives to accommodate the needs of individual learners. Learning strategy preference is thus a potentially important learner variable (Ausburn, 2004) that could be used by instructors to enhance students' learning experiences (Ausburn and Brown, 2006). Relevant to the present study, learning strategy preferences were not found to have been previously identified and discussed for working engineers.

Through a complex and lengthy process, Conti and his associates developed and validated the instrument known as ***A**ssessing **T**he **L**earning Strategies of **A**dult**S*** or ATLAS. This instrument was used in the present study to assess participants' preferred learning strategies. An important advantage of ATLAS is that it is simple to administer and is currently the generally-accepted method for measuring learning strategy preferences as defined by Conti (2009). The ATLAS instrument identifies three distinct learning strategy groups: Navigators, Problem Solvers, and Engagers. *Navigators* plan their learning and focus on completing the necessary activities to achieve their goals. Order and structure are important to these learners, who tend to be logical, objective, and perfectionists. They want clear objectives and expectations at the beginning of a course and in advance of activities, such as in an explicit and detailed syllabus and precisely defined assignments and learning activities. *Problem Solvers* are critical thinkers who like to explore multiple alternatives. For them, the learning process is important so they need flexibility in completing learning activities. They may have difficulty making decisions because finalizing activities means they have to make a choice among multiple viable alternatives and because the exploration process which they enjoy must come to an end. This may cause them to appear to procrastinate in making decisions because they do not want the process to end. *Engagers* are more affective learners who enjoy learning they perceive to be fun or personally beneficial. They are interested in building relationships with both instructors and fellow students during learning, which means they typically enjoy group activities. The emotional aspect of learning is important to Engagers, and this should be included in successfully facilitating and guiding their learning.

The three ATLAS strategy preferences in the general population have been consistently found to be relatively evenly distributed: 36.5% Navigators, 31.7% Problem Solvers, and 31.8% Engagers (Conti, 2009). However, research has also suggested that different professions may have different learning strategy preference profiles. For example, Birzer and Nolan (2002) found that law enforcement personnel had a distinctive profile compared to the general population in a comparison of known population norms to the preferred learning strategies of urban police in a Midwestern city. They also found there were some style profile differences between those who worked in community policing environments and those who did not, with police involved in community policing tending to be Problem Solvers. Ausburn and Brown (2006) studied career and technical education students in numerous occupations and found that most were Engagers. To date there have not been any studies to determine the ATLAS-defined learning strategy preferences of working engineers, the occupational group of interest here.

Verbal-Visual Preference

A major dimension of learning or cognitive style with implications for instructional design is the verbalizer-visualizer preference (Paivio, 1971; Riding, 2001). Unfortunately, there is no consensus on terminology for this dimension as it has been called a cognitive style, a learning style, and a learning preference (Plass et al., 1998). However, it has generally been accepted that "Visualizers tend to think more concretely, use imagery, and personalize information. While learning they prefer graphs, diagrams, or pictures added to text-based material. Verbalizers prefer to process information from words, either by reading or listening, rather than through images" (Jonassen and Grabowski, 1993, p. 191). Learners who have no strong preference for either verbal or visual processing are referred to as *flexible stylists*, also called *bimodal* or *mixed processors*. (Ong and Milcech, 2004). More visual learners may approach learning tasks with visual learning strategies, while more verbal learners may use more verbal strategies (Kirby et al., 1988). When given a choice, verbalizers tend to select more verbal content and visualizers tend to select more visual content (Riding and Watts, 1997). Individual preferences and strengths in the verbal-visual typology have significant implications for designing differentiated instruction and thus may be of considerable interest to those tasked with designing content for continuing engineering education.

Numerous instruments have been developed to measure the verbal-visual cognitive style. Richardson (1977) developed a 15-item questionnaire called VVQ (verbal and visual questions). His research showed 15 to 25% of people tested fell into what he called either habitual verbalizers or habitual visualizers, with the balance in between. He recommended using a distribution of 15% verbalizers and 15% visualizers with the balance in between for research purposes. Felder and Silverman (1988) wrote a frequently-cited paper on learning and teaching styles in engineering education. One of the five dimensions they discussed included visual-auditory. To facilitate verbal-visual style research, a research-based instrument was developed that is a self-scoring 44-item questionnaire called the Index of Learning Styles or ILS (Solomon and Felder, 1991). Montgomery (1995) used the ILS instrument to sample the learning styles of 143 students in an introductory sophomore-level chemical engineering class. She found that 69% were visual, 30% were verbal, and 1% were reported as none. Multimedia software was developed for the course, in part because multimedia software favors visual learners which were the strong majority of the students. Closer to the population of interest in the present study, Rosati (1999) used the ILS to sample a large group ($N = 858$) of engineering students at the University of Western Ontario and found that 80% were visual (89% of males were visual, 69% of females). The verbal-visual preference of the balance of the 20% of the participants was not reported. Kirkham, Farkas, and Lidstrom (2006) found that 85% of the University of Washington engineering students taking a particular class were visual as determined using the ILS. The verbal-visual preference of the balance of the 15% of the participants was not reported.

Despite the research history of other instruments, the verbalizer-visualizer preference was measured in the present study by the Verbal-Visual Learning Style Rating (VVLSR) established by Mayer and Massa (2003). This preference is reported to represent the *perceptual cognitive aspect of adult learning styles*, addressing individual choices in the preferred modality for perceiving and processing information. This instrument was validated against a number of other instruments and was used here because of its simplicity, i.e., a single question, for administration and interpretation. It was used in this study to examine possible relationships between perceptual/cognitive learning preferences and learner demographics.

A Study of Working Engineers' Learning Preferences

Analysis of individual learning preferences could be approached using many different learner variables. The study reported here used two specific learner preference characteristics: (1) preferred learning strategies and (2) preferred verbal-visual presentation media. Demographic variables of participants were also used to describe their learning strategy and verbal-visual preferences.

Study Design, Participants and Methodology

This study used a quantitative descriptive design based on survey methodology (Fowler, 2014), which uses instruments such as questionnaires to collect information from one or more groups of subjects (Ary et al., 2006). Participants ($N = 291$) were working engineers who were either employees from many organizations taking professional development courses at the John Zink Institute between September 2012 and October 2014 ($N = 176$) or employees of the John Zink Company ($N = 115$) participating in a 2014 dissertation research study at Oklahoma State University (Baukal, 2014). These engineers were questioned to determine their demographic characteristics, learning strategy preferences and verbal-visual cognitive styles. Three instruments were used: a demographics questionnaire created by the principal researcher, ATLAS, and the VVLSR. The demographics questionnaire was used to collect personal information including gender, age range, employer, ethnicity, native language, and native country. It should be noted that not all of the surveys used in this study included ethnicity, native language, and native country. Missing data were treated statistically as random and not systematic and are indicated in the tables below. The comprehensive survey incorporating all three individual instruments was given to subjects; it is shown in the Appendix. The surveys were completely voluntary and anonymous and all data were maintained with strict confidentiality and security by the principal researcher.

Table 1. Distribution of sample by gender and age ($N = 291$)

	n	%
Gender		
Female	52	17.9
Male	237	81.4
Missing	2	0.7
Total	291	100.0
Age Range		
<26	49	16.8
26-35	89	30.6
36-45	68	23.4
46-55	46	15.8
56-65	31	10.7
>65	4	1.4
Missing	4	1.4
Total	291	100.0

Table 2. Employers of the respondents ($N = 291$)

	n	%
Industry	288	99.0
Academia/Education	1	0.3
Government	2	0.7
Total	291	100.0

Table 3. Respondents' cultural attributes (ethnicity, native country, and native language) ($N = 291$)

	n	%
Ethnicity		
African American	0	0.0
Asian	24	29.6
Caucasian/White	45	55.5
Hispanic/Latino	7	8.6
Other/Multiple	5	6.2
Total	81	100.0
Missing	209	
Native Country		
U.S.A.	79	45.1
Other	96	54.9
Total	175	100.0
Missing	116	
Native Language		
English	107	61.1
Other	68	38.9
Total	175	100.0
Missing	116	

Paper copies of the survey were given to participants taking continuing education courses offered at the John Zink Institute (www.johnzinkinstitute.edu; now known as the Koch Engineered Solutions Institute). The surveys were distributed to all students in a given class and then collected by the lead researcher or by the Institute Administrator. Paper copies of the surveys were also given to John Zink employees and were collected by assistants rather than by the researchers. All data were then entered into the SPSS statistical program, version 22, for analysis using descriptive statistics and chi-square analyses.

The diversity and distribution of the study sample on the obtained demographic variables are presented in the descriptive tables above. Table 1 shows the gender and age distributions of the sample. Females represented 18% and males 81% of the total sample. Over half of the respondents were between 26 and 45 years of age. As shown in Table 2, nearly all the respondents worked in industry.

Table 3 presents some cultural information about the respondents including ethnicity, native country, and native language. As previously noted, this information was not requested on many of the surveys which is why so many were "missing."

Table 3 indicates diversity among the subjects. Most respondents were Caucasian/White with Asian as the next most reported. In the Other/Multiple category, of those that specified, students reported (1) Arab, (1) Jamaican, (1) Native American, and (1) West Indian. One respondent did not report their Other/Multiple ethnicity. The participants were approximately evenly split between those born in and outside of the U.S.A. In the Other category, students reported native countries of (1) Argentina, (1) Australia, (2) Belgium, (11) Brazil, (11) Canada, (3) China, (1) Columbia, (1) England, (3) France, (2) Germany, (6) India, (3) Italy, (1) Jamaica, (3) Japan, (2) Jordan, (1) Korea, (2) Mexico, (4) Netherlands, (1) Norway, (1) Pakistan, (1) Romania, (1) Saudi Arabia, (2) Singapore, (9) South Africa, (4) Thailand, (1) Trinidad, (3) United Kingdom, (5) Venezuela, (1) Venezuela/Germany, and (2) Vietnam. Six did not report what their Other native country was. For the majority of the participants, English was their primary language. In the Other language category, of those that specified, participants reported (2) Afrikaans, (1) Arabic, (1) Arabu, (4) Chinese, (6) Dutch, (4) French, (2) German, (1) Gujarati, (3) Hindi, (3) Italian, (3) Japanese, (1) Kannada, (3) Korean, (1) Norwegian, (1) Polish, (11) Portuguese, (1) Romanian, (10) Spanish, (4) Thai, (1) Urdu, (1) Vietnamese, and (1) Zulu. Three reporting Other for their native language did not specify what that language was.

Research Problem, Purpose and Questions

Addressing individual differences in how students learn and how to personalize learning options is not new. Many studies using the concept of learning or cognitive style are reported in the published literature. *Learning/cognitive styles*, also referred to as *psychological types* (McCaulley et al., 1983) and *learning strategies* (Conti, 2009), refers to how students prefer to receive and process information (Felder and Brent, 2005). This literature base has addressed a wide variety of populations and learning content, including numerous studies that considered learning styles for engineering students (e.g., Baukal et al., 2013). However, in a lengthy search of the published literature, no studies were found for working engineers. This lack of information and its potentially detrimental effects on lifelong learning for working engineers provided impetus and the research problem for this study. Therefore, the purpose of this study was to describe the learning strategy and verbal-visual preferences of *working engineers*, and the relationships of those preferences to selected demographic variables. To address this purpose, the following research questions guided this study:

1. What are the learning strategy and verbal-visual preference profiles for working engineers?
2. How do the learning strategy and verbal-visual preferences of working engineers compare to the norms for the general population?
3. What are the relationships of working engineers' learning strategy and verbal-visual preferences to the demographic variables of gender, age, employer, ethnicity, native country, and native language?

A final goal of the study was to discuss the findings relating to these questions in a framework of implications for further research and for instructional practice in continuing engineering education.

Results and Discussion

The study's research questions were addressed by the following analyses and findings.

Working Engineers' Learning Strategy Preferences

Table 4 shows data on the learning strategy preferences for the working engineers in this study.

Using chi-square tests of frequency distributions, several statistically significant results were observed for preferred learning strategies of the working engineers. The overall distribution of the three ATLAS strategy groups for the engineers was clearly significantly different ($\chi^2 = 99.172$, $df = 2$, $p = 0.000$) from the general population norms of approximately equal frequencies among the three style groups, established by a lengthy research history by Conti and his associates. The engineers had more problem solvers, fewer navigators, and even fewer engagers than the norms for the general population.

Table 4. Learning strategy preference for engineering sample in total and by gender ($N = 285$)

Gender	Navigator		Problem Solver		Engager		Total	
	n	%	*n*	%	*n*	%	*n*	%
Female	19	36.5	21	40.4	12	23.1	52	18.2
Male	58	24.9	146	62.7	29	12.4	233	81.8
Total	77	27.0	167	58.6	41	14.4	285	100.0
Missing							6	

There were also differences in learning strategy preference distributions by gender among the engineers in this study. The distribution of learning strategy preferences for males was significantly different from that for females ($\chi^2 = 9.074$, $df = 2$, $p = 0.011$). Additionally, the distribution of learning strategy preferences for males was strongly statistically different (χ^2 = 106.502, *df* = 2, *p* = 0.000) from the general population norms, while the distribution for females was not (χ^2 = 2.482, *df* = 2, *p* = 0.289). Thus, the distinctive occupational pattern of dominance of the problem solver style and minimization of the engager style observed among the engineers appeared to be gender-related, occurring more strongly among the males.

Table 5 shows learning strategy preference by age range for this study.

The distributions of learning strategy preferences by age ranges were not significantly different from each other (χ^2 = 8.780, *df* = 2, *p* = 0.553). However, most age ranges presented distributions that were different from general distribution norms for ATLAS. The distribution of learning strategy preferences for ages <26 was not quite statistically different (χ^2 = 5.788, *df* = 2, *p* = 0.055) compared to the general population distribution norms for ATLAS. The distribution of learning strategy preferences for ages 26-35 (χ^2 = 30.220, *df* = 2, *p* = 0.000); 36-45 (χ^2 = 21.210, *df* = 2, *p* = 0.000); 46-55 (χ^2 = 34.253, *df* = 2, *p* = 0.000); and 56-65 (χ^2 = 15.658, *df* = 2, *p* = 0.000) were strongly statistically different compared to the general population. Not enough data were collected for ages >65 to compare against the general population. Thus, it appeared that the occupational pattern of dominance of the problem solver style and minimization of the engager style was generally consistent across age ranges among the working engineers.

Table 6 shows a comparison of learning strategy distributions by ethnicity among the subset of engineers (*N* = 80) for whom this variable was measured.

Table 5. Learning strategy preference by age range (*N* = 283)

Age Range	**Navigator**		**Problem Solver**		**Engager**		**Total**	
	n	%	*n*	%	*n*	%	*n*	%
<26	16	32.7	23	46.9	10	20.4	49	17.3
26-35	27	31.0	50	57.5	10	11.5	87	30.7
36-45	18	26.5	39	57.4	11	16.2	68	24.0
46-55	7	15.9	32	72.7	5	11.4	44	15.5
56-65	7	22.6	20	64.5	4	12.9	31	11.0
>65	1	25.0	2	50.0	1	25.0	4	1.4
Total	76	26.9	166	58.7	41	14.5	283	100.0
Missing							8	

Table 6. Learning strategy preference by ethnicity (*N* = 80)

	Navigator		**Problem Solver**		**Engager**		**Total**	
	n	%	*n*	%	*n*	%	*n*	%
African American	0	0.0%	0	0.0%	0	0.0%	0	0.0
Asian	7	29.2%	11	45.8%	6	25.0%	24	30.0
Caucasian/White	18	40.9%	23	52.3%	3	6.8%	44	55.0
Hispanic/Latino	3	42.9%	3	42.9%	1	14.3%	7	8.8
Other/Multiple	1	20%	1	20%	3	60%	5	6.3
Total	29	36.3%	38	47.5%	13	16.3%	80	100.0
Missing							211	

Table 7. Learning strategy preference by native country and native language ($N = 172$)

	Navigator		Problem Solver		Engager		Total	
	n	%	*n*	%	*n*	%	*n*	%
Native Country								
U.S.A.	27	34.6%	39	50.0%	12	15.4%	78	45.9%
Other	21	22.6%	56	60.2%	16	17.2%	94	54.1%
Total	48	27.9%	95	55.2%	29	16.9%	172	100.0%
Missing							119	
Native Language								
English	35	33.3%	52	49.5%	18	17.1%	105	61.0%
Other	13	19.4%	43	64.2%	11	16.4%	67	39.0%
Total	48	27.9%	95	55.2%	29	16.9%	172	100%
Missing							119	

There were only enough participants in the Asian and Caucasian/White categories for a valid chi-square statistical analysis. There was no statistically significant difference ($\chi^2 = 4.590$, $df = 2$, $p = 0.101$) between those two categories. However, while there was no statistically significant difference ($\chi^2 = 2.215$, $df = 2$, $p = 0.330$) between Asians and the general population, there was a strongly significant difference ($\chi^2 = 14.744$, $df = 2$, $p = 0.001$) between Caucasian/White and the general population. Overall, the results for the issue of relationships of ethnicity to learning strategy preferences among the engineers were inconclusive.

Table 7 shows a comparison of learning strategy distributions by native country and native language among the subset of engineers for whom this variable was measured.

There was no statistically significant difference ($\chi^2 = 3.193$, $df = 2$, $p = 0.203$) between engineers born in the U.S.A. compared to those born outside the U.S.A. There was a strongly significant difference between both those engineers born in the U.S.A. ($\chi^2 = 14.926$, $df = 2$, $p = 0.001$) and those born outside the U.S.A. ($\chi^2 = 33.763$, $df = 2$, $p = 0.000$) compared to the general population ATLAS distribution norms. There was no significant difference ($\chi^2 = 0.241$, $df = 2$, $p = 0.886$) between those whose native language was English compared to those whose native language was not English. There was a strongly significant difference between both those engineers whose native language was English ($\chi^2 = 17.905$, $df = 2$, $p = 0.000$) and those whose native language was not English ($\chi^2 = 32.647$, $df = 2$, $p = 0.000$) compared to the general population. Thus, the occupational pattern of dominance of the problem solver style and mitigation of the engager style appeared to cross cultural boundaries and to be independent of the cultural factors of country of birth and native language.

Working Engineers' Verbal-Visual Preferences.

Table 8 shows the verbal-visual cognitive style preferences for the total sample and by gender for this study.

Overall, the verbal-visual distribution was strongly statistically different ($\chi^2 = 274.800$, $df = 2$, $p = 0.000$) from the one recommended by Richardson (1977) for the general population for research purposes (i.e., 15% verbalizers, 15% visualizers, 70% in between). The distribution for the engineers favored neutral and visual styles, with a very low percentage of verbalizers. Additionally, there was an unexpected statistically significant difference in the distribution of verbal-visual preferences of males compared to females ($\chi^2 = 13.236$, $df = 5$, $p = 0.021$), with females presenting the stronger visual preference. There was a statistically significant difference in the distribution of verbal-visual preferences of males (χ^2 = 185.138, *df* = 2, *p* = 0.000) compared to Richardson's recommended distribution for the general population. There were no females who reported More Verbal so a valid chi-squared analysis could not be done to compare against Richardson's distribution. Overall, it appeared the engineers demonstrated an occupational preference for visual and mixed cognitive styles, and visual preference was particularly evident among the females. This may suggest that females who choose to enter the engineering profession differ from the cognitive style patterns of the more general population of females who often present in available research greater verbal preference than males.

Table 9 shows learning strategy preference by age range for this study.

Table 8. Verbal-visual preference for engineering sample in total and by gender (*N* = 287)

Gender	More Verbal		Neutral		More Visual		Total	
	n	%	*n*	%	*n*	%	*n*	%
Female	0	0.0	19	36.5	33	63.5	52	18.1
Male	12	5.1	114	48.5	109	46.4	235	81.9
Total	12	4.2	133	46.3	142	49.5	287	100.0
Missing							4	

Table 9. Verbal-visual preference by age range (*N* = 285)

Age	More Verbal		Neutral		More Visual		Total	
	n	%	*n*	%	*n*	%	*n*	%
<26	2	4.1	18	36.7	29	59.2	49	17.2
26-35	2	2.3	44	50.0	42	47.7	88	30.9
36-45	5	7.4	35	51.5	28	41.2	68	23.9
46-55	3	6.7	17	37.8	25	55.6	45	15.8
56-65	0	0.0	15	48.4	16	51.6	31	10.9
>65	0	0.0	2	50.0	2	50.0	4	1.4
Total	12	4.2	131	46.0	142	49.8	285	100.0
Missing							6	

Table 10. Verbal-visual preference by ethnicity (*N* = 81)

	More Verbal		Neutral		More Visual		Total	
	n	%	*n*	%	*n*	%	*n*	%
African American	0	0.0	0	0.0	0	0.0	0	0.0
Asian	5	21.7	9	39.1	9	391	23	28.4
Caucasian/White	0	0.0	26	56.5	20	43.5	46	56.8
Hispanic/Latino	0	0.0	5	71.4	2	28.6	7	8.6
Other/Multiple	0	0.0	1	20.0	4	80.0	5	6.2
Total	5	6.2	41	50.6	35	43.2	81	100.0
Missing							210	

The distributions of verbal-visual preferences by age range were not statistically significantly different from each other (χ^2 = 9.155, *df* = 2, *p* = 0.517), demonstrating neutral and visual preference across age groups for the engineering occupation. The distribution of learning strategy preferences for ages <26 (χ^2 = 75.412, *df* = 2, *p* = 0.000); 26-35 (χ^2 = 77.368, *df* = 2, $p = 0.000$); 36-45 ($\chi^2 = 37.049$, $df = 2$, $p = 0.000$); and 46-55 ($\chi^2 = 58.101$, $df = 2$, $p = 0.000$) were all strongly statistically different compared to Richardson's (1977) profile, with minimal verbalizers. No analyses could be done for ages 56-65 and >65 as there were no participants who selected More Verbal.

Table 10 shows a comparison of the verbal-visual preference distributions by ethnicity.

The only groups with enough participants for a valid chi-square statistical analysis were the Asian and Caucasian/White groups which were not significantly different from each other ($\chi^2 = 5.597$, $df = 5$, $p = 0.347$). The Asians sampled were significantly different ($\chi^2 = 12.756$, $df = 2$, $p = 0.002$) than Richardson's profile. A chi-squared analysis could not be conducted on the Caucasian/White group because of the lack of participants selecting More Verbal. Thus, as for the gender demographic, the analysis for verbal-visual preference by ethnicity was inconclusive.

Table 11 shows a comparison of verbal-visual distributions by native country and native language.

There was no statistically significant difference ($\chi^2 = 3.846$, $df = 2$, $p = 0.572$) between those participants whose native country was the U.S.A. compared to those whose native country was not the U.S.A. There was a strong statistical difference for engineers both born in the U.S.A. ($\chi^2 = 57.334$, $df = 2$, $p = 0.000$) and outside the U.S.A. ($\chi^2 = 48.958$, $df = 2$, $p = 0.000$) compared to Richardson's (1977) profile. There was no significant difference ($\chi^2 = 0.888$, $df = 2$, $p = 0.971$) between those participants whose native language was English compared to those whose native language was not English. There was a strong statistical difference for engineers both whose native language is English ($\chi^2 = 58.209$, $df = 2$, $p = 0.000$) and is not English ($\chi^2 = 48.096$, $df = 2$, $p = 0.000$) compared to Richardson's profile. Thus, as for the gender demographic, an occupational preference for neutral and more visual styles appeared to cross cultural lines of nationality and language and to be independent of these cultural factors.

Table 11. Verbal-visual preference by native country and native language ($N = 173$)

	More Verbal		Neutral		More Visual		Total	
	n	%	*n*	%	*n*	%	*n*	%
Native Country								
U.S.A.	1	1.3	43	53.8	36	45.0	80	46.2
Other	6	6.5	50	53.8	37	39.8	93	53.8
Total	7	4.0	93	53.8	73	42.2	173	100.0
Missing							118	
Native Language								
English	4	3.8	59	55.7	43	40.6	106	61.3
Other	3	4.5	34	50.7	30	44.8	67	38.7
Total	7	4.0	93	53.8	73	42.2	173	100.0
Missing							118	

Conclusions and Recommendations

The results of this study point to two major findings. First, the preferred learning strategy profile of the working engineers in the study was significantly different from the general population norms established for ATLAS, and was somewhat different by gender. The profile presented was characterized by strong preference for the problem solver style and minimization of the engager style. This preference was particularly strong among the male engineers. While results for ethnicity were inconclusive due to sampling issues, the preference patterns were clearly observed across age groups and across the cultural lines of birth country and native language. The observed preference for the problem solver learning strategy in the working engineers was not surprising, for, as Sheppard et al. (2009) wrote, "Engineering practice is, in its essence, problem solving." With the exception of gender, the findings for the engineers support Conti's report of no statistically significant differences associated with any demographic variables such as gender or race (Conti, 2009).

Second, similar to the learning strategy profile of the working engineers, the verbal-visual cognitive style preferences presented clear patterns and some differences by gender. Mixed neutral and visual styles were preferred to more verbal ones, and this was most clearly present among the females. Also like the learning strategy profile, the neutral and more visual preferences were observed across age groups and across the cultural boundaries of birth nation and native language.

These findings of clear preference patterns of learning strategies and verbal-visual styles for the working engineers have an important implication for engineering as an occupation. The observed preferences appear to cross the lines of demographics and to characterize the occupation as a whole. Even the gender differences observed are ordinal rather than disordinal in nature, with the same preference patterns observed in both genders, just more strongly in one gender. This leads directly to the possibility that *the engineering occupation may have identifiable learning characteristics that set it apart as a profession and can guide its instructional designers*. The guiding principles may be attributable to the profession (engineering) rather than to most types of demographics. However, it is important that multiple learning strategy and verbal-visual cognitive style preferences were observed among the engineers, supporting a conclusion that preferences are not monolithic in the profession.

This conclusion would have major implications for the instructional strategies used to teach engineers, i.e., implications for the *how* to teach and not *what* to teach. Felder and Silverman (1988) recommended that teachers use techniques to address a range of learning styles to enhance learning. Similarly, Rutz and Westheider (2006) recommended that teachers use a variety of instructional methods to engage all learners. This study would support those recommendations. While the study identified some clear learning preferences among working engineers, it also found individual differences in both learning strategies and cognitive styles. This suggests that a range of techniques should be used with working engineers. It suggests that instructors might expect to have all three ATLAS groups (Navigators, Problem Solvers, and Engagers) in their classes and should understand the learning approaches favored by all three groups and be able to teach effectively to accommodate all of their students. Similarly, instructional designers and instructors for working engineers need to incorporate a full range of verbal-visual preferences to maximize learning in their classes and materials.

However, while variety in learning approaches and materials is both supported and recommended, this study suggests there may be occupational patterns in learning preferences among working engineers that can be used to guide and focus many instructional design decisions. It is recommended that designers and instructors become particularly familiar with the strategies that appeal to and facilitate learners with problem solving styles. These learners need to explore multiple resources and learning paths. They dislike forced choices and multiple choice alternatives, want to seek their own explorations, and may resist closure when they feel there are additional alternatives to investigate. Problem solvers enjoy the *process* of learning, like to work in their own time frames, and may avoid group work if they feel they are being confined by others. They may need the instructor's guidance to avoid wandering off-track and encouragement to seek closure decisions when necessary.

Given the observed preference for mixed and visual cognitive styles, it is also recommended that designers and instructors for working engineers develop their knowledge and skills with a variety of multimedia formats. These formats, including high resolution photography, video, animation, and virtual reality, offer many opportunities to incorporate more visual information into instructional materials. From a theoretical perspective, these highly visual media represent the more concrete/less abstract options proposed as beneficial by both Dale's iconic Cone of Experience and the newer Multimedia Cone of Abstraction presented early in this paper. Several factors relating to engineers' cognitive functioning and complex subject-matter requirements may contribute to this phenomenon, and further research is recommended to examine its details.

This research merely scratches the surface for the learner characteristics of engineers. More characteristics, such as specific multimedia preferences, should be measured in future studies. Nearly all of the participants in the present study worked in industry. More data should be collected for: engineers >55 years old, other employers such as academia/education and government, African Americans (none participated in the present study), and Hispanic/Latinos. Much research needs to be done to better understand the learner characteristics of working engineers, both as a professional group and as individuals; their content nature and complexity; and how to use media options to maximize the effectiveness and efficiency of their continuing education needs and requirements.

References

Ary, D., L.C. Jacobs, A. Razavieh, and C. Sorensen (2006), *Introduction to Research in Education*, Thomson Wadsworth, Belmont, CA.

Ausburn, L.J. (2004), Course Design Elements Most Valued by Adult Learners in Blended Online Education Environments: An American Perspective, *Educational Media International*, 41(4), 327-337.

Ausburn, L.J. and F. Ausburn (1978), Cognitive styles: Some information and implications for instructional design, *Educational Communication and Technology*, 26(4), 337-354.

Ausburn, L.J. and D. Brown (2006), Learning Strategy Patterns and Instructional Preferences of Career and Technical Education Students, *Journal of Industrial Teacher Education*, 43(4), 6-39.

Ausburn, L.J., Martens, J., Dotterer, G., & Calhoun, P. (2009). Avatars, pedagogical agents, and virtual environments: Cooperative learning online. i-manager's Journal of Educational Technology, 5(4), 1-13.

Baukal, C.E. (2012), Strategy for delivering continuing engineering education, *International Journal of Training and Development*, 16(4), 284-299.

Baukal, C.E. (2014), Learning Strategy Preferences, Verbal-Visual Cognitive Styles, and Multimedia Preferences for Continuing Education Instructional Design, Ed.D. dissertation, Oklahoma State University, Stillwater, Oklahoma.

Baukal, C.E. (2017), Learner Preferences for Continuing Engineering Education. Mauritius: Lambert.

Baukal, C. and L. Ausburn (2015), Learner Preferences and Continuing Engineering Education, presented at 2015 American Society for Engineering Annual Conference and Exhibition, Seattle, WA, June 16, 2015, paper 12223.

Baukal, C.E., L.J. Ausburn, J.E. Matsson, and G.L. Price (2013), Engineering Students' Learning Strategy Preferences, 2013 ASEE Midwest Section Conference, Kansas State University (Salina, KS), September 18-20, 2013.

Baukal, C., Olson, B., & Ernst, R. (2019, June). *Virtual reality for continuing professional development.* Paper presented at the American Society for Engineering Education, Tampa, FL.

Birzer, M.L. and R.E. Nolan (2002), Learning strategies of selected urban police related to community policing, *Policing*, *25*(2), 242-255.

Cagiltay, N.E. (2008), Using learning styles theory in engineering education, *European Journal of Engineering Education*, 33(4), 415-424.

Conti, G.J. (2009), Development of a user-friendly instrument for identifying the learning strategy preference of adults, *Teaching and Teacher Education*, *25*, 887-896.

Conti, G.J. and R.A. Fellenz (1991), *Assessing adult learning strategies*, retrieved from ERIC database (ED339847).

Conti, G.J. and R.C. Kolody (1995), The use of learning strategies: An international perspective, *Proceedings of the 36 Annual Adult Education Research Conference*, Edmonton, Alberta, Canada, 77-82.

Dale, E. (1969). *Audiovisual methods in teaching.* New York, NY: Dryden Press.

Dionne, R., Martens, J., Agnew, I., Baukal, Jr., C.E., Ausburn, L., & Ausburn, F. (2021, April). Influence of monitor screen size on immersion and presence in a virtual learning environment. Lightning Round Presentation/Session, Oklahoma Learning Innovations Summit (Online), Council for Online Learning Excellence (COLE), March 31-April 23, 2021.

Felder, R.M. and R. Brent (2005), Understanding Student Differences, *Journal of Engineering Education*, 94(1), 57-72.

Felder, R.M. and L.K. Silverman (1988), Learning and Teaching Styles in Engineering Education, *Engineering Education*, 78(7), 674-681.

Fellenz, R.A. and G.J. Conti (1993), *Self-knowledge inventory of lifelong learning strategies (SKILLS) manual*, Center for Adult Learning Research, Bozeman, MT.

Fowler, F.J. (2014), *Survey Research Methods* (5th ed.), Sage, Los Angeles.

Galloway, S. (1998), The professional body and continuing professional development: New directions in engineering, *Innovations in Education and Training International*, 35(3), 231-240.

Hargrove, S.K., J.A. Wheatland, D. Ding, and C.M. Brown (2008), The Effect of Individual Learning Styles on Student GPA in Engineering Education at Morgan State University, *Journal of STEM Education*, 9(3/4), 37-46.

Holvikivi, J. (2007), Learning styles in engineering education: the quest to improve didactic practices, *European Journal of Engineering Education*, 32(4), 401-408.

International Society for Presence Research (2019). *About presence.* Retrieved from http://ispr.info.

Johri, A. and B.M. Olds (2011), Situated engineering learning: Bridging engineering education research and the learning sciences, *Journal of Engineering Education*, 100(1), 151-185.

Jonassen, D.H. and B.L. Grabowski (1993), *Handbook of Individual Differences*, *Learning*, *and Instruction*, Lawrence Erlbaum, Hillsdale, NJ.

Kirby, J.R., P.J. Moore, and N.J. Schofield (1988), Verbal and visual learning styles, *Contemporary Educational Psychology*, 13(2), 169-184.

Kirkham, P., D.K. Farkas and M.E. Lidstrom (2006), Learning styles data and designing multimedia for engineers, Proceedings of the International Professional Communication Conference, 2006 IEEE, 57-67.

Kolb, D.A. (1984), *Experiential Learning*, Prentice-Hall, Englewood Cliffs, NJ.

Kolb, A. and D.A. Kolb (2005), The Kolb Learning Style Inventory – Version 3.1 2005 Technical Specifications, HayGroup (available at www.hayresourcesdirect.haygroup.com), 2005.

Kramer, C. (2002), *Success in On-Line Learning*, Delmar, New York.

Larkin-Hein, T. and D.D. Budny (2001), Research on Learning Style: Applications in the Physics and Engineering Classrooms, *IEEE Transactions on Education*, 44(3), 276-281.

Martens, L.J., J., Baukal, Jr., C.E., Agnew, I., Dionne, R., & Ausburn, F. (2019). Assembled Issue 2019. User characteristics, trait vs. state immersion, and presence in a first-person virtual world. *Journal of Virtual Worlds Research*, 12(3), 1-20. DOI: https://doi.org/10.4101/jvwr.v12i3.7366.

Mayer, R.E. and L.J. Massa (2003), Three facets of visual and verbal learners: Cognitive ability, cognitive style, and learning preference, *Journal of Educational Psychology*, 95(4), 833-846.

McCaulley, M.H., E.S. Godleski, C.F. Yokomoto, L. Harrisberger, and E.D. Sloan (1983), Applications of psychological type in engineering education, *Engineering Education*, 73(5), 394-400.

Montgomery, S.M. (1995), Addressing diverse learning styles through the use of multimedia, Frontiers in Education Conference, 1995 Proceedings, November, Atlanta, GA, pp. 3a2.13 - 3a2.21.

National Academy of Engineering, *Educating the Engineer of 2020: Adapting Engineering Education to the New Century*, The National Academies Press, Washington, DC, retrieved from http://www.nae.edu/Publications/Reports/25677.aspx, 2005.

Ong, Y.W. and D. Milcech (2004), Comparison of the Cognitive Styles Analysis and the Style of Processing Scale, *Perceptual and Motor Skills*, 99, 155-162.

Paivio, A. (1971), *Imagery and Verbal Processes*, Holt, Rinehart & Winston, New York.

Plass, J.L., D.M. Chun, R.E. Mayer, and D. Leutner (1998), Supporting visual and verbal learning preferences in a second-language multimedia learning environment. *Journal of Educational Psychology*, 90(1), 25-36.

Richardson, A. (1977), Verbalizer-visualizer: A cognitive style dimension, *Journal of Mental Imagery*, 1, 109-126.

Riding, R.J. (2001), The nature and effects of cognitive style. In R. J. Sternberg & L. Zhang (Eds.), *Perspectives on thinking, learning, and cognitive styles* (pp. 47-72), Erlbaum, Mahwah, NJ.

Riding, R.J. and M. Watts (1997), The effect of cognitive style on the preferred format of instructional material, *Educational Psychology*, 17(1-2), 179-183.

Rosati, P. (1999), Specific differences and similarities in the learning preferences of engineering students, Presented at the 29th ASEE/IEEE Frontiers in Education Conference, November, San Juan, Puerto Rico, pp. 12c1-17 – 12c1-22.

Rutz, E. and V. Westheider (2006), Learning Styles of Engineering & Engineering Technology Students – Similarities, Differences and Implications for Effective Pedagogy, paper 2006-419, *Proceedings of the American Society for Engineering Education Annual Conference & Exhibition*, Chicago, IL, June 18-21, 2006.

Sheppard, S.D., K. Macatangay, A. Colby, and W.M. Sullivan (2009), *Educating Engineers: Designing for the Future of the Field*, Jossey-Bass, San Francisco.

Smith, R.M. (1982), *Learning How to Learn: Applied Theory for Adults*, Prentice Hall, Englewood Cliffs, NJ.

Soloman, B.A. and R.M. Felder (1991), Index of Learning Styles Questionnaire. Online: http://www.engr.ncsu.edu/learningstyles/ilsweb.html. Accessed August 16, 2013.

Stice, J.E. (1987), Using Kolb's Learning Cycle to Improve Student Learning, *Engineering Education*, 77(5), 291-296.

Vanajakumari, M., Johnston, K., Lawrence, F., and Menon, R. (2015). An effective Teaching Methodology for Continuing Education Programs. *Journal of Engineering Technology*, 32(1), 18-29.

Appendix: Verbal-Visual & Learning Strategy Preferences Survey

Question	Circle Your Choice/Fill in the Blank
Employer	1 = Industry 2 = Academia/Education 3 = Government 4 = Non-Profit 5 = Other ____________________
Primary Position/Function/Role	1 = Engineer 2 = Operator/Technician 3 = Management/Administration 4 = Teacher/Educator/Professor 5 = Full-Time Student 6 = Other ____________________
Total engineering work experience (including teaching)	______ years
Highest earned engineering degree (e.g., Ph.D. in Mechanical Engineering)	
Gender	1 = Female 2 = Male
Age	1 = <25 2 = 26 – 35 3 = 36 – 45 4 = 46 – 55 5 = 56 – 65 6 = >65
Ethnicity	1 = African American 2 = Asian 3 = Caucasian/White 4 = Hispanic/Latino 5 = Other ____________________
Native Country	1 = U.S.A. 2 = Other ____________________
Native Language	1 = English 2 = Other ____________________
Length of time lived in the U.S.	______ years

Verbal-Visual Preference

In a learning situation, sometimes information is presented verbally (e.g., with printed or spoken words) and sometimes information is presented visually (e.g., with labeled illustrations, graphs, or narrated animations). Please place a check mark indicating your learning preference.

○	○	○	○	○	○	○
Strongly more verbal than visual	Moderately more verbal than visual	Slightly more verbal than visual	Equally verbal and visual	Slightly more visual than verbal	Moderately more visual than verbal	Strongly more visual than verbal

Learning Strategy Preference

Please complete the ATLAS assessment and then mark the result (Place ONE check mark next to your learning strategy preference):

○ **Navigator**	○ **Problem Solver**	○ **Engager**

The description of your learning strategy group in the ATLAS "Groups of Learners (page 2 of the ATLAS instrument) is reasonably accurate in describing you as a learner.

○	○	○	○	○	○	○
Strongly agree	Moderately agree	Slightly agree	Neither agree nor disagree	Slightly disagree	Moderately disagree	Strongly disagree

Chapter 4

Incorporating Learning Styles into Continuing Engineering Education: A Process Heater Case Study

Charles E. Baukal, Jr.[1,*] and Wesley Bussman[2]
[1]Koch Engineered Solutions Institute[SM], John Zink Hamworthy Combustion, Tulsa, Oklahoma, USA
[2]John Zink Hamworthy Combustion, Tulsa, Oklahoma, USA

Abstract

Lifelong learning is critical to engineers' success over the course of their career. Some topics are too specialized to be covered in the typical university engineering program and need to be learned while on the job. The specific example considered here is learning about process heaters used in the hydrocarbon and chemical processing industries (HPI/CPI) to heat hydrocarbon fluids that are being converted into fuels like gasoline and chemicals like ethylene. The Koch Engineered Solutions Institute[SM] (formerly the John Zink Institute[SM]) has been teaching a course on process burners used in those heaters for many years. However, until some time ago, relatively little consideration was given to adjusting the instructional methods to accommodate the various learning styles of the students. This chapter describes how multiple techniques addressing different learning styles are used to help explain a somewhat complicated, but critically important concept in that course, for properly operating process heaters. Failure to properly operate heaters can lead to incidents that can cause equipment damage and personnel injuries, so it is critically important to learn this subject.

Keywords: continuing engineering education, learning styles, process heater

Introduction

Continuing professional education is critical for graduate engineers because of rapid changes in technology (National Research Council, 1985, 1999; National Council of Examiners for Engineering and Surveying, 2004; National Academy of Engineering, 2005). This education may take a variety of forms ranging from short non-credit webinars to advanced graduate degree programs. This chapter focuses on technology-specific short courses that generally last one to five days, eight hours per day. Here, the short course under consideration is on the burners used in process heaters (see Figure 1). It is important to note that most undergraduate chemical and mechanical engineering programs have very little if any content on combustion. Therefore, most graduates who work with fired heater equipment learn the theory and

* Corresponding Author's Email: charles.baukal@kes.global.

applications in their field after graduation, through a combination of on-the-job training, magazines, journals, books, and courses. This chapter focuses on a course that uses a variety of techniques to teach heater control (Baukal and Bussman, 2009).

Figure 1. Photo of a typical process burners short course class.

The Koch Engineered Solutions Institute[SM] has been delivering training on process burners and heaters since 1950. Up until about the late 1990s, the predominant method of instruction was lecture using 35 mm slides. Students received a manual with black and white copies of the slides and a textbook entitled *Furnace Operations* written by the company's Technical Director (Reed, 1981). There was relatively little interaction, other than questions, between the instructor and the students. No instructive simulation models were available at that time. One of the more difficult concepts to understand in the course is how to set the excess O_2 and the draft level in a heater, both of which are critical for proper heater performance. Over the years, multiple techniques for different learning styles have been developed to teach this concept in the course. These are discussed here.

Process Heaters

Process heaters (see Figure 2) are used in the chemical and refining industries to heat hydrocarbon fluids flowing through tubes inside the radiant and convection sections. Most of these heaters are natural draft, which means no fans or blowers are used to supply the combustion air to the burners and to remove the combustion products from the heaters. Heater draft refers to the negative pressure that develops when hot gases rise inside a heater. Draft pulls in the air needed for combustion and pushes out the products of combustion such as CO_2, H_2O, O_2 and N_2. Excess O_2 refers to the amount of oxygen in those products of combustion. Ideally, this excess O_2 should come from the air being pulled through the burners, although some may also come from air leaking into the heater, which is called *tramp air* (Johnson and Baukal, 2021).

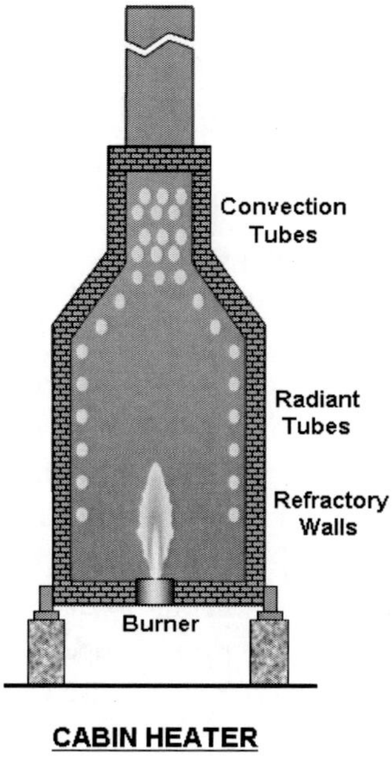

Figure 2. Schematic of a typical process heater.

It is very important that the draft and excess O_2 levels are closely controlled to ensure safe operations, maximize thermal efficiency and productivity, and minimize pollution emissions. Excess O_2 is directly related to the amount of extra air ("excess air") going to a heater beyond what is theoretically needed. Some amount of excess air is needed for complete combustion. If there is not enough air or too much air for combustion, a variety of problems arise. The terms *excess* O_2 and *excess air* are often used interchangeably, which can be confusing to new operators and engineers.

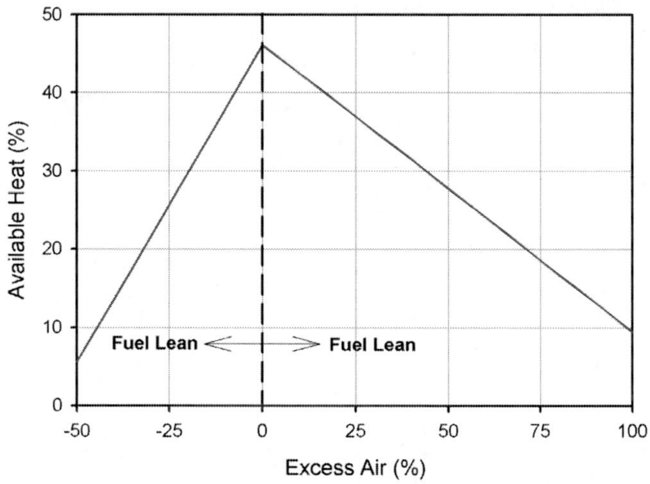

Figure 3. Available heat vs. excess O_2 for methane combusted with ambient air (2000°F flue gas temperature).

Figure 3 shows how thermal efficiency varies with excess air based on theoretical equilibrium calculations for air, simplified as 21% O_2 and 79% N_2 on a volume basis, reacting with methane. The exhaust products are assumed to be at 2000°F. Notice that the thermal efficiency decreases when excess air increases above 0%. This decrease occurs because of the added heat load from the extra O_2 and N_2 in the air that absorbs heat which is carried out with the flue gases. Also notice that the thermal efficiency decreases when the "excess" is below 0%. This decrease occurs because there is insufficient oxygen to completely combust all of the fuel, which means that not all of the chemical energy is converted into thermal energy. Insufficient air is also potentially unsafe because the unburned fuel could be ignited somewhere else inside of the heater. If a heater is significantly starved of combustion air, a condition called heater flooding may exist where there is too much fuel and not enough air. This is a dangerous condition that could lead to a serious incident (Baukal et al., 2020). Another example is tramp air leaking into the convection section, where the unburned fuel could mix with this air, leading to "afterburning." It is not the point of this chapter to discuss all the potential safety problems as it is the operator's responsibility to ensure safe operation. However, it is important that process heaters operate with sufficient oxygen to ensure complete combustion in a controlled manner.

The two most common pollutants regulated in process heaters are carbon monoxide (CO) and nitrogen oxides or NOx (Baukal, 2004b). These pollutants are sensitive to the amount of excess O_2. Figure 4 shows that with increasing excess air levels above 0%, the CO decreases and NOx increases. Ideally, the goal is to minimize both pollutants, which can be challenging since each reacts oppositely to excess air levels. Ultra low NOx burners use a variety of strategies to minimize both pollutants (Baukal, 2004a). High levels of CO indicate incomplete combustion and improper burner operation. High NOx emissions are a problem if they exceed permitted limits. Therefore, heater O_2 levels are typically controlled to levels as low as possible, without forming high levels of CO, to maximize thermal efficiency and minimize NOx emissions. Often, a manually-operated heater will be run with slightly more O_2 than is optimum to account for any variations in ambient air (e.g., temperature and humidity), fuel (e.g., temperature and composition), and process operating conditions (e.g., feed flow rate).

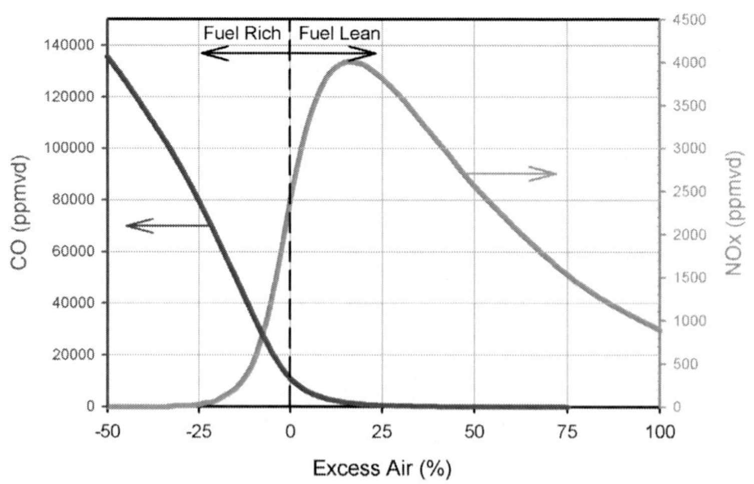

Figure 4. NOx and CO vs. excess O_2 for methane combusted with ambient air.

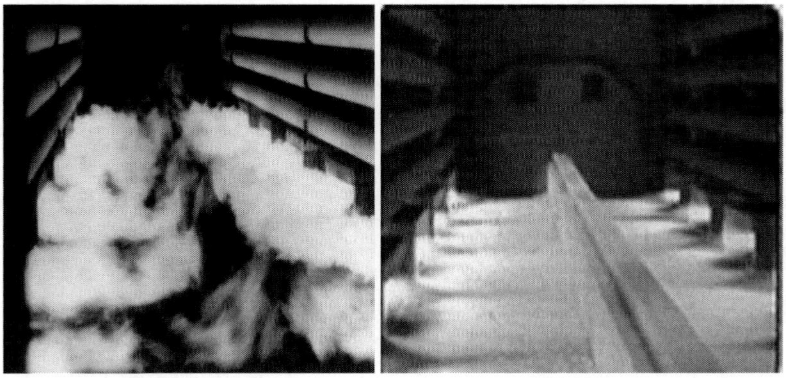

(a) (b)

Figure 5. (a) Before and (b) after adjusting the O_2 and draft in a process heater.

Figure 5 shows burners firing across the floor from both sides of a process heater. The figure on the left shows the heater before the O_2 and draft were properly adjusted. Notice how ill-defined the flames are and how relatively cold the heater is because of incomplete combustion. The figure on the right shows the flames after the O_2 and draft were properly adjusted. The fuel composition and firing rates are essentially the same in both photos. The flames in the right photo are now well-defined and the heater is hotter which typically means more throughput of hydrocarbon fluids. These heater adjustments are the responsibility of the plant operators and underscore the need for proper training and experience. The photos in Figure 5 show how proper adjustments improve flame quality. Training gives operators a clear physical understanding as to why these certain adjustments make the desired improvements. A better understanding of the why behind heater operation can help operators and engineers maintain peak performance over a broader range of operating conditions and allow them to better troubleshoot problems and plan for maintenance and equipment upgrades. In this case, pictures (Figure 5) are worth a thousand words because of the dramatic change with some relatively minor adjustments.

An operator can control the heater draft and excess O_2 by adjusting the damper on the heater exhaust stack, referred to as the stack damper, and on each burner, referred to as the burner damper. Although both the stack and burner dampers impact the draft and O_2, the stack damper is primarily used to control the heater draft, while the burner dampers are primarily used to control the excess O_2 in the heater.

To accommodate a variety of learning styles for teaching these important concepts, a combination of techniques has been developed to teach them to both engineers and operators working in chemical plants and refineries. These techniques include written descriptions, flowcharts, graphs, tables, everyday examples of similar phenomena, examples from actual process heaters, and a heater simulator. The techniques are categorized next according to learning style.

Learning Styles

Most education scholars agree that learners have an identifiable and preferred learning style, although they disagree about how to define or explain those styles (James and Galbraith, 1985).

Some concentrate on the cognitive aspects of learning, some on affective approaches to learning, some on the environmental aspects of learning, and still others on the perceptual modality which is the approach considered here. There are multiple elements that compose the perceptual modality: print, aural, interactive, visual, haptic, kinesthetic, and olfactory. A print-oriented learner learns best through reading and writing. An aurally-oriented person prefers to learn by listening. A learner who likes to talk and discuss ideas would be considered an interactive learner. A visually-oriented learner likes to see pictures, slides, graphs, videos, etc. Haptic learners are "hands on" and prefer to learn through touching and feeling. Kinesthetic learners prefer to move around while learning. Olfactory learners prefer to learn through smelling and tasting. No one style is better than another – they are based on learner preferences.

Some of these preferences are not always relevant in a particular learning situation. For example, haptic and olfactory are only applicable in those circumstances where the object of learning can be touched, tasted, and smelled. Kinesthetic may be limited in the typical classroom as it may be disruptive for learners to be walking around during class which could be distracting to non-kinesthetic learners. However, it may be possible for kinesthetic learners to use their hands or feet in ways that would not be as distracting.

The elements that are specifically considered here are print, aural, interactive, and visual. Each is discussed next. Haptic teaching methods are possible as pilot-scale test heaters equipped with burner and stack dampers are available (Erazo and Korb, 2013). However, all demonstrations in test heaters are conducted strictly by personnel certified to run the pilot heaters, to prevent unsafe conditions from being created.

Print Teaching Methods

Learners that prefer verbal content typically want training materials, such as student manuals and textbooks, that are in writing. In the course described here, students are provided with a copy of the *John Zink Hamworthy Combustion Handbook* (Baukal, 2013) which is a set of three textbooks in full color that provide a written description of excess O_2 and draft. Many of the course instructors have contributed materials to the books, which are primarily used as references for the students. Each student also receives a training manual containing the presentation slides in color. Plenty of space is provided for learners to make notes in the manual. Some of the slides also contain blanks designed to be filled in by the students to help encourage them to stay engaged and keep up with the instructor.

Table 1 is a print device designed to show learners how opening or closing the burner dampers or stack damper impacts the excess O_2 and draft. The color, direction, and size of the arrow indicate how the variables are relatively affected. The upward-pointing green arrow indicates an increase in the variable, while the downward-pointing red arrow indicates a decrease in the variable. The larger the arrow, the more it is impacted by the change. For example, opening the burner damper causes the excess O_2 to increase significantly and the draft to decrease slightly.

Table 1. Effects of burner and stack damper positions on O_2 and draft

Device	Position	O_2	Draft
burner damper	opening	↑	↓
	closing	↓	↑
stack damper	opening	↑	↑
	closing	↓	↓

Some of the slides have a written description, often accompanied by pictures. For example, Figure 6 shows the potential consequences of improper draft in a process heater. The students can read the information, which in this case also shows some pictures from actual heaters that relate to the information.

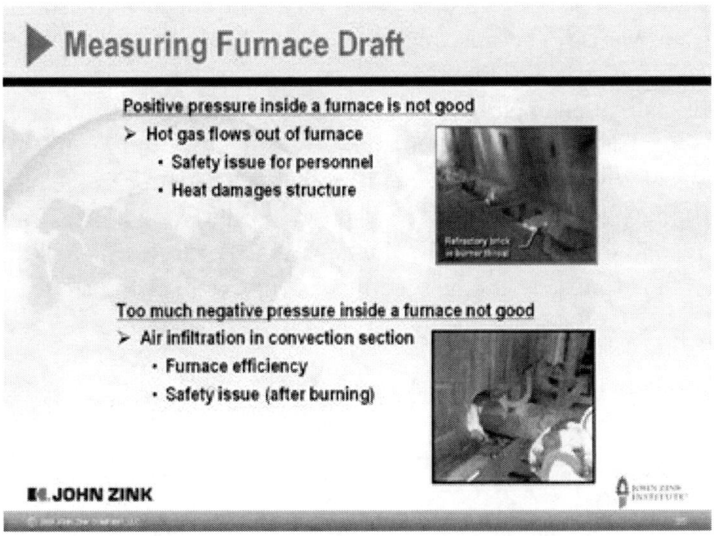

Figure 6. Slide showing the potential consequences of improper heater draft.

Aural Teaching Methods

Lecture is the primary style used by instructors who verbally explain the concepts, in conjunction with the presentation slides, while the students listen. This is the predominant method that most students have been exposed to during most of their previous learning experiences. While this may not be a particular learner's preferred learning style, it is one that all students need to be at least comfortable with because it is the preferred method used by most teachers. Despite its ubiquity, there is a wide range of lecture techniques. A team of instructors are used during this course, so students are exposed to a variety of voices and speaking styles. Some instructors prefer to remain behind a podium and speak with little voice inflection. Long periods of this can make it difficult for even aural learners to stay focused and engaged. Other instructors prefer to move around the classroom and vary their tone and pitch to make it easier for students to pay attention and learn. The latter is the preferred method to teach the concepts of excess O_2 and draft.

Another instructional method that has been incorporated which benefits aural learners is the use of video clips of actual operating equipment. Some of these clips demonstrate the various sounds that burners may make. For example, students can hear the sounds of burners that are not operating properly, such as burners that are flashing back or huffing. Recognizing these sounds can help operators quickly diagnose and troubleshoot problems.

Interactive Teaching Methods

The instructors teaching the concepts of excess O_2 and draft normally use the Socratic technique of asking students specific questions to determine their level of understanding and to promote interaction in the classroom. For example, an instructor may ask students which of the three heater configurations shown in Figure 7 would create the most draft and then ask them to explain why. The answer is (c) because higher temperatures and taller stacks create more draft.

There are also some problems that students are asked to solve to check their understanding of the class materials. An example is shown in Figure 8 where students are asked to discuss how they would calculate the draft that would be created at the bottom of a 100-foot tall stack with exhaust gases at 1000°F, using the chart provided. The graph is the calculated draft per foot for a specific flue gas composition. They simply need to look up the draft per foot for the given temperature, and then multiply it by the height of the stack which in this case is 100 ft. The main point of the exercise is not the specific answer, but to demonstrate that the higher the temperature and the taller the stack, the more draft that is created.

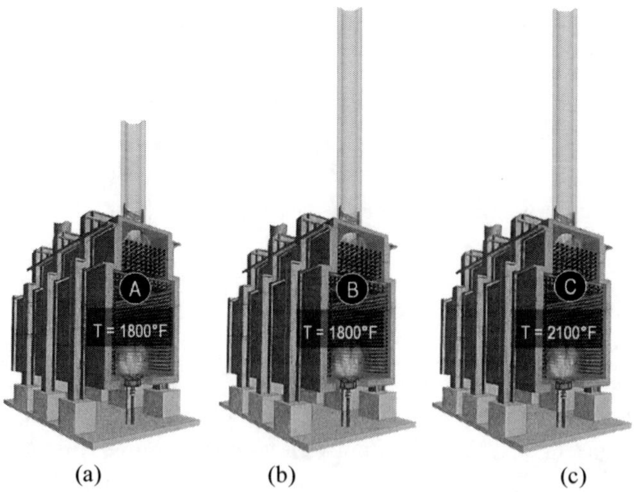

Figure 7. Process heaters: (a) at 1800°F with a short stack, (b) at 1800°F with a tall stack, and (c) at 2100°F with a tall stack.

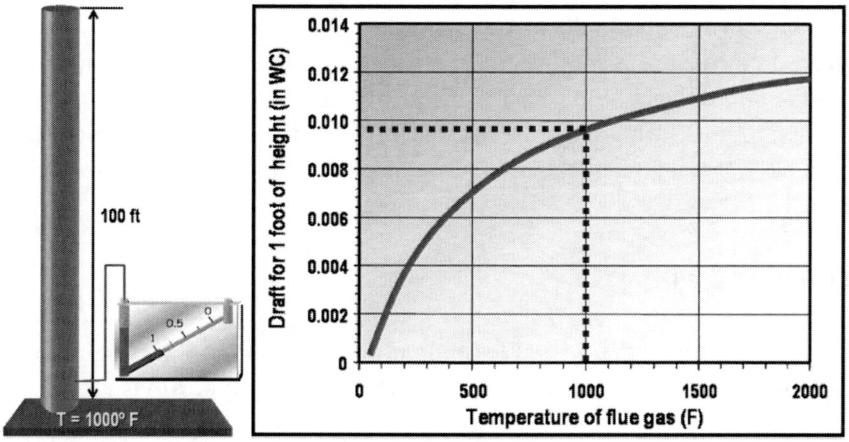

(a) (b)

Figure 8. Sample draft calculation for (a) a 100 ft stack with exhaust gases at 1000°F using (b) chart showing draft per foot of height for a specific flue gas composition at various temperatures.

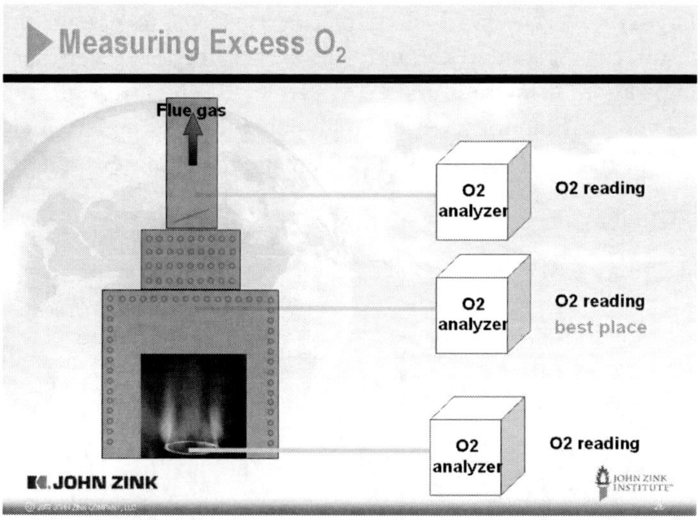

Figure 9. Slide showing multiple possible locations for measuring excess O_2 in a process heater, where the arch (top of the radiant section) is the best place.

Figure 9 shows a slide with a schematic of a process heater and three possible locations for measuring excess O_2. Students are asked to decide which is the best location and why. Measuring the excess O_2 at the floor is problematic because the combustion products may not be well-mixed at that location to give a representative reading. Measuring in the stack would include the tramp air leaking into the convection section (which tends to have high air leakage because of the tubes penetrating through the shell) and not the actual excess O_2 levels in the radiant section. The best location is at the top of the radiant section (often referred to as the arch or bridgewall) just before the convection section where the gases are well-mixed and where tramp air leakage should be minimal. This location is also preferred because it is where the draft should be measured as well so a single probe can be used. Students are also asked where excess O_2 and draft are measured in their heaters. Surprisingly, they are often being measured at some other location. They are instructed that their probes should be moved to the proper location, which again underscores the need for this training. Note that environmental pollution measurements are made in the exhaust stack, so the regulators know what is being emitted into the atmosphere. While that is the correct location to measure emissions, it is not the best location for measuring the draft and excess O_2 as previously discussed. This means the heater should have at least two locations for making important exhaust gas measurements.

Visual Teaching Methods

A study investigating the verbal-visual learning preferences of working engineers found them to be highly visual (Baukal, 2017). They strongly preferred any type of visual, such as pictures, drawings, videos, animations, and virtual reality simulations, compared to text. This suggests that training materials for engineers should be highly visual (see Chapter 3).

Many different visual illustrations are used to explain various aspects of draft and excess O_2. For example, Figure 10 shows a graph of how draft varies with elevation inside a typical process heater. The two most important locations to measure draft are at the floor (burner draft) and at the arch (target draft). The available burner draft is used to size the throat of the burner to ensure enough air can be pulled through the burner for the given operating conditions including fuel composition, firing rate, and excess air level. The lowest draft level in the heater is at the arch. It is even possible for the draft to go positive at the arch, which would cause hot gases to leak out of the heater. This could cause damage to equipment and injury to personnel.

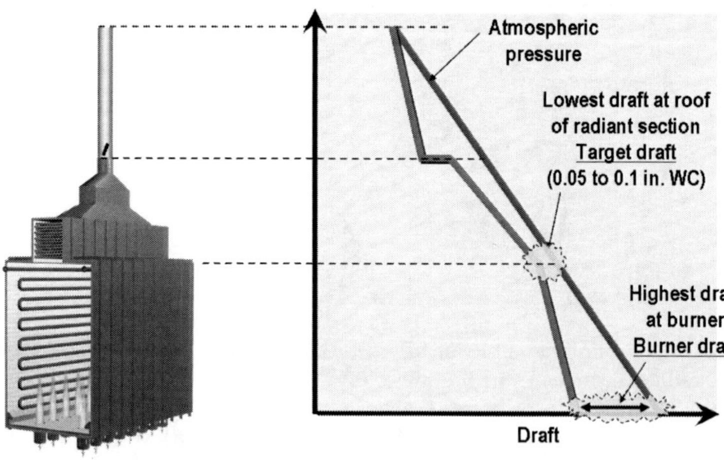

Figure 10. Draft variation with elevation in a process heater.

Figure 11 shows examples of high, normal, and low draft operations. High draft indicates there is too much negative pressure or suction inside the heater. A heater that operates with a high draft level is more susceptible to air leakage or tramp air as compared to one operating at the target draft level. Too much draft can lead to detrimental effects to the heater performance and structure. While the excess O_2 levels measured at the arch may be within acceptable limits, much of the air may not be coming through the burners due to high infiltration levels of tramp air. If this situation occurs, the burner flames may be unacceptable because not enough of the air for combustion is passing through the burners as designed. This could create high concentrations of CO and NOx, as well as extraordinarily long flames that could impinge on the process tubes resulting in a tube rupture. In more extreme cases, this could be dangerous as the flames could go unstable due to lack of combustion air going through the burners. Low draft indicates that the heater pressure may be positive at the arch, causing hot gases to leak out at that level. That could be dangerous for personnel opening a view port to look into the heater as hot gases could be blown into their faces. The graph visually shows what can happen if the draft is not properly set.

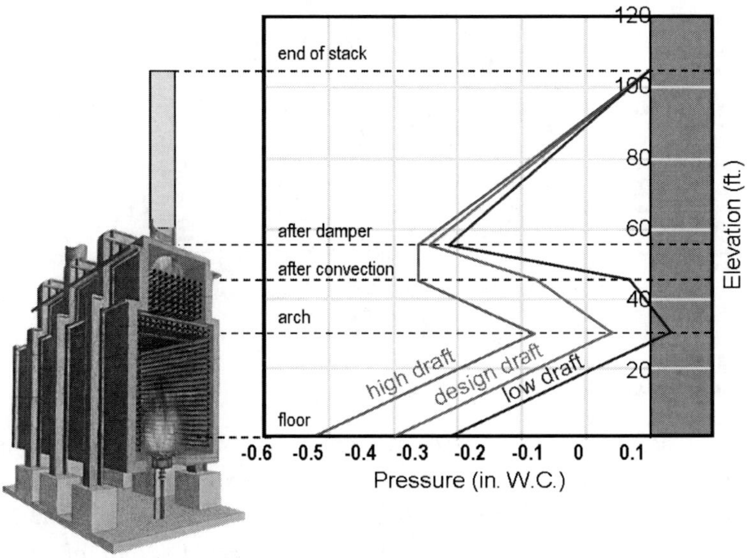

Figure 11. Draft variation with elevation in a process heater.

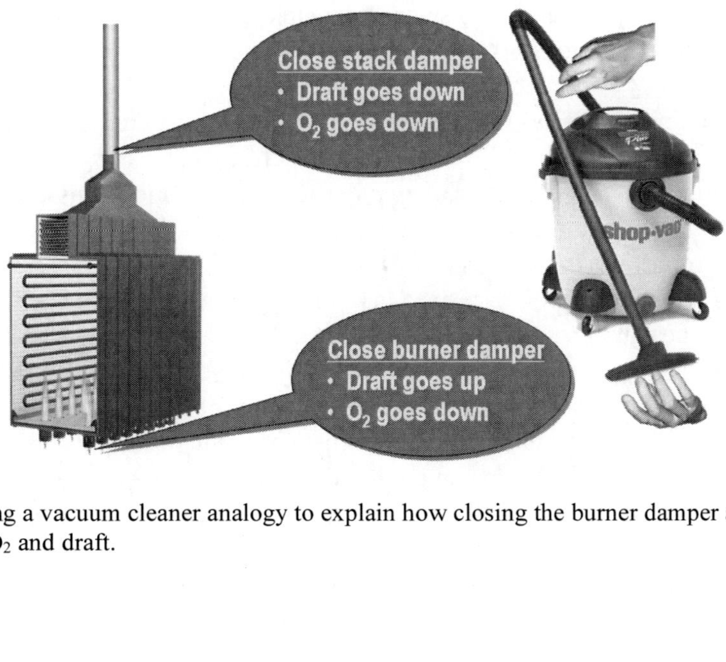

Figure 12. Using a vacuum cleaner analogy to explain how closing the burner damper and the stack damper affect O_2 and draft.

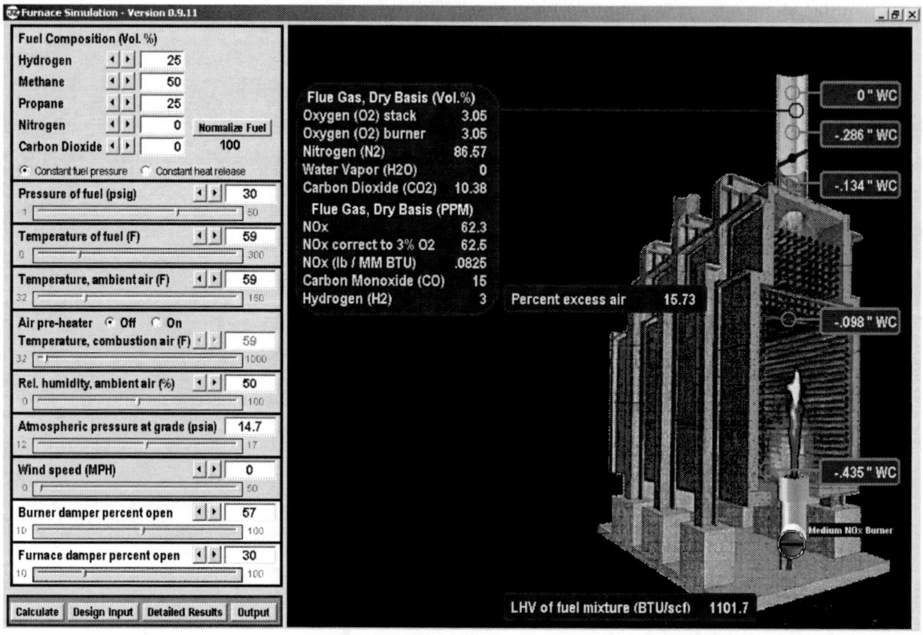

Figure 13. Example screen from heater simulator.

Sparkes (1999) discusses the importance of teaching scientific concepts in conjunction with everyday concepts. Figure 12 shows a common household example using a vacuum cleaner to help explain the effects of closing the burner or stack damper in a process heater. If the vacuum hose is pinched near the inlet to the canister, then both the suction and the flow are reduced. This simulates closing the stack damper, where both the draft and the excess O_2 decrease because the flow decreases. If a hand is placed over the inlet of a vacuum cleaner, the suction increases while the flow decreases. This simulates closing the burner damper which increases the draft while reducing the excess O_2.

Figure 13 shows an example screen from a process heater simulator that was developed to design equipment and which has been adapted for use as a teaching tool. The simulator has numerous inputs that can be varied to demonstrate the effects of a variety of parameters related to combustion, fluid flow, heat transfer, thermal efficiency, pollution emissions, and operating conditions. There are also numerous outputs, which have been limited in the screen shown to focus on specific topics – in this case excess O_2 and draft. The example shown is for a simulated refinery fuel consisting of 50% CH_4, 25% C_3H_8, and 25% H_2 at 30 psig supply pressure. The burner and heater design parameters were set on another screen. The default positions for the burner and stack dampers are wide open (100%). They are both adjusted so the excess O_2 and draft at the arch are approximately 3% (dry basis) and -0.1" water column, respectively. Students see how both variables are affected by changes in the positions of the dampers. If the O_2 level gets too low, a warning pops up on the screen stating that the heater is running out of air. Using a simulator is not only safer than making adjustments on an operating heater, it is also much faster and easier for instructional purposes. Real heaters sometimes take 15 minutes or more to react to changes in damper positions, while the simulator reacts immediately. Also, noise levels are often high around operating heaters which makes it difficult to communicate with a group of students. That is not the case using a simulator in a classroom. Students can safely learn how to tune a heater without affecting plant production. They also do not have to worry about causing an incident which could have severe consequences in the plant. A further benefit of the simulator is that operators can study conditions which may not happen very often which they could not easily do in an actual operating heater.

Conclusion

Continuing education for engineers in HPI/CPI plants often includes process heater operation. Excess O_2 and draft in a heater are critical variables that impact the safety, efficiency, productivity, and pollution emissions of the process. These variables are interrelated and are controlled using the burner and stack dampers. Teaching the how and why of controlling these variables can be challenging because they are not always intuitive and there are many other related factors that need to be considered.

Due to the range of preferred student learning styles, a variety of techniques have been developed to teach these concepts in a continuing education course. The four learning styles specifically considered here to teach those concepts include print, aural, interactive, and visual. For learners who prefer print, a textbook and student manual with color copies of the presentation slides are used. For aural learners, lecture and video clips are used. The Socratic method and classroom problems are used to engage students who prefer interactive learning. Illustrations, video clips, and analogies to common phenomena are used for more visual learners. This combination of approaches is much more effective for learning compared to the earlier methods that relied on almost exclusively on traditional lecture techniques.

References

Baukal, C., Ed. (2004a). *Industrial Burners Handbook*, CRC Press: Boca Raton, FL.

Baukal, C. (2004b). *Industrial Combustion Pollution and Control.* Marcel Dekker: New York.

Baukal, C., Ed. (2013). *John Zink Hamworthy Combustion Handbooks* (3 vols.). CRC Press: Boca Raton, FL.

Baukal, C. (2017). *Learner Preferences for Continuing Engineering Education*. Lambert: Mauritius.

Baukal, C. and W. Bussman (2009). Incorporating Learning Styles into Continuing Engineering Education: A Process Heater Case Study. Paper AC 2009-308, 2009 American Society of Engineering Education Annual Conference and Exhibition, Austin, TX, June 14-17, 2009.

Baukal, C., Pappe, M., Johnson, B., Weimer, B., and McGuire, J. (2020). The hazards of fired heater flooding. *Petroleum Technology Quarterly*, 25(5), 63-67.

Erazo, J. and Korb, T., (2013). Burner Testing. In The John Zink Hamworthy Combustion Handbook, Vol. 2: Design and Operations. C. Baukal (ed.). CRC Press, Boca Raton, FL, 191-201.

James, W. B. and Galbraith, M. W. (1985). Perceptual learning styles: implications and techniques for the practitioner. *Lifelong Learning*, 8(4), 20-23.

Johnson, B. and Baukal, C. (2021). The Problem with Tramp Air. *Hydrocarbon Engineering*, 26(10), 53-56.

National Academy of Engineering (2005). *Educating the Engineer of 2020*, National Academy Press: Washington, DC.

National Council of Examiners for Engineering and Surveying (2004). *Continuing Professional Competency Guidelines*. NCEES: Clemson, SC.

National Research Council (1985). *Engineering Education and Practice in the United States: Continuing Education of Engineers*. National Academy Press: Washington, DC.

National Research Council (1999). *Engineering Education Tasks for the New Century: Japanese and U.S. Perspectives*. National Academy Press: Washington, DC.

Reed, R. (1981). *Furnace Operations*, 3rd Edition. Gulf Publishing: Houston.

Sparkes, J. J. (1999). Learning-centred Teaching, *European Journal of Enginering Education*, 24 (2), 183-188.

Chapter 5

Selecting Multimedia for Continuing Engineering Education

Charles E. Baukal, Jr. [1,*] **and Lynna J. Ausburn**[2]
[1]Koch Engineered Solutions Institute[SM], John Zink Hamworthy Combustion, Tulsa, Oklahoma, USA
[2]Professor Emerita, Oklahoma State University, Stillwater, Oklahoma, USA

Abstract

The need for continuing engineering education (CEE) has been well-documented and is critical for working engineers because of the breadth of processes and equipment they design and use and because of rapid changes in technology. Research has shown working engineers have much different verbal-visual and learning strategy preferences compared to the general population, preferring more visual materials and more problem solving than most learners. They also have strong visual multimedia preferences related to learning. These preferences can be related to the engineers' prior knowledge of the subject being learned. As indicated by Dale's iconic Cone of Experience model, learners with high prior knowledge may prefer less concrete materials, while learners with low prior knowledge usually need more concrete materials. Selecting appropriate multimedia is an important element of effective training. This means CEE should typically be designed differently than traditional continuing professional development which is often highly verbal and less visual, with little problem solving. This chapter discusses the need for CEE and its unique challenges, engineers' learning preferences relating to verbal-visual, learning strategy, and multimedia, the importance of prior knowledge, and instructional design strategies for developing more effective training materials for working engineers.

Keywords: continuing professional development, working engineers, continuing engineering, learning strategy, problem solving, visual cognitive style

Introduction

Mayer (2021, p. 6) defines *multimedia* as "a technology for presenting material in both visual and verbal formats – words and pictures." Mayer has done extensive experimental research into the use of multimedia for learning. He lists three views of multimedia: delivery media (two or more delivery devices), presentation mode (verbal and pictorial representations), and sensory modality (auditory and visual senses). He believes there are two approaches to multimedia design: technology-centered and learner-centered. The goals of each are to provide access to

* Corresponding Author's Email: charles.baukal@kes.global.

information and to aid human cognition, respectively. He suggests two goals of multimedia learning: remembering (ability to reproduce or recognize presented material) and understanding (ability to use presented material in novel situations). He lists three kinds of multimedia outcomes: no learning, rote learning, and meaningful learning. Obviously the goal of CEE should be meaningful learning. He presents detailed theory for enhancing learning with multimedia. There are five representation forms in the cognitive theory of multimedia learning: words and pictures, acoustic and iconic representations, sound and images, verbal and pictorial models, and prior knowledge. He lists three goals of multimedia instructional design: reduce extraneous process, manage essential processing, and foster generative processing.

Mayer defines (p. 117) the *multimedia principle* as follows "People learn better from words and pictures than from words alone." He lists five ways to reduce extraneous processing:

- Coherence – delete extraneous words, sounds, or graphics
- Signaling – highlight essential words or graphics
- Redundancy – delete redundant captions from narrated animation
- Spatial contiguity – place essential words next to corresponding graphics on the screen or page
- Temporal contiguity – present corresponding words and pictures simultaneously

Three principles are given for managing essential processing:

- Segmenting – present lesson in user-paced segments rather than as a continuous unit
- Pre-training – precede lesson with pre-training in the names and characteristics of key components
- Modality – present lesson using pictures and spoken words rather than pictures and printed words

Four principles are suggested to foster generative processing:

- Personalization – present speech in conversational style rather than formal style
- Voice – present speech with a human voice rather than a machine voice
- Embodiment – present lesson by an instructor who uses human-like gestures and movement
- Generative activity – include prompts to engage in generative activities such as summarizing, mapping, drawing, imagining, self-testing, self-explaining, teaching, or enacting.

Applying these research-based principles will make instruction more effective.

Learning Preferences of Working Engineers

Vanajakumari et al. (2015, p. 18) wrote, "The key to the success of a teaching methodology is in understanding the learning behavior of the target audience." This suggests that developing effective CEE relies on understanding the learning behavior of working engineers. Research has shown working engineers have much different learning strategy and verbal-visual preferences compared to the general population, preferring more problem solving and visual materials than most learners (see Chapter 3; Baukal, 2017). This suggests continuing engineering education (CEE) should be designed differently than traditional continuing professional development (CPD) which is often highly verbal and less visual, with little problem solving. To support and facilitate this instructional design shift, this chapter discusses the multimedia learning preferences of working engineers and provides research-based recommendations for instructional designers and instructors for developing more effective training materials for working engineers.

Continuing Engineering Education: Importance, Challenges, Key Terms, and Principles

In many professions, lifelong learning is required to maintain proficiency and in many cases licensure (Baukal, 2012; Bennett & LeGrand, 1990). Mitigation of COVID-19 created a steep rise in the need for effective learning, both formal and informal, at-distance and often solitary. This situation resulted in increased interest in effective use of the cognitive and affective properties of emerging and already-popular visual multimedia such as high-resolution graphics and virtual reality (VR).

Saline (1983, p. 123) defined CEE as "professional technical or management education taken in relatively small doses throughout a career." A report sponsored by the U.S. National Academy of Engineering argued that lifelong learning for engineers is imperative for sustaining American competitiveness in the 21st century (Dutta, Patil, & Porter, 2012). The most recent accreditation requirements for engineering programs include an outcome that students are expected to achieve at the completion of their degree, "a recognition of the need for, and an ability to engage in life-long learning" (Engineering Accreditation Commission, 2017, p. 3).

Given this recognition of the importance of CEE, it is not surprising that its demand continues to increase. There are many reasons for this, including rapid technological advances, globalization, increasing environmental emphasis, and sociopolitical changes. Technical obsolescence (Ovesen, 1980) is a continuing problem due to the ever-increasing pace of changes in technology. Failure to continually learn after receiving an engineering degree leads to obsolescence because of the rapid changes in technology (Barton & Bommer, 1992). For example, the fields of nanotechnology and biotechnology are changing so rapidly that continuous training is required to maintain proficiency in those fields. As technology continuously changes engineering, CEE improves job performance and usually leads to increased compensation (Morris, 1978). It is recommended for quality assurance in a profession, and in some cases, it may be mandated to maintain licensure (Jeris, 2010). Licensure is particularly important in civil engineering because of the many public works projects done for governments where designs must be formally certified by a licensed engineer. Some argue that CEE is critical to enhancing the innovation (Keating, Stanford, Dunlap, Aherne, & Mendelson, 2001) that keeps a professional field viable and its workforce effective and competitive.

To meet their ongoing professional demands, engineers participate in many types of continuing education, including both informal and formal training (Cervero, Miller, & Dimmock, 1986). Employer-sponsored CEE dwarfs that offered by any other type of provider and possibly accounts for more than all other providers combined (Cervero, 2001; see Chapter 11). While much has been written, for example, on CEE offered by universities, relatively little has been written about employer-sponsored CEE (Baukal, 2012) despite its prominence in workforce education and training.

CEE courses include all types of technical (e.g., equipment, processes, software), management (e.g., leadership, employment law, and enhancing team performance), and soft skills (e.g., communication and presentation skills) courses (see Chapter 8). CEE courses combining engineering and management are the most popular types of courses (Beruvides & Ng, 2009). Cole, Moss, Gohs, Lacefield, Barfield, and Blythe (1984) broadly defined four categories of continuing education courses for engineers:

1. remediating and upgrading basic technical knowledge and skills,
2. extending and broadening previously-learned scientific and technical skills,
3. learning new concepts and skills to keep up with technology advancements, and
4. learning new knowledge and skills outside engineering.

CEE is one of the most efficient ways to transfer technology and improve productivity (Markkula, 1985). Besides facilitating maintenance of proficiency, CEE allows engineers to learn about new technologies, whether in their own organization or in others. This might be in the form of traditional classroom instruction, by reading trade journals, or by attending conferences. Productivity can be improved by applying new technologies or taking advantage of existing technologies, processes, or procedures that were not previously used.

The need for CEE has been well-documented (Baukal, 2010). Continuing education is critical for working engineers because of the breadth of processes and equipment they design and use and because of rapid changes in technology (Baukal, 2012). For example, plant engineers take courses to learn how to operate different types of equipment specific to their operations (Baukal & Crawford-Fanning, 2013; Valencia, Link, Baukal, & McGuire, 2008).

Unique Challenges of Continuing Engineering Education

There are some specific challenges unique to CEE. Continuing professional development is particularly important for engineers, compared to other professions, because of the rapid changes in technology (Evetts, 1998). The half-life of knowledge is the time it takes for half of one's knowledge to become obsolete. In some engineering fields, the half-life has become so short, due to rapid changes in technology, that some information is becoming obsolete even before completing an undergraduate education (Wulf & Fisher, 2002). However, it should be noted that the basics taught in many college engineering courses, such as the First Law of Thermodynamics, do not change with time.

Another unique challenge is globalization. New technology is being developed all over the world. It is no longer sufficient to keep up with advancements in one's own country. Today's engineers must be familiar with and be able to apply new technologies developed on a global scale. It is not practical for engineers to travel all over the world to learn about the latest advancements, especially in a pandemic environment. However, advances in learning applications such as group or individual online courses and seminars have made knowledge management easier than ever through the use of new and emerging multimedia, such as virtual reality, that can infuse realism and concreteness into the learning environment.

Learning new technologies can often be more challenging because of the need to use multiple senses to fully understand them. In some cases, this may mean hands-on instruction where students actually manipulate the new technology. For example, learning how to program a controller to automatically adjust a flow rate may require students to actually use the controller. In other cases, instruction may need advanced visualization and animation to aid in understanding the technology. This challenge is not usually faced by, for example, accountants or attorneys learning the latest developments in their fields which are largely text-based. The challenge of constantly-changing engineering technology is now, somewhat ironically, currently being addressed by similarly changing instructional technology. Bork (2004) believed computer-based distance education is the best way to provide continuing education on a global basis. Industry has recognized this challenge and has been a leader in using media-based continuing engineering education (Biedenbach, 1978).

The challenge addressed here is how to design effective CEE to maximize learning for working engineers. Of key importance in meeting the design challenge of CEE is an understanding of how engineers prefer to learn. What defines "good" instructional design for a specified audience is dependent on what defines the learning preferences of that audience. This design reality forces attention on what is known about how engineers prefer to receive and interact with learning content.

Engineers' Learning Preferences

Seaman and Fellenz (1989, p. 17) supported a direct relationship between instructional design and learner preferences and proposed that, "Those who develop or facilitate learning activities for adults should definitely pay attention to the personal preferences of the students...the teacher must select strategies that enable adults to achieve preferences for learning." *Supplantation* refers to customizing instructional design by using instructional modes preferred by learners to facilitate the requirements of specific learning tasks (Ausburn & Ausburn, 1978a,b; Ausburn & Ausburn, 2003; Salomon, 1970). This concept has also been referred to as the *meshing hypothesis* where instructional design is matched to a learner's preferred presentation style (Paschler et al., 2009). More specifically, *conciliatory supplantation* (Ausburn & Ausburn, 1978a) capitalizes on the strengths and preferences of learners. To be applied effectively using the conciliatory supplantation principle, those strengths and preferences must be known by the instructional designer. Three specific learning preference variables will be considered here: (1) learning strategy, (2) verbal-visual, and (3) multimedia.

Learning Strategy Preferences

Learning strategy preferences are important characteristics that vary among learners. Conti and Fellenz (1991, p. 1) defined learning strategies as "techniques or skills that an individual elects to use in order to accomplish a learning task." Through an intensive program of research, Conti and his associates developed and validated an instrument known as *A*ssessing *T*he *L*earning

Strategies of *A*dult*S* or ATLAS to determine learning strategy preferences (Conti, 2009). Three distinct groups were identified: Navigators, Problem Solvers, and Engagers. *Navigators* plan their learning and focus on completing the necessary activities to achieve their goals. *Problem Solvers* are critical thinkers who like to explore multiple alternatives. *Engagers* are more affective learners who enjoy learning they perceive to be fun or personally beneficial. The distribution of the three ATLAS strategy preferences in the general population was established as relatively even: 36.5% Navigators, 31.7% Problem Solvers, and 31.8% Engagers.

Baukal and Ausburn (2016a) studied the learning strategy preferences of working engineers at a medium-sized U.S. manufacturing company. As shown in Figure 1, they found the learning strategy preference profile for their sample was statistically significantly different than that for the general population. There were nearly double the number of Problem Solvers and less than half the number of Engagers in the working engineers' sample compared to the general population. This was a small sample ($N = 118$) which cannot be generalized to the entire population of working engineers, but in the absence of any other data the study suggests that working engineers may prefer problem solving in their CEE.

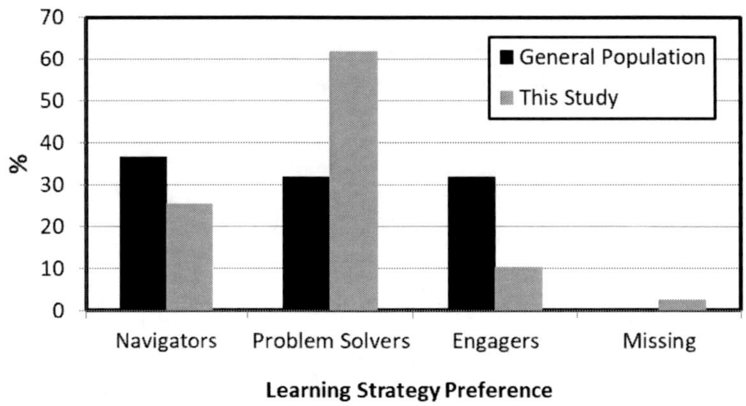

Figure 1. Learning strategy preferences of the general population and for the participants in Verbal-Visual Preferences.

A major dimension of cognitive or learning style is the verbalizer-visualizer dimension (Paivio, 1971; Riding, 2001). According to Jonassen and Grabowski (1993, p. 191), "Visualizers tend to think more concretely, use imagery, and personalize information. While learning, they prefer graphs, diagrams, or pictures added to text-based material. Verbalizers prefer to process information from words, either by reading or listening, rather than through images." Learners who have no strong preference for either verbal or visual processing are referred to as *flexible stylists*, also called *bimodal* or *mixed processors* (Ong & Milech, 2004). Richardson (1977) suggested 15% of the population are habitual verbalizers, 15% are habitual visualizers, and the 70% balance are considered bimodal. Leutner and Plass (1998) and Plass (2004) suggested one-third of the population are verbalizers, one-third are visualizers, and one-third are bimodal.

The *verbalizer-visualizer hypothesis*, sometimes called the *matching hypothesis*, maintains that verbalizers learn more effectively with verbal materials and visualizers with more visual material (Hayes & Allinson, 1996; Kollöffel, 2012; Plass, 2004; Riding & Watts, 1997). "The claim is that when students are matched with their preferred instructional mode, achievement and satisfaction with learning will be enhanced" (Hudak, 1985, p. 402). Plass, Chun, Mayer, and Leutner (1998) and Thomas and McKay (2010) experimentally found strong support for this hypothesis.

Mayer and Massa (2003) developed the Verbal-Visual Learning Style Rating (VVLSR) to measure a participant's verbal-visual preference. It consists of a single question: "In a learning situation, sometimes information is presented verbally (e.g., with printed or spoken words) and sometimes information is presented visually (e.g., with labeled illustrations, graphs, or narrated animations). Please place a check mark indicating your learning preference." This self-assessment measure is reported to represent the *perceptual cognitive aspect of adult learning styles.* This instrument compared favorably with other similar instruments containing many more questions and was used by Baukal and Ausburn (2016b) in a study of working engineers to determine the verbal-visual preferences of their sample. Figure 2 shows the measured profile for the sample in their study was strongly statistically different from Richardson's and Leutner and Plass' recommended profiles for the general population. This suggests working engineers may prefer much more visual CPD compared to the general population.

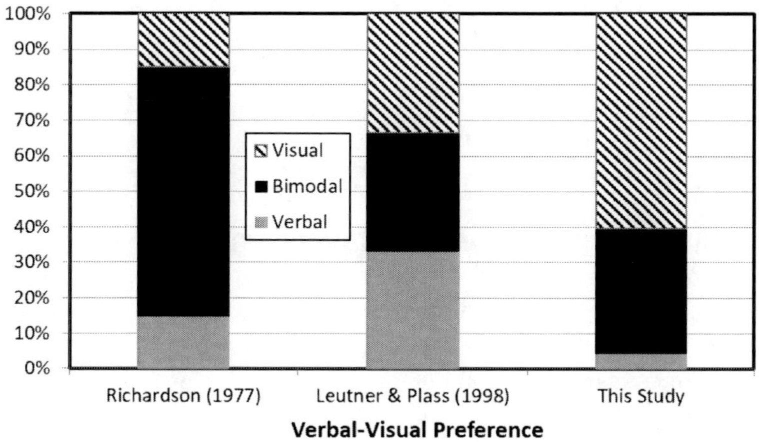

Figure 2. Predicted (from Richardson, 1977 and Leutner & Plass, 1998) verbal-visual preference profiles for the general population compared to the engineering sample studied (Baukal & Ausburn, 2016b).

Multimedia Preferences

There is no current consensus regarding what type of multimedia is best for learning (Baukal & Ausburn, 2016c). In fact, different learners and learning tasks may be better facilitated and supported by different media. In the absence of established guidelines for choosing media formats for learning, *learner preferences* should be an important factor as higher interest may motivate learners and increase learning, depending on learner characteristics and the learning context (Renninger, Hidi, & Krapp, 1992).

In a sample of working engineers, Baukal and Ausburn (2017) averaged five methods used to compare preferences among four multimedia types: text, drawings, animation, and simulated virtual reality (Baukal & Ausburn, 2017). These data, shown in Figure 3, indicate the drawing and simulated virtual reality (VR) multimedia types were approximately equally preferred, animation was only slightly less preferred, and text was much less preferred compared to the other three types. Working engineers clearly indicated a preference for graphic representations over text. This is consistent with the verbal-visual preferences discussed above that show working engineers are highly visual. This suggests CEE should be highly visual compared to many types of CPD which are highly verbal.

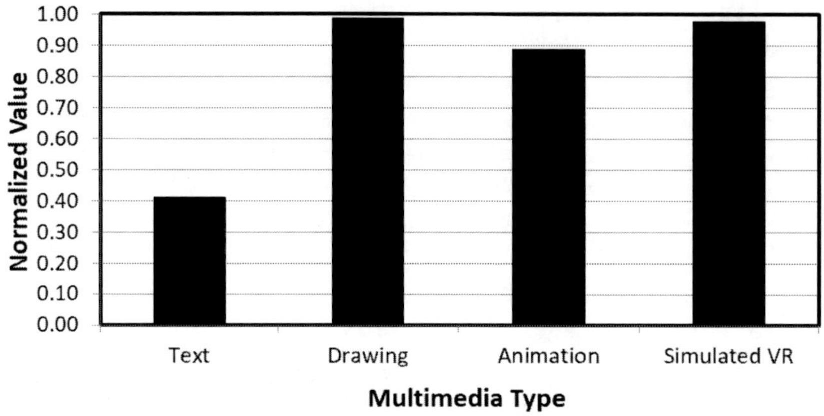

Figure 3. Averages of five normalized comparison methods for the preferences between text, drawing, animation, and simulated virtual reality (VR) multimedia types (Baukal & Ausburn, 2017).

Instructional Selection and Design Recommendations

Recommendations for selecting and designing CCE for working engineers can be gathered from two major sources: (1) empirical data and (2) applicable theory. Both are important in selection/design choices and are offered here for consideration by instructors and designers.

Include Problem Solving

While empirical research on the learning strategies preferred by working engineers has been largely unavailable, a study by Baukal and Ausburn (2016a; see also Chapter 3) reported relevant data on this variable. They found on the ATLAS test of learning strategy preference that working engineers strongly prefered problem solving as a learning strategy. This significantly differentiated the engineers from the general population, which has consistently presented an approximately equal distribution across three learning strategies. It is therefore recommended that effective instructional design for this group should include a significant amount of problem solving examples and activities. This recommendation is supported by additional empirical data from general brain research indicating that including problem solving in learning activities, sometimes referred to as *problem-based learning*, enhances the learning process (Wolfe, 2010).

Include Highly Visual Content

Like using problem solving, using visual content in CEE for working engineers has both empirical and theoretical support. The popular adage "a picture is worth a thousand words" (Larkin & Simon, 1987; Mayer & Gallini, 1990; Mayer & Sims, 1994; Ollerenshaw, Aidman, & Kidd, 1997) was supported by Baukal and Ausburn (2016b; see Chapter 5) in their study of verbal-visual preferences of working engineers. They found the engineers they sampled strongly preferred visual to verbal learning materials. This finding is consistent with other, more general, reported research. According to brain research, more information is taken in visually than by any of the other senses (Wolfe, 2010). One of Bowman's (2011) six learning principles is that images trump words where graphics are recommended over text. These reports suggest that highly graphic media should have a prominent place in CEE for working engineers.

There is also theoretical support for visual instruction. Dale's (1969) iconic Cone of Experience presents his theory graphically as a cone or pyramid of learning experiences (see Chapter 3) with actual real-world activities and items (i.e., realia) at the supporting base, the written word at the apex, and other forms of representation (films, photos, drawings, spoken words, etc.) in between. Dale's theory asserts that the various experiences shown on the cone represent decreasing realism, concreteness or abstraction from bottom to top. Thus, less abstract/more concrete experiences and media are positioned at the supporting base of the cone while more abstract/less concrete representations of reality are at the apex. Further, Dale suggests that the more concrete items, including highly visual media, can facilitate learning, particularly for complex concepts or information with which learners have little knowledge or experience. To update Dale's Cone to include new visual media such as high-resolution photography, realistic graphics, and virtual reality, Baukal, Ausburn, and Ausburn (2013) developed a proposed Multimedia Cone of Abstraction (see Chapter 3), applying Dale's theoretical propositions to new forms of multimedia as well as older ones.

While both empirical learner preference data and theory support using visual instructional materials in CEE for working engineers, this is not to say that only a single type of media should be used. However, as with most things in life, too much of any one thing is often not optimal or even desirable. As the saying goes, "Variety's the very spice of life" (Cowper, 1855, p. 74). This principle suggests a case for variety in multimedia choices in designing CEE. Kemp (1975) supported this point of view and recommended using a variety of multimedia types to meet both different instructional purposes and the various learning styles of students. Clark (1989) also recommended using a variety of instructional media as there is no one type that is inherently better than others. Hayes and Allinson (1996) recommended a variety of learning activities to meet trainees' range of learning style needs.

Using multiple representations can help learners develop deeper understanding (Ainsworth, 1999; McKenna & Agogino, 2004), which is particularly important for significant and challenging subjects. In engineering education, it is common to show an equation (a more abstract type of multimedia) along with a graph (a more concrete type of multimedia) to show how the variables in an equation are related. Concepts can be reinforced using different types of complementary multimedia. For example, a photo can show the outside of a device while a cutaway drawing can be used to show the inside. Using multiple types of multimedia for the same topic will also appeal to a wider range of student learning preferences.

Consider the Possible Effects of Learners' Prior Knowledge on Media Selection

A final concept with theoretical implications for selecting multimedia for CEE for engineers is the relationship between media choices and learners' prior content knowledge. This hypothesis is inherent in both Dale's Cone (Dale, 1969) and the newer Multimedia Cone of Abstraction (Baukal, Ausburn, & Ausburn, 2013). Learners' prior knowledge may be an important factor in designing training presentations, and CEE designers should consider the possibility that prior knowledge and preference variables such as learning strategies and verbal-visual affinity may interact.

In considering potential effects of prior knowledge on learning, it has been noted that novices and experts differ not only in their qualitative and quantitative knowledge of a subject, but also in how they construct knowledge (Litzinger, Lattuca, Hadgraft, & Newstetter, 2011) and that determining learners' prior knowledge is an important step in instructional design (Mayer, 2009). Instructional materials need to be tailored to learners' knowledge levels, which is referred to as the *prior knowledge principle* (Kalyuga, 2005; Naryanan & Hegarty, 2002). Learners with little prior knowledge in a subject area can be more easily cognitively overloaded compared to experts (Cook, 2006).

Baukal and Ausburn (2016d) empirically examined the relationship between prior knowledge and learning preferences in a sample of working engineers. While they found no statistically significant relationship between these variables, further research is recommended as it may continue to support a general recommendation that overall more concrete multimedia, such as pictures and video, are more appropriate for more novice learners, and more abstract multimedia, such as text, may be preferable for more expert learners (Baukal, Ausburn, & Ausburn, 2013).

Conclusion

Research on the learning preferences among working engineers and selecting multimedia for their CEE has to date been limited. However, from the limited amount of research available on learning and media for working engineers, some implications are clear. Available empirical evidence suggests these learners have strong learning strategy preferences, demonstrating affinity for problem solving learning strategy and visual multimedia presentation. Much more research needs to be done in this area to determine if engineers' learning preferences vary by demographics such as gender, age, engineering discipline, industry, and culture to determine if content needs to be customized accordingly.

Theoretical considerations suggest another important factor in choosing multimedia for engineering CEE may be the relative abstractness/concreteness of various media and the relationship between this characteristic and learners' previous content knowledge. Here too, further research is needed and recommended.

Through research and theory development, much may be discovered regarding working engineers' learning preferences and characteristics and the design or selection of appropriate multimedia. This knowledge can facilitate development of guidelines for designers and instructors to maximize learning for engineers through the CEE that is critical for their professional development and competence.

References

Ainsworth, S. (1999). The functions of multiple representations. *Computers & Education*, *33*(2-3), 131-152.

Ausburn, L. J., & Ausburn, F. B. (1978a). A supplantation model for instructional design: Investigation of a behavioural science approach. *Australian Journal of Education*, *22*(3), 277-294.

Ausburn, L. J., & Ausburn, F. B. (1978b). Cognitive styles: Some information and implications for instructional design. *Educational Communication and Technology*, *26*(4), 337-354.

Ausburn, L. J., & Ausburn, F. B. (2003). A comparison of simultaneous vs. sequential presentation of images in a visual location task to learners with visual and non-visual perceptual styles: A study of supplantation instructional design. *Journal of the Oklahoma Association of Teacher Educators*, *7*, 1-20.

Barton, B., & Bommer, M. (1992). Myths about engineers maintaining professional technical competence. *International Journal of Continuing Engineering Education and Lifelong Learning*, *2*(1), 22-30.

Baukal, C. E. (2010). Continuing engineering education through distance learning, *European Journal of Engineering Education*, *35*(2), 225-233.

Baukal, C. E. (2012). Strategy for delivering continuing engineering education. *International Journal of Training and Development*, *16*(4), 284-299.

Baukal, C. E. (2017). *Learner Preferences for Continuing Engineering Education*. Mauritius: Lambert.

Baukal, C. E., Ausburn, F., & Ausburn, L. J. (2013). A proposed multimedia Cone of Abstraction: Updating a classic instructional design theory, *Journal of Educational Technology*, *9*(4), 15-24.

Baukal, C., & Ausburn, L. (2016a). Learning strategy preferences of working engineers. *International Journal of Continuing Engineering Education and Lifelong Learning*, *6*(4), 52-69.

Baukal C., & Ausburn, L. (2016b). Verbal-visual preferences of working engineers. *European Journal of Engineering Education*, *41*(6), 660-677.

Baukal, C. & Ausburn, L. (2016c). Multimedia category preferences of working engineers, *European Journal of Engineering Education*, 41(5), 482-503.

Baukal, C., & Ausburn, L. (2016d). Relationship of prior knowledge and working engineers' learning preferences: Implications for designing effective instruction. *European Journal of Engineering Education*, 42(3), 302-322.

Baukal C., & Ausburn, L. (2017). Multimedia Type Preferences of Working Engineers, *Australasian Journal of Engineering Education*, 22(2), 63-72. doi: 10.1080/22054952.2017. 1392225, 2017.

Baukal, C. E., & Crawford-Fanning, M. N. (2013). Combustion training. In *The John Zink Hamworthy Combustion handbook: Fundamentals*, edited by C. E. Baukal, 513-550. Boca Raton, FL: CRC Press.

Bennett, N. L., & LeGrand, B. F. (1990). *Developing continuing professional education programs*. Urbana, IL: Illinois University.

Beruvides, M. G., & Ng, E-H. (2009). The continuing education of engineers and engineering managers. *International Journal of Continuing Engineering Education and Lifelong Learning*, *19*(1), 4-18.

Biedenbach, J. M. (1978). Media-based continuing engineering education. *Proceedings of the IEEE*, *66*(8), 961-968.

Bork, A. (2004). A new lifelong learning system for the world. *International Journal of Continuing Engineering Education*, *14*(1/2), 46-57.

Bowman, S. (2011). *Using brain science to make training stick: Six principles that trump traditional teaching*. Glenbrook, NV: Bowperson.

Cervero, R. M. (2001). Continuing professional education in transition, 1981-2000. *International Journal of Lifelong Education*, *20*(1/2), 16-30.

Cervero, R. M., Miller, J. D., & Dimmock, K. H. (1986). The formal & informal learning activities of practicing engineers. *Engineering Education*, *77*(2), 112-114.

Clark, R. C. (1989). *Developing technical training: A structured approach for the development of classroom and computer-based instructional materials*. Reading, MA: Addison-Wesley.

Cole, H. P., Moss, J., Gohs, F. X., Lacefield, W. E., Barfield, B. J., & Blythe, D. K. (1984). *Measuring learning in continuing education for engineers and scientists*. Phoenix: Oryx Press.

Conti, G. J. (2009). Development of a user-friendly instrument for identifying the learning strategy preference of adults. *Teaching and Teacher Education*, *25*, 887-896.

Conti, G. J., & Fellenz, R. A. (1991). *Assessing adult learning strategies*. Retrieved from ERIC database (ED339847).

Cook, M. P. (2006). Visual representations in science education: The influence of prior knowledge and cognitive load theory on instructional design principles. *Science Education*, *90*(6), 1073-1091.

Cowper, W. (1855). *The task*. London: James Nisbet and Company.

Dale, E. (1969) *Audiovisual Methods in Teaching.* New York: NY, Dryden Press.

Dutta, D., Patil, L., & Porter, J. B. (2012). Lifelong learning imperative in engineering: Sustaining American competitiveness in the 21st century. Washington, DC: National Academies Press.

Engineering Accreditation Commission. (2017). Criteria for accrediting engineering programs: Effective for reviews during the 2017-2018 accreditation cycle. Baltimore, MD: Accrediting Board for Engineering and Technology, http://www.abet.org/accreditation/accreditation-criteria/criteria-for-accrediting-engineering-programs-2017-2018/#outcomes [Accessed November 22, 2017].

Evetts, J. (1998). Continuing professional development for engineers: UK and European dynamics. *European Journal of Engineering Education*, *23*(4), 443-452.

Hayes, J., & Allinson, C. W. (1996). The implications of learning styles for training and development: A discussion of the matching hypotheses. *British Journal of Management*, *7*(1), 63-73.

Hudak, M. A. (1985). Learning styles inventory. In D. J. Keyser & R. C. Sweetland (Eds.), *Test critiques*. Kansas City: Test Corporation of America.

Jeris, L. H. (2010). Continuing professional education. In C. E. Kasworm, A. D. Rose, & J. M. Ross-Gordon (Eds.), *Handbook of adult and continuing education* (pp. 275-283). Los Angeles: Sage.

Jonassen, D. H., & Grabowski, B. L. (1993). *Handbook of individual differences, learning, and instruction*. Hillsdale, NJ: Lawrence Erlbaum.

Kalyuga, S. (2005). Prior knowledge principle in multimedia learning. In *Cambridge handbook of multimedia learning*, edited by R. Mayer, 325-337. New York: Cambridge University Press.

Keating, D. A., Stanford, T. G., Dunlap, D. D., Aherne, M. J. & Mendelson, M. I. (2001, June). Enhancing U.S. Technology Development Through Lifelong Education of Engineers and Technologists as Creative Professionals. Proceedings of the 2001 ASEE Annual Convention. Albuquerque, NM.

Kemp, J. E. (1975). *Planning and producing audiovisual materials* (3rd ed.). New York, NY: Thomas Y. Crowell.

Kollöffel, B. (2012). Exploring the relation between visualizer-verbalizer cognitive styles and performance with visual or verbal learning material. *Computers & Education*, 58, 697-706.

Larkin, J. H., & Simon, H. A. (1987). Why a diagram is (sometimes) worth ten thousand words. *Cognitive Science*, *11*(1), 65–99.

Leutner, D., & Plass, J. L. (1998). Measuring learning styles with questionnaires versus direct observation of preferential choice behavior in authentic learning situations: The Visualizer/Verbalizer Behavior Observation Scale (VV-BOS). *Computers in Human Behavior*, *14*(4), 543-557.

Litzinger, T. A., Lattuca, L. R., Hadgraft, R. G., & Newstetter, W. C. (2011). Engineering education and the development of expertise. *Journal of Engineering Education*, *100*(1), 123-150.

Markkula, M. (1985). Continuous studies as a basis for professional development. *European Journal of Engineering Education*, *10*(3/4), 221-227.

Mayer, R. E. (2009). Advances in specifying what is to be learned: Reflections on the themes in chapters 6-8. In *Development of professional expertise: Toward measurement of expert performance and design of optimal learning environments*, edited by K. A. Ericsson, 203-211. Cambridge, UK: Cambridge University Press.

Mayer, R. E. (2021). *Multimedia Learning*, 3rd ed. New York: Cambridge University Press.

Mayer, R. E., & Gallini, J. K. (1990). When is an illustration worth ten thousand words? *Journal of Educational Psychology*, *82*(4), 715-726.

Mayer R. E., & Massa, L. J. (2003). Three facets of visual and verbal learners: Cognitive ability, cognitive style, and learning preference. *Journal of Educational Psychology*, *95*(4), 833-846.

Mayer, R. E., & Sims, V. K. (1994). For whom is a picture worth a thousand words? Extensions of a dual-coding theory of multimedia learning. *Journal of Educational Psychology*, *86*, 389-401.

McKenna, A. F., & Agogino, A. M. (2004). Supporting mechanical reasoning with a representationally-rich learning environment. *Journal of Engineering Education*, *93*(2), 97-104.

Morris, A. J. (1978). *The return on investment in continuing education of engineers*. Palo Alto, CA: Genesys Systems, Inc.

Narayanan, N. H., & Hegarty, M. (2002). Multimedia design for communication of dynamic information. *International Journal of Human-Computer Studies*, *57*, 279-315.

Ollerenshaw, A., Aidman, E., & Kidd, G. (1997). Is an illustration always worth ten thousand words? Effects of prior knowledge, learning style and multimedia illustrations on text comprehension. *International Journal of Instructional Media*, *24*(3), 227-238.

Ong, Y. W., & Milcech, D. (2004). Comparison of the cognitive styles analysis and the Style of Processing Scale. *Perceptual and Motor Skills*, *99*, 155-162.

Ovesen, N. K. (1980). *Advances in the continuing education of engineers*. Paris: United Nations Educational, Scientific and Cultural Organization (UNESCO).

Paivio, A. (1971). *Imagery and verbal processes*. New York: Holt, Rinehart & Winston.

Paschler, H., McDaniel, M., Rohrer, D., & Bjork, R. (2009). Learning styles: Concepts and evidence. *Psychological Science in the Public Interest*, *9*(3), 105-119.

Plass, J. L. (2004). Learner preferences and cognitive abilities in multimedia learning. Presented at ED-MEDIA 2004, World Conference on Educational Multimedia, Hypermedia and Telecommunications, Lugano, Switzerland, June 21-26, 4255-4261.

Plass, J. L., Chun, D. M., Mayer, R. E., and Leutner, D. (1998). Supporting visual and verbal learning preferences in a second-language multimedia learning environment. *Journal of Educational Psychology*, *90*(1), 25-36.

Renninger, K. A., Hidi, S. & Krapp, A. (Eds.). (1992). *The role of interest in learning and development*. Hillsdale, NJ: Lawrence Erlbaum Associates.

Richardson, A. (1977). Verbalizer-visualizer: A cognitive style dimension. *Journal of Mental Imagery*, *1*, 109-126.

Riding, R. J. (2001). The nature and effects of cognitive style. In *Perspectives on thinking, learning, and cognitive styles*, edited by R. J. Sternberg and L. Zhang, 39-59. Mahwah, NJ: Erlbaum.

Riding, R. J., & Watts, M. (1997). The effect of cognitive style on the preferred format of instructional material. *Educational Psychology*, *17*(1-2), 179-183.

Saline, L. E. (1983). Continuing engineering education: One element of lifelong learning for engineers. *IEEE Transactions on Education*, *E-26*(4), 122-126.

Salomon, G. (1970). What does it do to Johnny? A cognitive-functionalistic view on media. *Viewpoints*, *46*(5), 33-62.

Seaman, D. F., & Fellenz, R. A. (1989). *Effective strategies for teaching adults*. New York: Merrill.

Thomas, P. R., & McKay, J. B. (2010). Cognitive styles and instructional design in university learning. *Learning and Individual Differences*, *20*, 197-202.

Valencia, R., Link, D., Baukal, C., & McGuire, J. (2008). Consider classroom training for plant operators. *Hydrocarbon Processing*, *87*(11), 55-59.

Vanajakumari, M., Johnston, K., Lawrence, F., and Menon, R. (2015). An effective Teaching Methodology for Continuing Education Programs. *Journal of Engineering Technology*, 32(1), 18-29.

Wolfe, P. (2010). *Brain matters: Translating research into classroom practice* (2nd ed.). Alexandria, VA: ASCD.

Wulf, W. A., & Fisher, G. M. C. (2002). A makeover for engineering education. *Issues in Science & Technology*, *18*(3), 35-39.

Chapter 6

Effective Instructional Practices in Continuing Engineering Education

Howard S. Kimmel[1,*] (Ret.), Ronald H. Rockland[2] (Ret.), John D. Carpinelli[3] and Gale T. Spak[4] (Ret.)

[1]Department of Chemical Engineering, New Jersey Institute of Technology, Newark, NJ, USA
[2]School of Applied Engineering and Technology, New Jersey Institute of Technology, Newark, NJ, USA
[3]Department of Electrical and Computing Engineering, New Jersey Institute of Technology, Newark, NJ, USA
[4]Division of Continuing Professional Development, New Jersey Institute of Technology, Newark, NJ, USA

Abstract

Currently, higher educational programs have been/are being shifted to mostly remote modes of learning either through live online (synchronous), asynchronous online, or hybrid, combining in-person with online instruction. Remotely delivered Continuing Engineering Education (CEE) can involve challenges that do not exist for in-person classes. Moving from in-person to remote instruction sometimes necessitates a redesign of content and assessment, as well as the mode of instruction. Regardless of the remote mode, effective instruction requires a common, consistent approach to the subject and strategies which engage learners in meaningful ways that overcome the socially isolated nature of remote learning. Another concurrent shift in higher education programming involves the 60-Year Curriculum construct as a grand strategy of particular relevance to CEE and to new instructional practices that involve remote learning as well as other trends and innovations.

This chapter is designed to address the practice of CEE, the design of effective instruction that provides opportunities for developing and improving the knowledge, skills, and engagement of learners, and their ability, when appropriate, to affect changes in practice in the workplace. A framework is developed that guides the design of content and assessment that maximizes learning, and is aligned with each other and with specified learning outcomes.

Assessments are designed to provide feedback as well as measure success. Synchronous and asynchronous engagement strategies are suggested for engagement activities that provide effective engagement between learners as well as between instructor and learners. Against this backdrop, instructional and DEI implications for faculty are explored arising from three components of the 60-year Curriculum. Alternative credentials, VR/AR/MX and engineering degree apprenticeships are higher education innovations that portent new effective instructional practices for faculty to consider as the academy

[*] Corresponding Author's Email: kimmel@njit.edu.

continues to produce an educated and upskilled CEE workforce that is diverse, equity-minded and inclusive and which has the ability to meet Industry 4.0 needs during careers spanning 60 years.

Keywords: continuing engineering education, distance education, instructional design, content, educational outcomes, assessments, learning performance, learner engagement, 60-year curriculum, industry 4.0, PCO units, DEI, credentials, VR/AR/XR, degree apprentices

Introduction

The future of education requires flexibility. It is generally accepted that an undergraduate engineering degree provides graduates with the foundation of knowledge and skills necessary to begin practice in their chosen field. However, it is also recognized that there will always be a continuous expansion of knowledge and an evolution of engineering practices over time. Hence, the first college degree should be just the start of a career-long education process, where continuing engineering education (CEE) becomes a significant part of the life-long learning process for the practicing engineer. The challenge for the CEE providers is to ensure that the nature, context, and programs keeps up with the changing base of knowledge and practices required for the working engineers. To this end, also explored are the instructional and Diversity, Equity and Inclusion (DEI) implications for faculty arising from alternative/micro credentials, Virtual/Augmented/Mixed Reality tools and engineering degree registered apprenticeships. These are three components within the 60-Year Curriculum grand concept associated with innovative continuing education programs.

The Challenges

Continuing professional education can be thought of as a process that tracks and documents the skills, knowledge and experience that are gained both formally and informally by participants. This chapter is designed to explore continuing engineering education (CEE) which is offered through Engineering, Engineering Technology, and Professional Continuing and Online (PCO) units of 4-year colleges and universities with a focus on effective approaches to improving the knowledge, skills and changes in practice in the workplace (Saline, 1983). Effective approaches require:

1. Identification of skills, knowledge and behavior required for the practicing engineer to do the job well.
2. Examination of the skills, knowledge and behavior that they currently have.
3. Comparing the actual with the "required" to identify the gaps between what they have and what they require to be successful in their work.

In addition, this chapter will examine the differential impact of learning in different environments, with a focus on the adaption of effective instructional practices from in-person classes to remote, blended, or hybrid environments; and "work and learn" and self-paced opportunities. Regardless of the mode of learning, effective instruction requires a common, consistent approach to the subject. The methodology for the creation of effective instruction that align specified learning outcomes with learning experiences and measures of student achievement should be adoptable for any mode of learning (Baukal, 2010).

As a final challenge, this chapter will focus on the implications for and illustrative innovative strategies to produce an engineering workforce with the wherewithal to meet Industry 4.0 needs during careers spanning 60 or more years.

A Framework

The framework discussed here is meant to provide the structure or system for the process of life-long learning from the earliest educational experience in the K-12 levels through the acquisition of a college degree and into the person's profession. Learning can be considered the act of acquiring skills, knowledge, or competencies through formal classroom instruction, study, and experience. After completing their post-secondary education, professionals must continue the learning process that should involve the continuing acquisition of new skills and knowledge and incorporation of changes in practice in their chosen field. The structure should also allow for the learning process to occur through multiple modes of learning, from in-person to various remote modes.

Engineering education can be thought of as a sequence of three levels of learning (Saline, 1983):

1. Provide background and generate maintain, and enhance the interests of K-12 students in the pursuit of an engineering career.
2. Provide a program that allows undergraduate students to acquire the skills and knowledge needed to succeed in their chosen field of engineering upon graduation (Stice, 1976).
3. Provide opportunities for engineering professionals to review, enhance, and acquire those competencies that allow them to extend and strengthen their knowledge and skills to an established level of accomplishment. These opportunities can include graduate courses/certificates/degrees; professional development short courses/ certificates/credentials/ workshops that yield Professional Development Hours (PDHs) and Continuing Education Units (CEUs), among others.

For each level, and with various degrees of emphasis, there are four phases that can overlap, repeat, or occur simultaneously:

1. Building a knowledge base.
2. Observing models and examples.
3. Reflecting on current practices.
4. Changing practices, where necessary.

Regardless of the level of learning, for many years the focus had always been on a "content-based" model of learning, where the emphasis was on the content presented in the curriculum and textbook, rather than a focus on the specific skills and knowledge to be acquired by the students (Manzoor, 2017). The content is important, but without specification of learning outcomes, the assessment of student learning becomes problematic.

Over 40 years ago, a change started in the K-12 level, when learning standards were developed, that provided teachers with the path to develop learning objectives/outcomes, and classroom experiences. Those experiences provided opportunities for students to achieve the outcomes and assessments that measure the acquisition of those skills and knowledge specified by the outcomes. However, while K-12 teachers began moving to the "outcome-based" model, higher education had not begun.

Such criteria for undergraduate education exist for different areas by accreditation agencies, such as the Accreditation Board for Engineering and Technology, ABET (ABET, 2010). For many years, criteria were usually just the counting of courses required for a degree, until about 20 years ago when ABET began revising their criteria to requiring evidence of student learning, not only in their classes, but also in their courses and programs. Thus, ABET began moving the focus to program outcomes that specified those skills and knowledge acquired by their students and needed for their graduates to succeed as they pursued their post-graduate careers (McGourty, et. al., 1999; Felder & Brent, 2003). This became part of a continuous process for promoting student achievement and program improvement for students earning undergraduate degrees (Kimmel, et. al., 2011).

Life-long learning can be considered as the practice of continuing to learn throughout one's entire life, especially after the completion of an engineering student's formal undergraduate education (Marra, 1999). The process of life-long learning should be continuous so that learning occurs on a regular basis of new skills and knowledge, and of incorporation of changes in practice. Recognizing that the first degree should be just the start of a career-long education process, the change from content-based learning to outcomes-based learning must also be part of the process of life-long learning. The challenge for CEE is to make this shift from content-based learning to outcomes-based learning.

The framework should guide the design of content and assessment for the CEE that aligns the learning experience and assessment with each other and with specified learning outcomes, so that the assessments can provide feedback to instructor as well as learner. The assessment should evaluate whether the professional can successfully transfer the knowledge and skills gained in the classroom to various contexts in their careers.

Regardless of the mode of learning, effective CEE requires a common, consistent approach to the subject. The methodology for the creation of effective instruction that align specified learning outcomes with learning experiences and measures of achievement should be adaptable for any mode of learning (Felder, et. al., 2000).

The Instructional Issues

One purpose of this chapter is to provide an understanding of the instructional and assessment strategies that are most effective in the online learning environment. Strategies are available for providing effective instructional practices in the online environment as well as face-to-face, that allow the use of a variety of instructional and assessment methods. Online assessment strategies can include a variety of assignments that can provide meaningful and timely feedback to students as well as instructors. Effective assessment techniques include projects, portfolios, self-assessments, and regular assignments with immediate feedback. The role of meaningful feedback is critical.

It is assumed that the undergraduate degree earned by the professional has provided the person with the basic foundation of technical skills and knowledge in preparation for entry into the workforce. The CEE should then be able to meet their more specialized needs on a regular basis, keeping up with the increasing knowledge and changes in practice in their chosen field.

Effective Instruction

While the development of learning standards for the K-12 sector and revised accreditation criteria from ABET provided instructors with a path to developing outcomes-based instruction, many still rely on the content-based model, where the emphasis is on the content presented in the curriculum and textbook, rather than a focus on the specific skills and knowledge to be acquired by the students. These instructors tend to formulate specifications of skills and knowledge to be acquired by students that fit existing course materials, such as textbooks, with little or no connection to desired course objectives and learning outcomes. The use of content-based models of instruction may be even more prevalent at the post-undergraduate level, serving professionals enrolled in graduate courses/certificates/degrees programs, and/ professional development short courses-certificates-credentials-workshops that yield PDHs and CEUs, as part of their life-long learning endeavors.

The content is important, but without specification of learning outcomes, the assessment of student learning becomes problematic (Besterfield-Sacre, et al., 2000). The content-based model emphasizes what is presented, whereas the outcome-based model focuses on what students can actually learn and do. The utilization of an outcome-based model, aligned with accreditation criteria or learning standards, can provide instructors with the path to develop learning objectives/outcomes, and classroom experiences. Those experiences provide opportunities for students to achieve the outcomes and assessments that measure the acquisition of those skills and knowledge specified by the outcomes. Thus, the outcomes become linked to course topics and assessments of assignments, tests, student work, etc. The continuity of life-long learning by engineering professional should include the model of learning utilized across all educational levels including CEE (Manzoor, et. al., 2017).

Regardless of educational level, the quality of engineering education must include specification of clear and unambiguous learning objectives that are aligned with learning outcomes, instructional practice, and assessment methods. Learning Objectives should focus on the learner and contain action verbs that describe measurable behaviors. Learning objectives should specify those skills and knowledge that the learner should acquire as a result of their learning experience, that is, the outcomes. The objectives and outcomes should be aligned with the learning experience, and the assessment that measures success, and provides the information to both the instructor and the learner. Learning objectives should provide a basis for designing course material, since, while not all topics need to have learning objectives, the learning objectives should be embedded in at least one of the topics taught in the course (Felder & Brent, 1997, 2003; Mager, 1997; Gronlund, 2000).

Bloom's Taxonomy of Measurable Verbs is a classification of measurable verbs that describe and classify observable knowledge, and skills acquired by students as a result of classroom instruction (Felder & Brent, 1997; Chandio, et. al., 2016)). Learning objectives use measurable verbs to indicate explicitly what the student must do in order to demonstrate learning. In order to write learning objectives, one must start by identify what skill/knowledge the instructor wishes for the students to acquire. The learning objectives should be used to identify where each one is applied in the course content, i.e., the list of topics to be covered in the course/workshop. While other topics may be considered in the course, it is critical that there are topics that are relevant to the learning objectives. Then it is necessary to determine the nature of the activity that will allow the students to acquire the specified skill/knowledge. Finally, the instructor must specify how the skill/knowledge has been acquired by the students. This constitutes the learning outcome and assessment, which provides feedback to both instructor and students.

Faculty in undergraduate engineering education has been making progress in the transition from the content-based model to the outcomes-based model of instruction. However, such does not appear to be the case in CEE in graduate courses/certificates/degrees; professional development short courses/certificates/credentials/ workshops that yield PDHs and CEUs, among others. These programs either have no sets of learning objectives/learning outcomes, or they are poor objectives/outcomes that are not measurable. A major issue appears to be the use of three verbs that should not be used in learning objectives: know, learn, and understand. These actions are not observable or measurable, and thus cannot be used in a learning objective.

Let's look at some examples of what is contained in specific courses. An example of a short course on Project Management: The syllabus lists 8 objectives:

1. Know the characteristics…
2. Have the ability to define…
3. Understand how to…
4. Understand …
5. Understand how to…
6. Understand how to…
7. Know how to…
8. Learn about…

The terms "understand", "know", and "learn" are considered very weak action verbs as they are not measurable. Except for the first two objectives, verbs like "describe" "explain", or "compare" are measurable action verbs. For the first one on the list, "Identify" or "List" would be a more suitable verb. For the second verb, "Have the ability is not appropriate and, nor is it necessary. Define is an acceptable verb. All learning objectives are included in the list of topics for the course.

An example of a graduate course is entitled "Computer Systems Architecture", which has 14 learning objectives, an extraordinary number for a course. "Understand" is used 4 times, "An ability to" is used 6 times, "Learn is used twice, and "Improve is used once". This is a very poor set of learning objectives for a course.

It would be a very difficult task to translate such courses from a classroom environment to an online environment.

Assessment Issues

Assessment of student performance is a fundamental aspect of teaching and learning (Shaeowitz, 1998). While seat time (clock or credit hours) are commonly used as an indicator of quality, the length of a certificate program is considered a poor predictor of its outcomes. Comparing certificate programs of varied lengths does not reveal a clear or consistent pattern. In several cases, shorter programs lead to better outcomes than longer programs. A framework is needed to guide the design of content and assessment that aligns the learning experience and assessment with each other and with specified learning outcomes (Caruana, 2021). Assessments should demonstrate whether the student/professional can successfully transfer the knowledge and skills gained in the classroom to various contexts, or scenarios, as well as provide feedback to instructor and learner as well as measure success. The measurement of the acquisition of the knowledge, skills, and competencies that program graduates (i.e., students and professionals) have acquired should also address how student level of attainment is assessed, and how assessment results are used to improve instruction and learning (Carter, et. al.,2001). Assessment should also promote learning as well as monitoring it. Assessment practices should be both formative and summative. The utilization of formative assessment during instruction should provide feedback that allows instructors to adjust their practice leading to improved students' achievement of learning outcomes.

The outcomes-based model aligns with assessment that provides a measure of what and how much learning is occurring and whether a student/professional knows and can do what his/her degree or other credential implies, so that they are well prepared for success in their workplace (Stephanchick & Karim, 1999)). Assessment must be more than written tests. Assessments can include open-ended assignments, critical thinking exercises, and simulation and problem-solving activities (Lorenzetti, 2014). Outcomes are assessed in terms of acquired knowledge, skills, and competencies. Knowledge should include facts, principles, theories and practices; that is, the integration of information through learning. Skill refers to the ability to apply knowledge and use it for solving problems. Competence refers to the ability to practice knowledge and skills in work and professional development.

Transfer of Effective Instruction from Face-to-Face Mode to Remote Mode of Learning

Distance learning as a mode for CEE could be traced back over 35 years to the National Technological University, NTU (Eurich, (1985, Fowler, 1986). Initially, NTU courses were delivered via videotape, until the delivery system for the courses was converted to the utilization of a telecommunication satellite system that broadcast videos of classroom lectures to professionals at corporate and government sites that were members of the NTU network. The professionals were enrolled in selected graduate degree programs, the cost of which was incurred by the corporations and government agencies. Professionals viewed the videos and course materials at the corporate sights and most assessments consisted of submitted reports of assigned projects and proctored exams at these sites. Hence there were no problems adapting the in-person classes to remote learning. NTU also provided a regular schedule of non-credit short courses that enabled professionals to update their skills and knowledge or acquire new ones due to changing technologies.

Delivering educational services through a telecommunication satellite system was so expensive that NTU became a consortium of engineering universities across the country that provided the instruction and corporations and government agencies that received the instruction. But by the year 2000, the internet became sufficiently well-developed, universal, and almost free, that the participating universities were able to establish their own individual educational networks, which led to the decline and closing of NTU.

At the start of the pandemic with the move to remote instruction, many instructors believed that instruction could not be easily transferrable from the face-to-face environment to a remote environment. This is not really a true statement. Effective instructional practices that include learning objectives/outcomes, aligned with experiences that allow student to achieve the outcomes, and assessment of student achievement can be adapted to a remote instructional environment, or any mix of in-person and remote learning, with appropriate modifications (Gowda, & Dean, 2017). Redesign of content and assessment can be related to the work experiences of the professional. Outcomes do not require specific course material or curriculum, since they can be adapted to different modes of instruction. Effective in-person instructional practices, including hands-on experiences can be adapted for online learning, either totally remote or hybrid. Such instructional practices would include learning outcomes, classroom experiences that allow students to achieve the outcomes, and assessment that demonstrates the student achievement. The different modes of instruction can include total face-to-face, total remote, and hybrid, which can be synchronous or asynchronous modes of learning. Adapted instructional practices can maintain the continuity of the learning process by integrating videos, instructions on using simulations, and other digital resources (Blaha, 2021). In any environment formulating a set of learning outcomes that professionals have acquired should result from their acquisition of learning objectives, addressing learning outcomes, and how their level of attainment is assessed. In addition, regular interactions between instructor and students, as well as between students, is essential

Remote learning requires the design of meaningful assessments of learning (Robles & Broaathen, 2002). Assessments can draw from multiple ways of demonstrating knowledge and competencies, including application to the work environment of the professional. Approaches to online assessment strategies should include a wide variety of clear assignments that could include projects, portfolios (Rogers & Williams, 1999), papers (Felder & Brent, 1992), and regular assignments that are related to the job experience of the professional. Learning experiences and assessment methods should be closely connected in well-designed courses, whether in-person, or remote environments. When combined with valid learning objectives/outcomes, assessment techniques should demonstrate the acquisition skills, knowledge, and competencies (Sewell, et. al., 2010).

Since doing assessment remotely is a challenge, we can look at an example of an assessment for an online graduate course or a short course. This could involve dividing the content of the course into weekly units, one unit per week, where each unit is assessed at the end of each week. (Note that the period of time for each individual unit would be the decision of the instructor. One unit per week is one possibility.) Breaking the content down into these manageable sections can be important to keep students from becoming overwhelmed with the topic. Then, for each unit, students can be provided with a list of concepts and learning objectives/outcomes that they are expected to learn, leading to acquisition of relevant skills and knowledge. Students can be provided with online supplemental materials, textbook readings, and mini-lecture videos produced by the instructor. Assessments are then aligned with the outcomes and learning experience.

For online and hybrid environments, these units can be referred to as "learning objects" (Ritzhaupt, 2010). A learning object may be defined as an entity, either digital or non-digital, that can be used for learning, education, or training. Since learning objects focus on small snippets of information, they are quick and easy to consume, meaning the engineer can still learn alongside their other day to day tasks, without minimum disruption. They can fit it around their busy schedules and complete modules at a time that suits them, from wherever they are. Or adopt a fluid learning schedule by choosing which topics they want to tackle when and setting their own pace. This flexibility can help to keep them engaged and support different learning preferences. Also, learning objects can focus solely on practical, action-based content that users may need to perform their job more effectively.

Thus, learning objects are any grouping of materials that is structured in a meaningful way and tied to a learning objective or outcome. They should contain the objective/outcome, a learning activity, and an assessment (Rockland, et. al., 2013). Learning objects can vary in terms of size and scope, content, and design. They are usually small units that can be fitted together in different and meaningful ways to provide a customized learning experience. "Fitting together in a meaningful way" implies that the units are related and are arranged in a logical order. The learning experiences provided by the learning objects can be documents, pictures, simulations, movies, etc. As an example, a learning object can be created that describes the process for the manufacturing of a certain product. This learning object can have a learning outcome, "Identify the key elements in the manufacturing process." The assessment can involve writing a one-page paper that describes the steps of the project presented in the learning object.

Remote

Remote learning, which also can be termed online learning, can be implemented in two distinct modalities – synchronous and asynchronous. Synchronous courses run in real time, with students and instructors attending together from different locations. Instructors lecture in the same fashion as if they were in a face-to-face class. An asynchronous course is a general term used to describe forms of education, instruction, and learning that do not occur in the same place or at the same time. It uses resources that facilitate information sharing outside the constraints of time and place among a network of people and uses learning management systems such as Canvas or Blackboard. Typically, this type of delivery uses pre-recorded lectures.

However, one should look at all modalities of teaching as a continuous spectrum. On one end of the spectrum is the traditional face-to-face lecture style, while at the other end is the asynchronous method. In between the two extremes would be synchronous online learning, which would be similar to face-to face learning, but online.

You can also combine these modalities. For example, one of the authors lectured face to face, and at the same time, presented online videos of the lecture that could be viewed asynchronously to complement the student learning process. The same could be done in an online synchronous modality, where additional information, separate from the actual lecture, would be available to the students in an asynchronous mode. As pointed out by Maloney and

Kim (2020), "there are now multiple ways to combine and deploy these two major categories" There are several advantages of both modalities.

Synchronous

Synchronous online class meetings and well-structured small group meetings can help students feel a stronger sense of connection to their peers and instructor and stay engaged with course activities, while being online. Studies have found that while engaged in synchronous learning when compared to asynchronous learning, participants (a) find a stable means of communication, (b) tend to stay on task, (c) feel a larger sense of participation, and (d) tend to experience better task/course completion rates (Hrastinski, 2010). During a prolonged shutdown, such as the pandemic, the advantage of online learning became more apparent and even in non-pandemic times, there is still an advantage for the students for synchronous online learning.

A face-to-face class can be weather dependent, and students can miss one or more classes. In a synchronous online class, unless there is a technology failure, students can still attend class and the instructor can still give a class from anywhere. Students can also choose a hybrid form of synchronous learning, where one week they are meeting with the instructor face-to-face and another week, perhaps due to their job, they meet online synchronously. This has been used in the pandemic times to reduce the number of students in a classroom (alternating half in class and half online).

Asynchronous

Asynchronous learning, on the other hand, has several advantages, especially in the workforce where being tied down to a fixed time for a lecture could present problems. Materials, in the forms of links and videos are used to provide the week-by-week learning. While asynchronous learning is does not have a fixed lecture time, as well as a time for student to review the material, usually learning modules in this modality are structured within a week-by-week basis.

A good method of providing the information is with learning objects (Bradley, & Boyle, 2004), which provides "learning chunks of information". Typically, this would be a short, ten minute or so video on one topic of the material for that week, and the instructor would break down individual lectures to two or more learning objects. Learners want short information, as opposed to just videoing the entire lecture and putting it on a learning management system, such as Canvas or Moodle.

Garrison and Cleveland-Innes (2005) found through a multi-case comparison study of asynchronous courses that participant interaction alone does not instill a shared feeling of social presence or engagement in an online course. They found that participants of asynchronous online courses need structures placed by the instructor/designer or participants themselves to help them engage in meaningful learning activities.

Challenges for Both Modalities

Synchronous online learning can have issues in terms of equipment use, since the instructor is trying to combine online lecturing, presentation of visual materials online, and monitoring the remote class in terms of questions (either from the instructor or from the students). It requires a lot of monitoring by the instructor, and if there are technical issues with the instructor during the class, then that lesson might have to be postponed. If there are technical issues for individual students, then only those students would be kept from participating.

For asynchronous learning, the challenges are much greater, in terms of the time for material creation, monitoring the student progress, and examinations (which is also a challenge for synchronous online learning). One of the authors has an asynchronous circuit analysis course where there are over 70 learning object videos. That takes a lot of time to prepare and was done initially as additional learning for a face-to-face class. Each video, in terms of designing the content, publishing it and editing it, took between 5-10 hours. This is a serious time commitment before even teaching the course.

The next challenge is the organization of the material for an asynchronous course. One of the authors found that having a consistent week-by-week organization helped the students. The course, using Canvas as a learning platform, is structured each week with the following subheadings and links. The next section, Best Practices, is a description of some of the key elements in an asynchronous course.

1. What the lesson is about
2. Weekly learning objects that, when combined, would be the lecture for that week.
3. Quick Guides
4. Challenge Problems
5. Additional Links and Resources
6. Assignment uploads
7. Discussion Forum for that week.

Best Practices for Asynchronous Online Class

The structure above lists many of the best practices for asynchronous online class.

1. What the lesson is about – a good practice for any instruction is to give a brief introduction at the beginning of the lesson as to what that lesson will cover and how it relates both to material prior to the lesson and following the lesson. This is accomplished by a simple text document, although one could have a short video.
2. Weekly learning object – one of the longest times for creating an asynchronous course would be to go over the material (from prior notes and from the text) for that week, identify 2-3 main subtopics that are distinct, and then create a short (less than 10 minute) video on that topic. The total of all the videos that week should encompass all the key learning outcomes for that lesson.
3. Relating to the videos should be emphasized to the students that they should not only watch and listen to the videos but also take notes. It has been found (Bohay, et. al., 2011) that the process of taking notes from the videos enhances learning.

4. Quick Guides – a Quick Guide is a written document that summarizes (in steps) how to analyze or solve a specific group of problems. It can either be used before reading the text and/or video, to give the student a sense of the specific topic, or could be used after the student has reviewed the material, to reinforce the key steps.
5. Challenge Problems – they are problems related to the week's work that are more difficult than then the typical homework assignment. The problem and the solution are on video, and students are asked, after the problem is introduced, to stop the video, and try to solve it. If the student has difficulty, the video can be replayed and then the student can once again solve the problem.
6. Additional Links and Resources – there is a wealth of information on the Internet, some very good and some very poor. The links provided to the students have been thoroughly vetted in terms of their relevancy to the topic and to the quality of the material (either a video or webpage).
7. Assignment Uploads – In an online course, there needs to be a way to upload assignments and grade them so the students can see the assignments and get feedback.
8. Discussion forums – since an asynchronous class is not meeting on any regular schedule, there is need to have some communication between instructor and students and student to student. Email is an effective way for instructor-student communication, but it does not involve the rest of the class. A weekly discussion forum, where students can post a question, and have both fellow students as well as the instructor post responses that everyone can see is a more preferred method of communication. The instructor can "seed" the weekly discussion forum with one or more questions, engaging multiple students as well as encouraging students to use this means of communication.

An additional best practice regarding homework revolves around student self-assessment (Beumann, S and Wegner, SA. 2018). A typical method in a face-to-face class, or any online class, would be for the instructor to grade the homework and comment on what parts of it was incorrect. A better way is to divide the grading of the homework into two parts. The first part is the completeness – did the student attempt to solve all the problems, whether they were incorrect or not. The second part, which would be worth the same number of points, would be the student's self-assessment of what they did correctly, what errors they made, and why do they think they made those errors. Either a video or Word document is provided to the students the day after the deadline for homework (homework is not accepted after that deadline) of the answers and the method to arrive at the correct answer. Students then have two days to review the video/document, mark up their original homework (as if they were the instructor) and then upload that corrected homework. The grading for the second part is based on the completeness of that self-assessment.

Hybrid, Converged, and Hyflex Delivery Modes

Within the context of remote learning, there can be several different modes of online instruction. This section examines three of these instructional modes: hybrid, converged, and hyflex modes. Since definitions for these modes can vary from institution to institution, this section first specifies the attributes of each mode as used at the authors' home university.

Subsequent discussion is based on these definitions. The advantages and disadvantages of each mode, as well as best practices, are then discussed. Finally, this section looks at laboratory courses and remote experimentation, which present unique challenges for online instruction.

Definitions

For the purposes of this paper, the following definitions of these three modes of instruction are used (NJIT, 2021). These definitions are aligned with those of the Online Learning Consortium (Online Learning Consortium, 2017).

Hybrid Mode

Unlike the other two modes of instruction discussed in this section, hybrid mode courses cannot be completed entirely online. Some face-to-face instruction is required. Typically, some of the courses in person instruction is replaced with online instruction. The online component can be scheduled at regular intervals, for example, one of two class meetings is held online every week, or the class meets online every other week, or on some other predetermined schedule. Online components can be synchronous, meeting at specific times, or asynchronous. In the latter case, students would view pre-recorded classroom presentations no later than a given deadline. This mode is sometimes referred to as blended learning.

Converged Mode

This mode of instruction is independent of location, but not of time. Students can choose to participate in each class session either in person or online. Students are not restricted to a single mode of participation; they may choose to participate in person some days and online other days. The class is broadcast in real time using a videoconferencing system such as Webex or Zoom, so all students participate at the same time, regardless of location. Although the authors' university does not require any in-person attendance, some universities may require students to attend some class sessions in person, for example, on days that tests are given. Courses with no in-person requirements may still require proctoring for tests that may be given online.

Hyflex Mode

This mode of instruction combines converged mode and asynchronous course delivery. The instructor holds class at regularly scheduled times, and students can choose to participate in person or online in real time. Students also have the option to view the class sessions at another time as their schedules permit. All class sessions are recorded and made available to all students once the class session ends. Asynchronous does not mean self-paced; students are expected to adhere to a schedule, typically week by week, that is given at the beginning of the course. As with converged courses, students can vary how they participate in the course from one class session to the next. Instructors can require all students to participate at fixed times for some class meetings, whether in person online, usually to take tests.

Advantages and Disadvantages

As with many things, it isn't possible to say that there is a single best mode of instruction. Each mode has advantages and disadvantages. This section examines the pros and cons of the three instructional modes.

Hybrid Mode

The hybrid mode of instruction offers several advantages to instructors, students, and institutions. For hybrid courses that include asynchronous instruction, this component offers students a degree of flexibility. Although they are required to adhere to a course schedule, they can choose to view online videos and complete online assignments at times that work best for their own schedules within a given week or other set block of time. Some students may not be able to attend a fully in person course but can participate in a hybrid course.

With this flexibility, however, comes the need for stronger organizational skills (Carlton, 2021). It is easier to put off working on online, asynchronous material than it is to delay material in an in-person class. Such delays can lead to lower grades or greater withdrawal rates from students who fall behind in their work.

There are also some logistical advantages. For instructors and students who commute to campus, online class sessions, whether synchronous or asynchronous, can be accessed anywhere an Internet connection is available. For some, the savings in commuting time can be significant. Institutions need to schedule fewer classroom hours for hybrid courses than for face-to-face courses, freeing up rooms for additional classes.

The technology needs and costs for hybrid classes are greater than those for fully in- person courses. Students and instructors need computers and high-speed Internet connections. Universities must provide the infrastructure to support remote class delivery for synchronous class sessions. For asynchronous class sessions, they must provide technology and support services for both the creation and dissemination of pre-recorded materials. All of these come at a cost to the university.

Converged Mode

The converged mode offers some of the same advantages and disadvantages as the hybrid mode, as well as some that are not shared with that mode of instruction. The converged mode does offer students flexibility in terms of location, but not of time. Classes still meet at scheduled times, whether students attend in person or online. For students lacking organizational skills, this can be advantageous. Students who have jobs and travel for work may still be able to attend classes online while on business trips. Students who are sick, or quarantined due to covid-19 exposure, can still attend class online without jeopardizing the health of other students, as long as their own health permits. One of the authors had a student who was unable to enter the United States due to covid-19 travel restrictions, but was able to take a full schedule of classes in converged mode. The student attended classes for part of the semester from eastern Europe and for another part from western Africa. Other than the difference in time zones, there was no significant difference between this student and others who attended remotely from much less further away. Technology costs are similar to those encountered in hybrid mode, except that it is not necessary to record and stream class lectures online.

From an instructional standpoint, it can be difficult to monitor in person and online students simultaneously. Invariably, instructors' focus will be drawn toward one group, often the in-person students, leaving the other group feeling disconnected from the rest of the class.

Hyflex Mode

The hyflex mode offers the flexibility of converged mode and more. Since all class sessions are recorded and made available to students, they do not need to participate at regularly scheduled times, except possibly for tests. This can benefit students with work commitments that vary from week-to-week, as well as students who might miss a class due to illness, car trouble, or other reasons. Universities must provide the same technology as is used in converged learning. In addition, the university must also provide the ability to record lectures and servers to make them available to students on demand.

Best Practices

Regardless of the mode of instruction, there are strategies instructors and universities can use to benefit students. This section presents some best practices that are applicable to one, two, or all three modes of instruction.

Call in the Pros

Designing a course with an online component requires more than just recording lectures and posting the videos online. Although the content may be the same as a face-to-face course, it must be prepared and presented differently for different course modes to maximize student learning. This is especially true for asynchronous online instruction that is sometimes incorporated into hybrid mode.

Instructional designers can help instructors design and deliver their courses to improve student learning in any course mode. Instructional designers are typically staff members, not faculty members, with expertise in educational technology and pedagogy that many faculty members lack. Together, the faculty member and instructional designer can create course content best suited to the desired instructional mode.

Call in Some Backup

As mentioned earlier, many faculty members find it difficult to teach both in person and online students simultaneously. When teaching in converged or hyflex mode (only for students participating synchronously), it is helpful to have assistance monitoring the online platform. For example, a teaching assistant can monitor the online system for student questions posted while the instructor teaches "live" to all students. Students typically post questions in the online chat; the teaching assistant can aggregate questions for the instructor, ensuring all students have the same opportunity to ask about material that is unclear to them.

Keep Them Active

The effectiveness of active learning has been well documented; see for example, (Felder and Brent, 2016). While it is more difficult to incorporate active learning into courses with online instruction, it still can be done. Many online systems, including Webex and Zoom, support breakout rooms. The instructor can create a brief activity, perhaps requiring only a few minutes, and use the breakout feature to create separate areas for different groups of students to collaborate as they work on the activity. The instructor can join and leave groups at will to see how students are doing with the activity and to answer questions. At the end of the allotted time, the class resumes and groups can report on their results.

This works well when all students are online. When some students are in person and some are online, this can be implemented in a couple of ways. One is to form groups consisting of all in person or all online students. Another is to have all in-person students also connect to the online system and to use it when breakouts are in progress. From a technology standpoint, this has become easier to implement as major online platforms have developed apps that run on cell phones, which just about every student has.

In hyflex mode, this only works for students participating in real time. Students attending the class asynchronously cannot participate in breakout rooms. Nevertheless, there are other strategies to incorporate active learning for these students. One strategy is to prepare collaborative assignments that require students to work in small groups. The students can decide how to collaborate in the best way for themselves, with possibilities ranging from scheduling in person or online meetings to discussing via email, discussion forms, or another asynchronous mechanism. Another strategy is to post online discussion questions with each student required to post their answer and also to reply to a few other students' answers.

Laboratory Courses

How can you run a hands-on laboratory course when some or all of your students are remote? Well, you can't at least not in the same way you would if all the students were present in the lab. His section examines several ways to provide a meaningful laboratory experience in these instructional modes.

Converged Labs

Converged labs are useful when only some students can be physically present in the lab. For example, when covid-19 capacity restrictions were in place, one of the author's laboratory courses was limited to having half the class physically in the lab room. The instructor set up a schedule with one student from each two-person lab group present in the lab each week, and the other student participating online. The students alternated weeks being physically present so that each student received some hands-on experience in the course.

The breakout room feature of the online system is essential for converged labs. It allows the instructor to set up a separate breakout room for each lab group. The instructor can join individual breakout rooms to check students' progress with the experiment and to answer questions. Students can ask the instructor to come to their breakout room to ask questions or to demonstrate successful designs.

Instructor-Led Experiments

When all students are remote, whether by design or not, it may not be possible for students to "perform" the experiment, especially if the experiment requires equipment that cannot be accessed outside of the lab room. This strategy addresses this situation.

In this method, the instructor or a teaching assistant performs the actual experiment (Gonsar & Berger, 2020). The students participate remotely in real time, watching the instructor perform the experiment. The instructor may take measurements and give them to the students, or simply show the measuring equipment and have students read the measurements themselves. In either case, students perform the same data analysis as they would if they were present in the lab.

By performing the data analysis, students complete at least part of the original experiment, even though they do not actually set up and run the experiment. This preserves much of the original learning experience even when students cannot physically participate in the experiment.

Remote Instrumentation

During the past several years, the availability of equipment that can be accessed via the Internet has increased significantly (Hariton, et. al., 2020). It is possible to "run" an experiment remotely, using Internet-connected instruments to record measurements in real time. As with instructor-led experiments, however, this method requires an instructor or assistant to physically set up the experiment to be run. As the experiment is being run, the remote student has partial or complete control of the experiment. This is a costlier option than others discussed so far. Depending on the availability of equipment, it may not be possible for more than one lab group to perform the experiment at a time.

PC-Based Instrumentation

Several manufacturers offer equipment that interfaces directly to a personal computer, usually via a USB port (Hantek 2021). Although this equipment does not have all the capabilities of equipment found in a typical undergraduate lab, it often has sufficient functionality for undergraduate laboratory experiments. If students have the parts they need to perform the experiment remotely, and it is safe to do so, this instrumentation can allow them to complete the experiment outside of the lab.

Costs for PC-based equipment have dropped during the past few years, but there still is a cost to students, the university, or both. This strategy is more amenable to some disciplines than others. For example, a lower division circuits lab in an electrical and computer engineering program could be performed remotely by students with the necessary components, a protoboard, a power supply, and a PC-based multimeter or oscilloscope. A chemical engineering experiment that requires chemicals and a fume hood, however, could not be performed remotely regardless of the PC-based instrumentation available.

Simulation

Computer-based simulation is a viable alternative for hybrid, converged, and hyflex mode lab courses. Students set up and run their experiments in the simulator, which generates results for students to analyze. Simulation is used widely in engineering practice to reduce the time needed for design and testing. Simulation requires students or the university to purchase software licenses, though some companies offer student versions of their simulators at greatly reduced cost or at no cost. These versions may have limited features, but are often sufficient for undergraduate laboratory experiments.

Simulators are not always perfect. One of the authors taught an electronics laboratory experiment that required students to measure the propagation delay of a component under varying output loads. The model for the component in the simulator, however, had a fixed delay for the component, regardless of load, meaning that the simulator could not provide valid data for this experiment.

Flipping the Classroom

In a traditional face-to-face class, the typical structure for a student is to read the textbook, listen and take notes on the lecture, complete the relevant homework, and take the required exams. One question is whether students who read the textbook before the lecture come into the classroom setting with some prior knowledge of the topic or topics to be covered. Typically, many students do not read the material prior to the lecture, and therefore their first introduction to the topic is during a lecture. The problem with this traditional class approach is that since a lecture deals with the weekly required material, the amount of time students can do problems, case studies or other active learning concepts in class is very limited. In the case of STEM related courses, the ability to do multiple problems is key to learning the material.

Flipping the classroom was developed to overcome this problem, by providing more in-class time for active learning (Barrett, 2012). The concept is that students gain their first exposure to new material outside of class, usually via reading, lecture videos or other related web related links, and then use class time to do the harder work of utilizing that knowledge, perhaps through problem-solving, case studies discussion, or debate, meaning that flipping the classroom enables more active learning time (Fulton 2014). The flipped classroom aims at creating a student-centered environment where the emphasis is on the provision of meaningful activities that enable students to gain higher order thinking skills (Arslani, 2020).

In addition to providing more time for active learning in the class, another advantage is that by having materials in a video format prior to the lecture, rather than listening to the instructor during class, students can replay the video several times. Another advantage of this concept is that topics that would aid the students' knowledge but would normally not be covered in a class discussion can be added for the student's review without spending classroom time.

Flipping the classroom can be thought of as blended learning (Rutherford, 2013), since it involves both face-to-face interaction with other students and the instructor during the class, as well as online learning. Other terms that have been used for this concept include backwards classroom, reverse teaching, and reverse instruction.

There are some issues with flipping the classroom. The first issue is that students learn the material prior to the class, but how do you assess that learning prior to class? A second issue is if this is the first time that students are involved with a flipped classroom, there would be resistance by the students to learn the material, as opposed to the traditional way of sitting through a lecture. A related issue is that since this concept is applied to face-to-face instruction, students are not expecting to have part of the instruction online and feel that they are not getting the full instructional value. This was the experience by one of the authors when trying to implement a flipped classroom for the first time.

Since one of the main goals of flipping the classroom is to allow more time for problem solving, one of the authors tried a different approach. Rather than flipping the classroom, he created videos of additional problems, called challenge problems since they were at a more difficult level than the homework problems. The challenge problem videos introduced the problem, and then students were instructed to stop the video, and try to solve the problem. Then, the students were able to compare their answer and method to the instructor's answer and method, and possibly try it again. This concept appeared to be more acceptable to students, since they continued to ask for additional challenge problem videos.

Looking Ahead: New Faculty Roles and Perspectives

Thus far specific effective instructional practices have been examined for remote modes of learning and for content and assessment that maximizes education and is aligned with learning outcomes. Against this backdrop, it also is prudent to look ahead to more macro-level ways in which CEE faculty can begin to keep pace with advances in educational technologies and teaching models as well as changes in the engineering profession which is becoming more digitized, more global and more mobile and comprised of engineers who are likely to be working for 60 years or more. Using the framework of the 60-Year Curriculum and three of its components, new faculty instructional practices and perspectives will be examined next.

The 60-Year Curriculum (60YC) and Upskilling Engineering Professionals

The 60-year curriculum (60YC) is a new way of thinking about higher education. No longer are engineers likely to follow a straight and uninterpreted pathway from Bachelors, to Masters, and even to PhDs, but rather their learning will be stretching over the six decades which parallels how long today's engineers are expected to work over their lifetimes; and with many starts and stops, concurrently require new knowledge and skills to remain relevant and productive (Tugend, 2019).

Not so much a concrete program, the 60YC takes as its goal a focus on developing a higher education system that is much more nimble and able to respond quickly to the needs of what is being called Industry 4.0 or the revolution caused by automation, robotics, mechatronics, artificial intelligence, etc. In a 4.0 world of rapidly evolving industries, it is clear that engineers, in particular, must possess complimentary new knowledge. The 60YC has a wide range of implications for successful CEE instruction by faculty.

In engineering and computing, the breadth of developments in the world of work is astonishing and still unpredictable given the exigencies imposed by technological change, globalization and climate change on the speed needed for innovative solutions. As one example, Boeing Executive Leader, Michael Fors recently divided specific new STEM areas into three categories (Fors, 2021). The categories are Biological, Digital and Physical. See Table 1 below for the topics that are involved.

Taking this emerging STEM content into account, no wonder Huntington D. Lambert, Dean of the Division of Continuing Education and University Extension at Harvard University, who is a leader in the 60YC movement has remarked, "The real driver of the 60-year curriculum is the job market and length of life" (Tugend, 2019). And according to his fellow evangelist, Gary Matkin, Dean of the Division of Continuing Education and Vice Provost of the Division of Career Pathways at University of California at Irvine, "While the 60-year curriculum may sound like a new name for lifelong learning, the difference is that it focuses on how an institution can provide formal education courses and programs….The 60-year curriculum is really the organizing principle behind a lot of different trends in higher education, including the trend of being more accountable and very relevant to the workforce needs of a particular region" (Tugend, 2019).

Table 1. Illustrative STEM Categories

Biological	Digital	Physical
Optogenetics	Digital twins	Next generation batteries
Next generation genomics	Augmented & Virtual Reality	Advanced materials
Engineering	Blockchain	Nanomaterials
Bioinformatics	Cloud technology	Mechatronics
Neurotechnology	Artificial Intelligence	Autonomous vehicles
Biological	Digital	Physical
Stem cells	Machine learning	Additive manufacturing/3D printing
Symbiotic biology	Robotics	Nanodevices
Personalized medicine	Internet of things/sensors	Wearable technology
Nanoparticles	Cybersecurity	Micro/nano satellites
	Quantum computing	Organ microchips
		Smart cities/buildings
		Advanced energies

As suggested by the growing literature, the educational trends commonly associated with the 60YC are shown in Table Two (Dede, 2018; Johnson, D.P.. 2021; Tugend, 2019; Warner, 2021). (Note may be made that the trend of remote instruction has already been analyzed in detail above.) Among the other trends or components, *three are demarked in italic typeface* to indicate each will be discussed in further detail below.

The Broad Sweep of Components in the 60YC includes:

- Greater variety of ways for students to attend classes — in person, via video calls, hybrid or entirely online and so as to provide more flexibility for busy adult learners.
- *Micro-credentials are mini-degrees in specific competency areas, which involve short courses that emphasize job skills — both hard skills, for example, associated with Industry 4.0 (see Table 1) and soft skills to engage in difficult conversations — using real-life problems and case studies.*
- Portable transcripts also known as lifelong transcripts of degrees and credentials that move with the student rather than stay with the institution.
- *Greater use of educational technology tools such as Virtual Reality (AR), Augmented Reality (AR) and eXtended Reality (XR) to stimulate attention and engagement to increase knowledge retention.*
- More deliberate connections between continuing education and a university's undergraduate and graduate programs as well as connections with external entities such as companies, and third party educational providers.
- Greater support over the long haul and new roles not just from academic advisers, financial aid and career services professionals but also from academic faculty, because learners may come back, albeit in fits and starts, to their alma mater to acquire new content.

- *More experimentations in new academic programs and co-curricular tools which for engineering include makerspaces and apprenticeships each of which are designed to help students to develop practical skills in emerging technologies needed for a life-wide world of work.*
- Utilization of data mining and AI integration to support holistic credential tracking.
- Expansion of Student Management Data Systems (e.g., Banner) beyond support for fully matriculated degree seekers to give equal relevance to supporting lifelong learners studying part time in academic programs as well as in professional development or non-credit/non-degree training.
- Innovative programs that allow for acknowledging the academic credit worthiness of non-credit programs.
- New sources of financial aid for adult learners to pay for their continuous learning.

This may seem like a concept kaleidoscope, which to some extent it is, but for advocates, "The goal is to make this a thoughtful evolution rather than doing it piecemeal" (Tugend, 2019).

60YC, Professional, Continuing and Online (PCO) Units and Empowering Engineers for Industry 4.0

To be clear, the 60YC is more of an evolving and aspirational model than one specific program and according to experts, it is primarily being forged in the Professional, Continuing and Online (PCO) arms of universities (Gazi & Baker, 2021; Tugend, 2019). Or as Georgia Institute of Technology's Gazi and Nelson Baker (2021) have put it in their exploration of the shifting role of higher education in response to changing engineering workforce needs, "new knowledge, especially in STEM fields is being created at an increasing rate. Accordingly STEM professionals…. "need to continuously engage in learning to stay employed. This renewed need for lifetime professional, continuing and online education moves PCO operations and capacities from the fringes of higher education to center stage…. so professionals can be provided with…. ongoing preparation for career-long learning journeys by the engineering faculties.

There is a reason for this movement of PCO units to more center stage in university function and organization, especially at a time when higher education is being challenged to better prepare both emerging and seasoned engineers for the increased sophistication of the advanced technologies of AI, robotics, mechatronics, etc. At its essence, the 60YC is an expression of the belief that PCO arms--whether they are organized centrally or are located specifically within Engineering schools/colleges—are emerging as the educational entity especially well-equipped and positioned to carry out what is needed to prepare students and professionals for life-wide learning. It isn't that traditional academic engineering departments are unable to do this over time—and indeed Covid-19 has led to unexpectedly fast pivots especially about online learning—but the magnitude and speed of changes happening in engineering work are requiring a partner in higher education whose primary mission is to get in front of the changes happening outside the ivy-covered walls of academe. That is, it might be said that PCO units, by the nature of how they operate and are organized, are in unique positions to know how to design and implement the required pivots in learning, content and outcome assessment especially for seasoned engineers who are "adult learners/working learners" or the population of students who were once referred to as "non-traditional."

According to the editors of a new book, *The Great Skills Gap: Optimizing Talent for the Future of Work*, "within the changing employment landscape, there are questions about how well the current higher education model is positioned to continue launching its graduates into productive careers. Indeed, the future of work creates an imperative for radically rethinking the purpose and current approach to employment readiness by higher ed institutions" (Wingard & Parrugia, 2021). Of course, many higher education leaders would dispute this assertion and argue they are doing well at educating emerging engineers for employment and seasoned professionals to stay employed. But still in all, there are empirical data which have begun to question this belief even about emerging engineers. For example, the Strada Gallup 2017 College Student Survey found that while nearly all (96%) of chief academic officers believe that their institution is effective in preparing future workers, just 11% of business leaders strongly agree (Strada & Gallop, 2017).

To cap, one can think of the 60YC as lifelong learning "on steroids" or an umbrella concept covering the many different STEM topics in Table 1, which are being widely discussed in the academy today. Clearly, several, but certainly not all, of these trends or components are likely to have significant implications for effective CEE instruction by academic and PCO faculty. Yet three, shown in boldface in the list above, not only have such direct consequences for CEE instruction; but, when implemented, also have a significant role to play in advancing Diversity, Equity and Inclusion (DEI) goals.

Defining DEI

Before delving into the CEE-relevant 60YC topics and how they might maximize the potential to ameliorate the DEI challenge, prudence and clarity calls for some basic definitions. The WICHE Cooperative for Educational Technologies (WCET) has been undertaking such a task (Downs, 2021).

Diversity

Diversity is "a synonym for variety. A diversity focus emphasizes 'how many of these' we have in the room, organization, etc. Diversity programs and cultural celebrations/education programs are not equivalent to racial justice or inclusion. It is possible to name, acknowledge, and celebrate diversity without doing anything to transform the institutional or structural systems that produce, and maintain, racialized injustices in our communities."

Equity

Equity is "the effort to provide different levels of support based on an individual's or group's needs in order to achieve fairness in outcomes. Working to achieve equity acknowledges unequal starting places and the need to correct the imbalance."

Inclusion

Inclusion refers to "a state of belonging, when persons of different backgrounds and identities are valued, integrated, and welcomed equitably as decision-makers and collaborators. Inclusion involves people being given the opportunity to grow and feel/know they belong. Diversity efforts alone do not create inclusive environments. Inclusion involves a sense of coming as you are and being accepted, rather than feeling the need to assimilate."

The DEI Challenge within CEE Instruction

DEI goals are an especially thorny issue in CEE given the historic male-domination of the field and the threshold requirement of possession of an undergraduate degree. In other words, the instructors who are shaping both emerging and seasoned engineers for Industry 4.0 increasingly are committing to finding new tactics and strategies--many of which coincidentally are also 60YC-related--to enlarge the ranks of students in their undergraduate classes and thus the profession with more women, Black, Indigenous, and People of Color (BIPOC) populations as well as other disadvantaged and underrepresented groups/ demographics. (Taken together these groupings will be referred to below as BIPOC+ populations.)

DEI programs and activities are becoming ubiquitous throughout higher education today as cross-cutting and deliberate initiatives emerge. Less discussed in the literature, but elaborated further here, are how three CEE-relevant 60YC components have the potential to yield ancillary and constructive contributions to solving the vexing problem of latent and unintentional institutional and instructional biases which limit undergraduate educational opportunities for BIPOC+ populations.

Academic engineering faculty may play less of a direct role in addressing "diversity" among emerging engineers since undergraduate admission decisions, in particular, are often outside their purview, but they and PCO non-credit faculty certainly have a new role to play regarding Equity and Inclusion both for emerging and seasoned engineers. This is surely one of the reasons that many universities today are appointing new DEI-specific officials, both at the institutional and departmental level, to help faculty examine factors such as whether their syllabi and assignments are biased through their use of inequitable language and pedagogy (Kelly, 2021). These and other new resources are helping professors to become "equity-minded" for example about ways to deploy educational technologies, curriculum, and course design strategies, and teaching practices that increase the likelihood for inclusive instruction appropriate for well- and BIPOC+ populations alike (Teich, 2021).

Three 60YC components particularly relevant to addressing the DEI challenge are:

Academic and PCO professors' development and conduct of micro and stackable STEM credentials.
  - Because non-credit alternative credentials, for the most part, are open to all, do not require a long-term commitment and cost far less per credit than typical academic degrees, they are more accessible to a wider range of learners.

Academic and PCO professors' incorporation of digital instructional tools such as VR/AR/XR (see Chapter 7) in both degree and non-credit classes.
  - Because these AI-driven tools, by their very nature, are customizable, they can be deliberately designed to be bias-free and so better engage and speak to the sensibilities of diverse populations.

Academic professor's support for "work and learn" models such as "Degree Registered Apprentices" in high skill STEM and engineering areas.

- Because Degree Registered Apprentices, in particular, involve hands-on learning and overcome the prohibitive cost of higher education, they remove recognized obstacles to enrollment by underserved populations.

Growing Impact of Alternative Credentials for CEE Upskilling: Effective Instructional Practices and Dei Amelioration

Current engineering undergraduates, newly-minted engineering graduates, as well as mid-career and seasoned engineers already know that the typical shelf-life of their technology knowledge is around five years; and they also are coming to realize they have a shortened amount of time to devote to staying relevant; or in other words, they require upskilling (Gazi & Baker, 2021). There are growing data and consensus that an increasingly preferred method for upskilling appears to be through credential programs which are referred to by many different labels, such as micro-, nano, mini-certificates, badges, alternative credentials, non-degrees, etc. For working adults, theses shorter and often non-credit programs versus masters and doctorates, in particular, "provide more opportunities to develop skills and advance their knowledge in smaller blocks of time and lower cost" (All Campus Briefing, 2021).

The phenomena of micro and stackable credentials have become pervasive in higher education across all disciplines, but for STEM and engineering it may be even more imperative due to the acceleration of digital innovation in Industry 4.0. Some are beginning to suggest that micro or short-term credentials, many of which are stackable, are becoming the new "currency" of education versus degrees and credit offerings. Or put another way, "The new currency of learning will no longer be measured exclusively through degrees and credit offerings, but rather through demonstrated skills and competencies completed through more stackable pathways of higher education" (Fong, et. al, 2021).

Workcred, an affiliate of ANSI, has spent considerable time in attempting to untangle what most see as the "wild west of certifications" with a mushrooming growth in the number and types of programs that are available (Forte, 2020). Beyond the classic credentials from colleges and universities of associate's, bachelor's, master's, and doctoral degrees, there are the credentials and certifications that meet the varying needs of industry and/or learners. They also vary in length, format, modality, and start times. Workcred has illustrated just how complex the question is in the abridged version of their chart, shown as Table 2 below (Janzow, 2021):

An important fact to understand is that there is a difference between a certificate and certification as column one and column two outline, and that university faculty are directly involved only with issuing certificates (and, of course, degrees) versus certifications. About certification credentials, which are not the province of universities, the big difference is that they are offered by associations or industry groups, and are based on standardized exams that measure a person's knowledge of relevant job skills. Also, they are valid for only a limited time period, can be revoked for incompetency or ethical lapses, and are not awarded by the same organization that provided the education or training.

Complicating the situation more is that universities are not the only entities offering certificates, as other actors, such as for-profit boot camps, educational vendors, associations, companies and governments compete to upskill working learners. Credential Engine has been cataloging these programs and of the 967,734 credentials they have identified (as of February

2021), only 359,713 (37.2%) were offered by postsecondary educational institutions. "Not only are colleges and universities not alone in offering such programs, but they are also outnumbered" (Hadsell & Poulin, 2021).

Table 2. How Do Credentials Differ?

	Certificates	Certification	Degrees	Licenses
Awarded By	Colleges & External Training Providers, Unions, & Industry Associations	Industry certification bodies	Colleges & Universities	Government agencies
Awarded For	Exam at the end of training or education course or one-time assessment	Third-party, independent competency assessment	Course of study	Meeting requirements of an occupation
Indicates	Education/knowledge/skills	Skill mastery/competencies	Education, successfully passed courses	Legal permission
Time to Complete	Variable, generally less than 2 years	Variable	Variable, generally 2 years or more	Variable
Time & Renewal Require-ments	Often none	Time-limited, includes re-certification	None	Time-limited, renewal generally required
Revocation Process	Cannot be revoked	Can be revoked for incompetence or unethical behavior	Cannot be revoked	Can be revoked for incompetence or unethical behavior
Example	CNC Machinist	CompTIA Cybersecurity Analyst	B.S. in Engineering Technology	Professional Engineer
Standard for Accred-itation	ANSI/ASTM/E2659, globally recognized US National Std.	ANSI/ISO/IEC 17024:2012, International & National Std.	Regional Educational Accreditation Board/ABET	State law

And even university-issued certificates are rather complicated concepts as the content of the 2018 SUNY Micro-Credentialing Task Force Report makes abundantly clear (Johnson, K., 2018).

Adapted from this Report, Certificate Credentials:

1. Are issued by a higher education institution in recognition of the completion of a curriculum.
2. Developed through established faculty governance processes and designed to be meaningful and high quality.
3. Represent the content of credit or non-credit study; when it is for-credit, the courses must be applicable towards a degree program.
4. Can be offered online, on-campus, or a hybrid of both.

5. When credit-bearing, are awarded in recognition of the completion of a curriculum that represents a smaller domain of knowledge than established degrees and must be approved by the higher education institution and registered with the appropriate State Education Department.
6. When non-credit, needs no external approval, (though accreditation commissions are now getting involved) and must be identified as such but requires similar faculty participation as that needed for credit-bearing programs.
7. May be stackable or sequenced together and accumulated over time to build up an individual's qualifications and help that individual move along a career pathway and further education. When this occurs, it is referred to a "Stackable Credential" or Career Pathway.

Implications for Effective Instructional Practices and DEI

There is one fundamentally important aspect of a university certificate conducted by faculty for adult learners which distinguishes it from the offerings of all the other types of providers. As more universities begin to offer alternative credentials, this role revolves around the fact that it is faculty who develop and/or approve the certificate's curriculum and who teach its content following standard governance protocols as the SUNY Task Force Report makes clear and utilize the kinds of effective instructional practices examined earlier in this chapter. Moreover, the vetting process which colleges and universities employ to invite new members to join their ranks is unparalleled for its emphasis on subject matter expertise in research and teaching and for the candidate's discernible commitment to quality teaching and successful student outcomes. In the long run, this may give the CEE credentialing programs offered from colleges and universities an edge in what is becoming a race to upskill working adults.

Moreover, another section of the SUNY Task Force Report has an important implication for why university certificates have a special nexus to DEI goals. Individual campuses within the SUNY system are explicitly encouraged to develop criteria for a process that will permit existing non-credit certificates to be articulated into credit courses or in other words, a process for granting credit worthiness to non-credit classes that comprise a certificate (Johnson, K., 2018).

Just as there is a difference between the terms certificate and certification, there is a difference in the admission or eligibility standards germane to credit-bearing certificates versus non-credit certificates. Typically, in engineering, the former requires possession of an undergraduate degree with a specific grade point average from an ABET-accredited engineering school whereas admission or eligibility standards for entry into non-credit certificates is to a large extent "open access."

Therein lies one reason why this component of the 60YC offers a new opportunity to attract BIPOC+ populations. Not only are certificates less expensive and less time-consuming to acquire, but admission to non-credit certificates is available to a greater breadth of the population than those who typically qualify for admission into degree programs. Historically, BIPOC+ groups constitute this latter population. That is, PCO units, from where non-credit certificates largely originate, are perfectly fine with populating the ranks of their programs with men and women from the kinds of diverse backgrounds of concern to DEI goals meaning that they have always been able to accept learners with varying levels of education beyond high school.

Furthermore, if faculties from other universities adapt provisions like those found in the SUNY Task Force Report for articulating non-credit certificates with degree programs, this suggests that non-credit programs, which are deemed by faculty to have conveyed credit-worthy content to enrollees, may become a new mechanism through which BIPOC+ groups can demonstrate, with credits-in-hand, that they are eligible for degree matriculation. Existing on a continuum from foundational to advanced, CEE non-credit training certificates may yield the unanticipated benefit of being able to address the challenge of lifting the lives and prospects of BIPOC+ workers of all stripes and providing this segment of American society with a financial safety net in new high paying engineering-related jobs they can do in this decade and going forward. One might argue, then, that an effective instructional practice to ameliorate DEI issues in CEE revolves around a new role for faculty in standardizing the criteria that permits academic credit to be conferred on learners who successfully complete non-credit training programs.

Growing Impact of Virtual Reality (VR), Augmented Reality (AR) and Extended Reality (XR) in Effective CEE Instruction and DEI Amerioration

Artificial Intelligence (AI)-based technologies have the potential to become important tools to provide personalized learning and in so doing also to promote DEI goals. Among the most promising are the uses of VR/AR/XR in curriculum whether embedded in courses as segments or used as options for providing engineering students with hands-on experiences in labs, fieldwork, and practicums. The importance of these tools became especially apparent during Covid especially for hands-on aspects of the curriculum but they are expected to become given more ubiquitous after the pandemic has been contained. Following a term coined by Global Venture Solutions (GVS), the latter time period will be referred here as AD for "After Disease" (Moe & Rajendum 2020).

To state the obvious, the digital transformation that most readily comes to mind is the 60YC component and covid-19 begot explosion of online learning. Indeed, according to many predictions, the explosion of live online learning via Zoom, WebEx, Microsoft Teams, etc. and anytime/anyplace or asynchronous online learning will not be going away when (and if) the AD age emerges (Moe & Rajendum 2020). But this prediction poses a particularly acute dilemma for many universities which teach engineering with its vital lab and hands-on foci as has been examined earlier in this chapter. The challenge for now is to find a way to simulate for engineering students as closely as possible distinctive and valuable faculty-led experiences that are just as meaningful and just as easy as handling equipment, doing research in labs, and speaking face-to-face with professors.

One solution to this challenge is that VR/AR/XR will begin being adopted by faculty both in digital and face-to-face courses. These AI-powered derivatives hold the potential to help students learn faster and retain knowledge by making the overall student class experience more personal, engaging, collaborative and more connected to the outside world albeit virtually and to maximize a student's sense of inclusion a la the DEI initiative. These effects can occur even though no or limited physical bonding may be occurring among classmates and with faculty members, and even should AD predictions prove wrong and in-person CEE classroom instruction becomes the norm again.

Defining VR/AR/XR

VR is a simulated experience that can be similar to or completely different from the real world. An VR experience requires goggles that cover the eyes, making wearers feel as though they are immersed in a different universe. One description of the experience is that VR has the ability to sever the learners from his/her physical environment, allowing for the kind of real-world experience that wouldn't otherwise be possible and certainly during Covid was not; or in other words, "it is possible to replicate the past, envision a future or experience the present without doing anything more than donning a headset" (Edelman, 2017). Often avatars are used which by their nature can be exemplars of DEI sensitivity in humanoid figures, especially if intentionally and deliberately, the forms represented are from BIPOC+ populations. The googles are bulky (but becoming lighter and less expensive) but when learners put them on, they can see only the alternative reality being displayed and they lose any sense of their actual surroundings. The experience is so real that heart rates change, and senses are engaged—except of course at least for now—the senses of taste and smell. So immersive is the experience, that it is impossible for learners to take notes while the googles are on. Some faculty may see this as a positive because learner's foci can be only on the content being taught versus, as is so often the case, itchy student finger's drift to non-class related websites and materials.

AR is a little different and is a distinct type of AI-style technology which is a variation of the sensation created by the VR googles. However, it doesn't require users to wear any headgear. AR layers digital information in real time using various sensors, cameras, and projectors to augment or enhance reality and it utilizes technology to layer these superimposed images, sounds or text on the existing world around a user (Verizon White Paper, 2021). Unlike VR, AR builds upon the user's actual environment. Instead, users just aim their smartphones or tablet's cameras (and in some cases, specialized eyewear) toward an object and instantly a display pops up on the smartphone providing the viewer with details about whatever is being looked at. XR, sometimes referred to as MR for Mixed Reality, is a combination of VR and AR techniques powered by AI-coordinated algorithms.

In the AD age, analysts are predicting that colleges will realize that increased implementation of AI-powered technology will become a necessity; and in fact, over and above AI's use in instruction, non-academic administrators are already designing in-house programs or contracting with vendors for AI driven-products that increase recruitment, retention and graduation rates and increase overall operational efficiencies. But the reason for using VR/AR/XR in instruction which is often cited especially by venture capitalists is that it can help students learn faster and retain more information (Moe & Rajendum 2020).

There are psychological underpinnings for why. They derive from "context-dependent memory" theories that show individuals have improved recall of specific episodes or information when contextual cues relating to the environment are the same during memory encoding and retrieval. Think of when a faculty member is in her classroom and she thinks of something she wants to get from her office; but on route back to her office, she inevitably thinks about other things so by the time she reaches it, she has completely forgotten why she came there. So what does she do to recover the memory? She either re-imagines being in the classroom, or better, she goes back to it. Why? And why did she intuitively know this? When she had the thought in the classroom to go to the office to fetch something, the classroom environment is (unconsciously) being coded into having the thought. So, when she reimagines the classroom, or goes back into it, the environment serves to cue the memory. The classic study of this phenomenon was undertaken by British psychologists Godden and Baddeley (Godden & Baddeley, 1975). They asked 18 deep-sea divers to memorize 36 unrelated words of two or three syllables and discovered that recall for subjects who had learned words under water was better performed underwater than on land, and the reverse for those who learned the words on land. Since then, this idea of learning as context-dependent has been studied in many forms.

The implication of context-dependent memory in terms of VR/AR/XR is that immersive learning in simulated real-world contexts comes closer to harnessing learner's cognitive tendencies and more of their senses. In the 2D world, student's eyes and ears are used but in the 3D world, it is their eyes, ears, emotions and brains. The idea of environmental context dependent memory predates contemporary developments in VR/AR/XR, but it provides strong support for considering the learning advantages of these exponential technologies. What's more, environmental context-dependent memory provides backing for the use of these tools to overcome the problem that many employers point to where engineering graduates have mastered theoretical knowledge but that they do not know how to apply it in real-world situations. Providing simulated real-world contexts "embodies" the knowledge, the way in the example above, the faculty member's classroom "embodies" the thought of going to her office to retrieve something.

Implications for Effective Instructional Practices and DEI

The importance of adopting a student-centric focus has been gaining acceptance in the academy for the last many years and is related to why VR/AR/XR is likely to become more frequently used by faculty in their teaching. To start, it is that students will demand it, having had their expectations raised of how a class can operate during covid-19. That is, as has been examined earlier in this chapter, no longer will students find it acceptable for a professor to stand in a large lecture hall and talk for 45 minutes and more. Students simply will no longer stay engaged with this traditional format, often referred to as the "Sage on the Stage." They will require more collaboration among themselves and independent learning that has become associated by necessity with many of the effective instructional practices discussed above in live online, hybrid and asynchronous online teaching especially because they are already familiar with such AR games as Pokemon Go; and, if predictions prove correct, they will shortly become enamored with Facebook's recently announced new bet on VR which is powering their Conference Room product tool (Isaac, M. (2021). These experiences undoubtedly allow students to understand that there are newer ways to teach engagingly if only these same tools were to be adopted by faculty for use in their classes. And faculty are coming to understand too that these tools can strengthen a student's learning environment providing the opportunity to interact with and visualize an abstract concept that is otherwise tough to access or, as is sometimes the case in engineering, difficult to comprehend even when taught by the best of faculty.

Moreover, because these tools stimulate attention and engagement typically in non-biased ways, they can contribute to reaching DEI goals. Speaking at a 2021 conference, Michael Crow, President of Arizona State University (ASU), said, "People have talked for a long time about entertainment—how do you take something that is entertaining that activates certain parts of the mental process and ties it into education in a meaningful way… What we're after is the connection to emotional engagement" (Young, J.R., (2021). Which is why an ASU lab unit has been redesigned for an introductory Biology course that through the use of AR will help students form emotional connections with fictional "frog-cats" or in other words DEI-bias free constructs.

Of course, not all engineering content is expected to benefit from embedment of VR/AR/XR in instruction. According to Gaertner and referring to how VR is being reevaluated for workforce training, "VR works best for jobs that require empathy-based decision making as well as jobs that carry a physical risk such as working on oil rigs or manufacturing plants" (Combs (2021). One can expect CEE faculty will develop criteria for deciding the most appropriate context in which to teach content which uses 3-D simulations rather than putting their students in any danger which is sometimes inherent in lab and fieldwork.

Further, besides the ASU example, experimentation in STEM and CEE classes is already underway where learners are being offered hyper-immersive experiences that cater to an alternative learning environment through purely digitized VR/AR/XR images, fostering teaching and learning that closely mimics in-person interaction.

Examples include:

- Pennsylvania State University's Engineering Professor Nicholas Meisel (2021) uses AR in his classes teaching additive manufacturing.
- University of Michigan has their engineering students use an AR track to test autonomous vehicles in simulated traffic (Carney, 2018).
- Colorado State University's 2,500 sq. ft. VR lab which is equipped with 100 VR headsets/googles is helping neurology and anatomy students to look at structural relationships as they really exist, in a collaborative, immersive environment while interacting will data and with each other versus in the past only being able to see (and try to interpret) single plane images (Lang, B. (2019).
- Columbia University's Professor Brent Stockwell VR-driven biochemistry class taught over Zoom allows learners to "touch" a protein in 3D as they learn about it (Stockwell, 2021).

However, before wholesale adaptation of these AI-derived tools can happen in instruction either by simulating full environments with VR, or blending digital and real world elements in AR and XR, there inevitably will be new activities to access the learning outcomes from their use. Are the learners subject to these tools successfully transferring the knowledge and skills gained in the classes and is this transfer more successful than traditional methods of teaching? According to a University of California at Berkeley study, the answer is "Yes", specifically 80% of information presented to learners via VR was retained compared with 20% when students listened to a lecture alone (Verizon White Paper, 2021). But can these tools improve CEE classroom orchestration or help faculty organize for longer-term, more complex, more collaborative problem-solving? Will VR/AR/XR platforms generate avatars capable of communicating with learners through more personalized and DEI-sensitized speech, facial expressions, gestures, gaze and posture as educational scenarios are build, thus helping ameliorate DEI issues that are needed to stimulate the attention of BIPOC+ groups from the get-go so that they stay engaged and retain the content which their professors are teaching them?

Time will tell, especially because there are good reasons to be uneasy about AI, but if assessments yield positive results, then these AI-derived technologies are poised to affect CEE education at all levels and for learners of all backgrounds, having a particularly profound effect both on how emerging and seasoned engineers are upskilled during 60 or more years of their careers; and in the AD era, this will likely be the case regardless whether CEE learners study in classrooms or entirely online.

Work and Learn Models in Effective Instructional Practices and Dei Amelioration

In his chapter in *The Great Skills Gap*, Williams relates conversations he has held with numerous executives, professors and deans about threats to higher education from the changing world economy and from all the non-academic providers of micro credentials, and concludes that "the resounding belief is that higher education will remain as relevant as ever, but it is going to take significant—and occasionally painful—overhauls to make that possible. Faculty will need to rethink the way they interact with students, the populations they aim to target, what courses they offer and the medium in which students take courses" (Williams, 2021).

Perhaps one of the most important changes that can happen in engineering schools and departments refers to a growing acceptance not just of certificates but also of "work and learn" models, one of which is exemplified by what is termed the "Degree Registered Apprentice." This is an innovation explored by New America's Mary Alice McCarthy (2021). "The challenge of educating a much larger share of Americans for a longer amount of time, even as they raise families and work full time, is forcing colleges to reevaluate how they design, deliver, and fund higher education and skills training". Not just the fact that colleges were able to make the rapid pivot to ubiquitous online learning due to Covid-19 demonstrates that higher education today can move faster than the proverbial glacier, but so too do other innovations in the past two decades, such as MOOCs and certificates. But, as McCarthy points out, even with all this movement and change, which no doubt has broadened access to higher education, particularly for middle class working learners, none thus far "have moved the needle significantly on postsecondary attainment or disparities in degree attainment" among BIPOC+ populations.

The reasons are numerous and complex for why many new approaches fall short for BIPOC+ populations. But, according to McCarthy, one under-appreciated limitation in general to their impact is that they are all built on the assumption that formal education is an activity that "takes place outside of work, on the learner's own money and their own time, rather than as part of his or her job." However, these particular constraints are not associated with Registered Apprenticeship as an educational strategy which has begun to be applied in engineering.

More than 100-years old, the Apprentice model is a teaching and learning system which usually is associated with training tradespeople. In the United States, the system is strongly codified in federal legislation and regulated and is implemented by the U.S. Department of Labor. Akin to but not identical to the German model, the U.S. apprentice system is employer-driven and is a model that integrates formal academic instruction, known as "Required Technical Instruction" (RTI), with structured, paid, "On-The-Job Learning" (OTL). RTI and OTL are essential components. Companies must first hire full-time employees who are 18 and older as Registered Apprentices in a specific occupational title (e.g., mechatronics technician) among the thousands already developed (and with new titles added all the time to match the needs of the emerging work in Industry 4.0 industries). Each occupational title has its own employer-driven standardized curriculum that has been developed by members of aligned professional and union associations versus academic faculty. The mantra is "earn (or work) and learn" since Registered Apprentices receive a paycheck from day one when OTL commences along with RTI which often is conducted by faculty of colleges. Wages from paychecks can be used to cover the cost of RTI, but many employers offer "tuition remission" as an employment benefit to their Apprentices. Every occupational title in this program has a defined program length varying from one to six years; and, when, completed, results in the participant being awarded a portable and stackable credential by the U.S. Department of Labor.

Implications for Effective Instructional Practices and DEI

No doubt, apprenticeships are more commonly associated with producing tradespeople than engineering professionals, but experimentation has begun to change this equation. One example is the active engagement of The American Council on Education (ACE) which has been playing the role of an aggregator or integrator between the US Department of Labor and higher education (American Council on Education, 2020). That is, ACE is currently implementing an Apprenticeship Pathways project designed to expand the range of Registered Apprenticeships which are eligible for college credit including in engineering fields. Specifically, ACE has begun to evaluate three different U.S. Department of Labor approved Software Engineering Apprenticeship programs which are offered by IBM, Microsoft Leap and Techtonic to their employees for the academic creditworthiness of the learning experience. Concurrently, through ACE's efforts, Bismarck State College, Rowan University and California State University at Bernardino are among colleges which *a priori* have agreed to accept ACE's credit recommendations from students who have completed these company-led apprenticeship programs so that should these students enroll in the participating college's matching academic degree programs, they start out with a pool of academic credits already on their transcripts.

With the burden of paying for higher education largely covered when a learner is in an apprenticeship program, accessibility to learning is opened for BIPOC+ populations like never before. But will it be really opened in engineering? Entry into jobs in the engineering profession requires, at minimum, a bachelor's degree in an engineering discipline and this requirement is not likely to ever change. So for Apprentice programs to become vehicles through which BIPOC+ populations can achieve this specific degree, support for and changes by faculty within engineering departments will be needed.

As mentioned above, all Registered Apprenticeship programs--even in higher level STEM areas--result in the award of a national credential issued by the U.S. Department of Labor. This is where experimentation and faculty practices are coming into play with development of the "Degree Registered Apprentice." Daniel Kuchn (2021) has hypothesized precisely how apprentices can concurrently earn both the US Labor Department credential and a Bachelor of Science degree in an engineering discipline. His thinking, expressed in a piece, entitled, "Bachelor's Level Registered Apprenticeship for Engineers" imagines how an engineering department at a university could pursue the radical idea of using apprenticeships to educate individuals for STEM jobs and in so doing attract a more diverse student body because, according to McCarthy, the apprenticeship model can solve some of the complex problems preventing BIPOC+ groups from enrolling, which is the concern of DEI initiatives.

The apprentice system, McCarthy (2021) concludes:

> …addresses the three biggest challenges that keep adults from staying in or going back to school: cost, time and relevance. Because apprentices are paid to learn, apprenticeships make education more affordable. Because apprentices learn on the job, they do not have to squeeze training into time outside work. And because their learning is tied directly to the occupation for which they are being trained, they don't have to worry about whether the program is teaching them relevant knowledge and skills.

Further Kuchn believes that "the engineering discipline is the most logical place to begin this re-imagination." Of key importance, he begins his analysis by taking ABET requirements for engineering programs as a given and then shows how these requirements can guide adherence to the "Code of Federal Regulations" governing the award of the Registered Apprenticeship credential. Then he breaks down how any bachelor's degree program meeting ABET accreditation requirements can readily meet the Apprentice system's RTI requirement; and, with the adoption of a substantial coop component in a university's ABET approved engineering degree program, also the Apprenticeship's OTL requirement.

Clearly, for the Degree Apprentice model to become a reality, it will require new thinking from engineering faculty and increased importance, understanding and assessment of outcomes from learning that occurs on-the-job through OTL. It would also require compromise and innovation. But should consideration be given to this model, there is little doubt that the field of engineering would become more accessible to BIPOC+ populations.

Conclusion

A model is presented for effective instruction at all levels of life-long learning from elementary and secondary grade levels, through post-secondary education, and continuing into the person's profession, when professionals must continue the learning process that involve the continuing acquisition of new skills and knowledge and incorporation of changes in practice in their chosen field. The practice of effective instruction utilizes an outcome-based model that aligns learning objectives/outcomes and classroom experiences. Those experiences provide opportunities for the professional to achieve the outcomes and assessments that measure the acquisition of those skills and knowledge specified by the outcomes which are linked to course topics. The assessment should also evaluate whether the professional can successfully transfer the knowledge and skills gained in the classroom to various contexts in their careers. This process of life-long learning is designed to ensure that the content and practice is relevant, and that the activities authentically match their needs in their workplace. The continuity of life-long learning by engineering professional should include the model of learning utilized across all educational levels including CEE.

It is also seen that effective instruction allows for the learning process to occur through multiple modes of learning, from in-person to various remote modes. Each mode offers its own advantages and disadvantages, and there are best practices in place that are applicable to one or more of these modes. Laboratory instruction presents its own challenges that are quite different from those associated with a typical lecture course, and there are several ways to offer a meaningful laboratory experience via remote instructional modes. Regardless of the differences between the different modes of learning, these professionals are able to achieve the specified skills and knowledge to be acquired. For remote or online learning, two modalities were discussed, synchronous and asynchronous, either as individual modalities or in combination with traditional face to face learning. Both challenges with each modality as well as best practices for the asynchronous modality were identified. In addition, the concept of flipping the classroom, advantages and issues with this concept, and an alternative to flipping the classroom were presented.

In a transition from discussing direct effective instructional practices to that of innovations that can help emerging and seasoned engineers remain vital and current in the Industry 4.0 world, examination is made of the concept of a 60YC as a grand strategy where some, but certainly not all, of its multitude of components will be realized. There are many hurdles to be overcome stemming not just from resistance to or lack of internal faculty buy-in, or from the burgeoning requirements that faculty address diversity, equity and inclusion issues in their instruction, but also from a lack of internal infrastructure, systems, and resources. Nevertheless, three components of the 60YC were highlighted. Micro-credentials, VR/AR/XR tools that faculty can use to stimulate attention and engagement to increase knowledge retention and registered apprenticeships represent a sampling of new 60YC approaches in CEE, each of which in their own way has the potential to contribute to the production of an educated and upskilled engineering workforce that is diverse, equity-minded, inclusive and responsive to Industry 4.0 needs.

Across these approaches is one common denominator. It is the hope and expectation that effective instructional practices are adopted by academic and PCO faculty who teach degree and certificates programs, digitally embed new immersive tools into their courses, and experiment with work and learn models as well as other innovations that will be emerging in the coming years. If the 60YC can be thought of as a macro-level "sum" of parts, then the instructional denominator, no doubt, is its key micro-level part.

References

ABET (Accreditation Board for Engineering and Technology). (2010). Criteria for accrediting engineering programs: Available online at: https://www.abet.org/wp-content/uploads/2015/04/abet-eac-criteria-2011-2012.pdf.

All Campus Briefing. (2021). How Universities Can Lead Upskilling: The Next Revolution in Learning. August 4. allcampus.com. Chicago, Il. Available online at: Allcampus.com/about-all-campus/press-llcampus.com/about-all-campus/press-insights/upskilling-in-higher-education/?cid=2624&utm_source=upceaemail&utm_medium=email&utm_campaign=upcea email.

American Council on Education. (2020). ACE Apprenticeship Pathways Program. Washington D.C. Available online at: https://www.acenet.edu/Programs-Services/Pages/Credit-Transcripts/ Apprenticeship-Pathways.aspx.

Baukal, C.E. (2010) Continuing Engineering Education Through Distance Learning, *European Journal of Engineering Education*, 35 (2), 225-233.

Besterfield-Sacre, M.E., et al. (2000). Defining the outcomes: A framework for EC 2000. *IEEE Transactions on Engineering Education.* 43(2): 100–110.

Beumann, S., Wegner, SA. (2018). An outlook on self-assessment of homework assignments in higher mathematics education. *Journal of STEM Education*, 5, 55.

Blaha, K. (2021). Instructional Practices that Support Virtual and Online Learning Environments. *Faculty Focus,* September 3. Available online at: https://www.facultyfocus.com/ articles/online-education/online-course-delivery-and-instruction/instructional-practices-that-support-virtual-and-online-learning-environments/.

Bohay, Mark, Blakely, Daniel P, Tamplin, Andrea K. and Radvansky, Gabriel A. (2011) Note Taking, Review, Memory and Comprehension, *The American Journal of Psychology*, 124 (1), 63-73.

Bradley, C. & Boyle, T. (2004, December). Students' Use of Learning Objects. *Interactive Multimedia Electronic Journal of Computer-Enhanced Learning*. Volume 6, Number 2.

Bradley, B. (2021). New Institute Backed by National Science Foundation to Explore AI's Role in Education. August 9. *Edweek Market Brief.* Bethesda, MD. Available online at: https://marketbrief.edweek.org/marketplace-k-12/new-institute-backed-national-science-foundation-explore-ais-role-education/.

Carlton, G. (2021). Hybrid Classes: What Are They and Pros and Cons. Available online at: https://www.thebestschools.org/magazine/hybrid-classes-pros-cons.

Carney, S. (2018). Augmented Reality at U-M Improves Driverless Vehicle Testing. *The University of Michigan Record.* November 18. Available online at: https://record.umich.edu/articles/ augmented-reality-u-m-improves-driverless-vehicle-testing/.

Carter, M., Brent, R., and Rajala, S. (2001). EC 2000 Criterion 2: A procedure for creating, assessing, and documenting program educational objectives. *Proceedings 2001 ASEE Annual Conference.* Albuquer-que, New Mexico.

Caruana, V. (2021). The Power of Significant Learning during Course Development. *FACULTY FOCUS*, June 28. Available online at: https://www.facultyfocus.com/articles/course-design-ideas/the-power-of-significant-learning-during-course-development/

Chandio, M., Tufail, P. Saima, M., & Iqbal, R. (2016). Bloom's Taxonomy: Improving Assessment and Teaching-Learning Process. *Journal of Education and Educational Development.* 3(2), December.

Combs. V. (2021). VR Training Expands to Make Collaborative Education Relevant to All Workers and for All Skills. July 8. Techrepublic.com. Available online at: echrepublic.com/article/vr-training-expands-to-make-collaborative-education-relevant-and-available-to-all-workers-and-for-all-skills/.

Dede, C. (2018). The 60 Year Curriculum: Developing New Educational Models to Serve the Agile Labor Market. November 19. The Evolllution. Available online at: https://evolllution.com /revenue-streams/professional_development/the-60-year-curriculum-developing-new-educational-models-to-serve-the-agile-labor-market/.

Downs, L.R. (2021). Diversity, Equity, and Inclusion and Digital Learning. *WCET Frontiers.* September 3. Available online at: https://wcetfrontiers.org/2021/08/05/dei-and-digital-learning/.

Edelman, R. (2017). *The Truth about Your Future: The Money Guide You Need now, Later and Much Later. Simon and Schuster.* New York. NY. 8-9.

Eurich, N.P. (1985). *Corporate Classrooms: The Learning Business, Carnegie Foundation for the Advancement of Teaching Special Report.* Princeton University Press.

Felder, R M. & Brent, R. (1992). Writing assignments—pathways to connections, clarity, creativity. *College Teaching, 40*(1), 43–47 (1992).

Felder, R.M., & Brent, R. (1997). Objectively speaking. *Chemical Engineering Education.* 31(3): 178–179. Available online at _http://www.ncsu.edu/felder-public/Columns/Objectives.html.

Felder, R.M., & Brent, R. (2003). Designing and Teaching Courses to Satisfy the ABET Engineering Criteria. *Journal of Engineering Education*, 7-25, January, 2003.

Felder, R.M. & Brent, R. (2004). The ABC'S of engineering education: abet, bloom's taxonomy, cooperative learning, and so on. *Proceedings 2004 Annual ASEE Conference*. Salt Lake City, Utah.

Felder, R.M. & Brent, R. (2016). *Teaching and Learning STEM: A Practical Guide.* Jossey-Bass, San Francisco, CA.

Felder, R.M., Woods, D.R. Stice, J.E. & Rugarcia, A. (2000). The future of engineering education. 2. Teaching methods that work. *Chemical Engineering Education*. 34(1): 26–39. Available on-line at: http://www.ncsu.edu/felder-public/Papers/Quartet2.pdf.

Fong, J., Ahluwalia, A., & Erter, B. (2021). *Shifting Paradigms: Understanding Institutional Perspectives on Microcredentialing: UPCEA & The Evolllution: A Modern Campus Illumination.* Available online at: https://moderncampus.com/research-paper/shifting-paradigms-understanding-institutional-perspectives-on-microcredentialing.html.

Forbes, A., & Reamer, A. (2021). *Counting U.S. Postsecondary and Secondary Credentials.* Credential Engine (www.credentialengine.org). February. Washington D.C.

Fors, M. (2021). The Future of Work and Learning from a Corporate Perspective—Boeing. *IACEE World Conference* (virtual). May 24-26.

Forte, J. (2020). *Understanding Certifications. December 10. Workcred: An Affiliate of ANSI Blog.* Available online at: https://blog.ansi.org/workcred/understanding-certifications/.

Fowler, E.M. (1986) Engineers Take Classes by Satellite. *New York Times,* September 6.

Garrison, D.R., & Cleveland-Innes, M. (2005). Facilitating cognitive presence in online learning: Interaction Is not enough. *American Journal of Distance Education*, 19, 133–148.

Gazi, Y. & Baker. N. (2021). Higher Education's Changing Face. In Wingard, J. & Parrugia, C., (Ed.), *The Great Skills Gap: Optimizing Talent for the Future of Work.* Stanford University Press, Stanford, CA. 85-91.

Godden, D.R. & Baddeley, A.D. (1975). Context-dependent Memory in Two Natural Environments: On Land and Underwater. *British Journal of Psychology*. 66(3). 325-331.

Gosnar, S. and Berger, T. (2020). *5 Smart Ways to Run Science labs when you're Teaching Remotely.* Available online at: https://www.edutopia.org/article/5-smart-ways-run-science-labs-when-youre-teaching-remotely.

Gowda, R.S., & Dean, S.V. (2017). A Comparative Analysis of Traditional Education System Vs. e-Learning. *International Conference on Innovative Mechanisms for Industry Applications* (ICIMIA 2017).

Gronlund, N.E., *How to Write and Use Instructional Objectives.* Merrill–Prentice Hall, 6th ed., (2000).

Hadsell, J. & Poulin, R. (2021). New Post: The Bigger Shift: Student ROI for Higher Education and The Role of Digital Learning. *WCET Frontiers*. September 29. Available online at: https://wcetfrontiers.org/2021/09/29/student-roi-for-higher-ed-the-role-of-digital-learning/.

Hantek. (2021). *Hantek6004EU Series.* Available online at: http://hantek.com/products/detail/12169.

Hariton, A., Zet, C., Vrerama, E., and Foselau, C. (2020). Remote Laboratory – Study of Digital Oscilloscope. Proc. *International Conference and Exposition on Electrical and Power Engineering (EPE)*, 495-500, October.

Hrastinski, S. (2010). How do e-learners participate in synchronous online discussions? Evolutionary and social psychological perspectives. In N. Kock (Ed.*), Evolutionary psychology and information systems research* (pp. 119–147). US: Springer.

Isaac, M. (2021). Facebook's New Bet on Virtual Reality: Conference Rooms. *New York Times*. August 19. Available online at: https://www.nytimes.com/2021/08/19/technology/facebooks-new-bet-on-virtual-reality-conference-rooms.html.

Janzow, P., (2021). Webcast Moderator. *ASEE Industry 4.0 Workforce Summit Series #5.* Available online at: https://workforcesummit.asee.org/video-archives/. September 10.

Johnson, D.P. (2021). Webcast Moderator. *The 60 Year Degree: Learning Without Boundaries.* January 21. Available online at: https://www.youtube.com/watch?v=t655pXMZ_7A.

Johnson, K. (2018). *SUNY Micro-Credentialing Task Force: Report and Recommendations.* Available online at: http://www.nysed.gov/common/nysed/files/principal-project-phase-2-suny-microcredentials.pdf. January 23. 1-34.

Kelly, B.T. (2021). *Beyond Inclusion: Transforming Syllabi and Assignments for Equity webcast.* (September 13). EducationAdminWebAdvisor.com.

Kimmel, H., Perna, A, Klotzkin, S., Carpinelli, J.D., & Tomkins, R., P. (June, 2011). Student Learning and the Continuous Program Improvement Process in a Chemical Engineering Program, *Proceedings ASEE Annual Conference*, Vancouver, Canada.

Kuehn, D. (2021). Bachelor's-Level Registered Apprenticeship for Engineers: Possibilities and Challenges. In Wingard, J. & Parrugia, C., (Ed.). *The Great Skills Gap: Optimizing Talent for the Future of Work.* Stanford University Press, Stanford, CA. 194-202.

Lang, B. (2019). Colorado State University has Deployed a 100 Headset VR Lab for Biomedical Education. October 10. *RoadTovr Newsletter*. Available online at: https://www.roadtovr.com/colorado-state-university-immersive-reality-training-lab-vr/.

Mager, R.F. (1997). *Preparing Instructional Objectives; A Critical Tool in the Development of Effective Instruction.* Center for Effective Performance, Inc., 3rd ed.

Maloney, E.J. & Kim, J. (2020). *The Low-Density University: 15 Scenarios for Higher Education.* Johns Hopkins University Press, Baltimore, MD:

Manzoor, A., Aziz, H., Jahanzaib, M., Wasim, A., & Hussain, S. (2017). Transformational Model for Engineering Education From Content-Based to Outcome-Based Education. *International Journal of Continuing Engineering Education and Life-Long Learning*, 27 (4), 266-286.

Marra, R.M., Camplese, K.Z., & Litzinger. T.A. (1999). Lifelong learning: A preliminary look at the literature in view of EC 2000. *Proceedings. 1999 Frontiers in Education Conference*. San Juan, Puerto Rico.

McCarthy, M.A. (2021). Past as Prologue: Apprenticeship and the Future of Work. In Wingard, J. & Parrugia, C., (Ed.), *The Great Skills Gap: Optimizing Talent for the Future of Work.* Stanford University Press, Stanford, CA. 184-193.

McGourty, J., Besterfield-Sacre, M., & Shuman, L. (1999). ABET's eleven student learning outcomes (a–k): Have we considered the implications? *Proceedings, 1999 Annual ASEE Conference,* Houston, Texas.

Meisel, N. (2021). Augmented Reality Computer-aided Design Education: Tool to Improve Student Engagement. *Proceedings ASEE Annual Conference* (virtual). July 26 – 29.

Moe, M. & Rajendum, V. (2020). *The Dawn of the Age of Digital Learning: An Acceleration of Trends that have been Building for Years.* (May 6). Medium Open Platform/Blog. San Francisco, CA. Available online at: https://medium.com/gsv-ventures/dawn-of-the-age-of-digital-learning-4c4e38784226.

National Fund For Workforce Solutions. (2021). *The Words of the Workforce: A Field Guide to the Terms and Ideas Shaping the World of Work—A Collaborative Effort of a Diverse Group of Non-Profit Organizations Focused on Workforce Policy and Equity Issues.* September 21. Washington D.C. Available online at: https://workingnation.com/lets-talk-about-work/?emci=7475f40f-0118-ec11-981f-501ac57ba3ed&emdi=34773d5b-fd1a-ec11-981f-501ac57ba3ed&ceid=2392157.

NJIT. (2021). *Modes of Instructional Delivery.* Available online at: https://www.njit.edu/registrar/instructionaldelivery.

Online Learning Consortium. (2017). *Updated E-Learning Definitions.* Available online at: https://www.onlinelearningconsortium.org/Updated-e-learning-definitions-2.

Rafe, G., & Manley, J.H. (1999). Assessing Quality Issues in Interactive Video Teleconferencing-based Graduate Level Engineering Courses. *Journal of Engineering Education*, 88 (1), 119-128.

Ritzhaupt, A.D. (2010). Learning Object Systems and Strategy: A Description and Discussion. *Interdisciplinary Journal of E-Learning and Learning Objects*, 6 (1):217-238. Available online at: https://www.researchgate.net/publication/288964268_Learning_Object_Systems_and_Strategy_A_Description_and_Discussion.

Roberson, G. (2021). How the Future of Work Impacts the Workforce of Technical Organizations. In Wingard, J. & Parrugia, C., (Ed.), *The Great Skills Gap: Optimizing Talent for the Future of Work.* Stanford University Press, Stanford, CA. 59-66.

Robles, M. & Broathen, S. (2002). Online Assessment Techniques. *The Delta Pi Epsilon Journal*, 44 (1), 39-49.

Rockland, R.H., Burr-Alexander, L., Kimmel, H.S., Carpinelli, J.D., Hirsch, L.S., & Tylutki, T.M. (2013). Integration of Medicine and Robotics – an Online Teacher Professional Development Program for STEM Education. *Proceedings 2013 Annual ASEE Conference*, Atlanta, GA.

Rogers, G., and Williams J.M. (1999). Asynchronous assessment: Using electronic portfolios to assess student outcomes. *Proceedings, 1999 Annual ASEE Conference*, Houston, Texas.

Saline, Lindon, E, (1983). Continuing Engineering Education: One Element of Lifelong Learning for Engineers. *IEEE Transactions on Education*, E 26 (4), 122-126.

Sewell, J.P., Frith, K.H., & Colvin, M.M. (2010). Online Assessment Strategies: A Primer. *MERLOT Journal of Online Learn.ing and Teaching*. 6 (1).

Shaeowitz, J.A. (1998). Classroom assessment. *Journal of Engineering Education*. 87(2): 179–183.

Stephanchick, P., & Karim, A. (1999). Outcomes-based Program Assessment: A Practical Approach. FIE'99 Frontiers in Education. *29th Annual Frontiers in Education Conference*, San Juan, Puerto, Rico.

Stice, J.E. (1976). A First Step Toward Improved Teaching. *Engineering Education*, 66 (5), 394-398.

Spak, G. (2013). US Advanced Manufacturing Skills Gap: Innovative Educational Solutions. Paper and Presentation at International Conference on New Horizons in Education. Rome, Italy. June 25 – 27 (published in *Procedia Social and Behavioral Science Journal*).

Stockwell, B. (2021). *Meet the Columbia Professor Who is Pioneering VR to Teach Biochemistry in 3D.* September 29. Columbia University YouTube Channel. Available online at: https://www.youtube.com/watch?v=7cXND8QC4kE.

Strada Education Network/Gallup. (2017). *College Student Survey: A Nationally Representative Survey of Currently Enrolled Students.* Washington, DC. Gallup.

Teich, A.G., (2021). *Webcast Moderator. Using Digital Tools to Create Equity in Higher Education.* July 8. Available online at: https://www.fierceeducation.com/best-practices/using-digital-tools-to-create-equity-higher-education?mkt_tok=Mjk0LU1RRi0wNTYAAAF_0KJZvxw2ZQN1SpUDhAhcGKWeNieQYWtUZpco8XWjvclofis77TNt6znr04oWOwUab3TdSCGtYSOB4NqUEUNc7QxbiL8Hepb7aelD18Dm7iNOOdZ4s0M&mrkid=12969654.

Tener, R.K. (1999). Outcomes Assessment and the Faculty Culture: Conflict or Congruence? *Journal of Engineering Education* Volume 88, Issue 1, 65-71.

Trevisan, M.S., Davis, D.C., Calkins, D.E., and Kenneth L. Gentili, K.L. (1999). Designing sound scoring criteria for assessing student performance. *Journal of Engineering Education*. 88(1), 79–85.

Tugend, A. (2019). *60 Years of Higher Ed – Really? New York Times,* October 10. Available online at: https://www.nytimes.com/2019/10/10/education/learning/60-year-curriculum-higher-education.html

Verizon White Paper. (2021). *Using Augmented Reality to Enhance College Learning: How Higher Education Can Benefit from Implementing Augmented Reality.* September 28. Available online at: https://www.techrepublic.com/article/vr-training-expands-to-make-collaborative-education-relevant-and-available-to-all-workers-and-for-all-skills/.

Warner, E. (2021). *Taking the Long View: Postsecondary Institutions and the 60-Year Curriculum. April 1. The Evolution.* Available online at: https://evolllution.com/revenue-streams/extending_lifelong_learning/taking-the-long-view-postsecondary-institutions-and-the-60-year-curriculum/.

Williams, J. (2021). Introduction-Part II/Higher Education Still the Solution for a Workforce in Flux? In Wingard, J. & Parrugia, C., (Ed.), *The Great Skills Gap: Optimizing Talent for the Future of Work.* Stanford University Press, Stanford, CA. 75-79.

Wingard, J. & Parrugia, C., (Editors), (2021). *The Great Skills Gap: Optimizing Talent for the Future of Work.* Stanford University Press, Stanford, CA. 1-7.

Yamagata-Lynch, Lisa C. (2014), Blending Online Asynchronous and Synchronous Learning. *International Review of Research in Open and Distributed Learning*, 15(2, April 2014, 189–212.

Young, J.R., (2021). *Can a Hollywood-Backed VR Company Make Science Labs More Effective?.* August 19. Edsurge.com. Available online at: https://www.edsurge.com/news/2021-08-19-can-a-hollywood-backed-vr-company-make-science-labs-more-effective.

Chapter 7

Virtual and Augmented Realities - Continuing Engineering Education

Dane Laughlin[1,*] and Brian Laughlin[2]
[1]INVISTA- Koch Industries, Wichita, KS, USA
[2]The Boeing Company, Wichita, KS, USA

Abstract

This chapter serves as an overview for current extended realities (XR) tools, history, applications, design, and Human Factors theories underpinning this rapidly evolving area of technology. The chapter begins with a brief history of XR, along with functional definitions and examples of applications of the sub "Realities." The Milgram Reality-Virtuality continuum is used to provide conceptual spectrum across the Reality types with particular attention paid to the two most commonly used subtypes of XR; Augmented Reality (AR), and Virtual Reality (VR).

Given the highly engaging nature of XR that makes it such a powerful tool for use in User Interfaces, extra attention has been paid to the Human Factors and cognitive psychological principles, foundational to these technologies. The goal of which is to not only describe how these technologies function and how they are applied, but to also explain underlying mechanisms we use to ingest information about the world and how the appropriate use of XR can enhance our abilities to better process information and perform meaningful tasks. In the XR space, there is a lot of focus on technology and hardware. This is important, as developments in the hardware and technology drive the expansion of capability and use case, but a keen focus needs to be placed on the real needs of the end user. Simply stated "Nobody who ever bought a drill wanted a drill… they needed a hole." The synergies between effective design of the user interface and application, paired with the improvement of technology and hardware over time, will result in maximum value capture for both the end user and XR ecosystem.

The chapter closes out by walking through a number of real-world, use case examples that demonstrate the successful application of a myriad of XR technologies in industry. This is intended to be an XR reference guide, providing examples spanning from consumer to industrial applications.

Introduction

It is well documented that the predominant sense that human beings use to navigate and orchestrate their environment is sight. The vision systems of the body help with everything from balance to perception, and are integral to the way that external stimulus is translated within

* Corresponding Author's Email: dane.b.laughlin@invista.com.

the brain. In other words, a majority of the human experience happens through the perception of what people see. This is important to note when discussing the implications of using extended reality (XR) tools in the context of engineering.

In order to discuss XR, it is important to first define what XR is and the vernacular of the space. Fortunately, Paul Milgram came up with a framework called the reality-virtuality continuum in 1994 that breaks XR down into 4 categories: Real Environment, Augmented Reality, Augmented Virtuality, and Virtual reality.

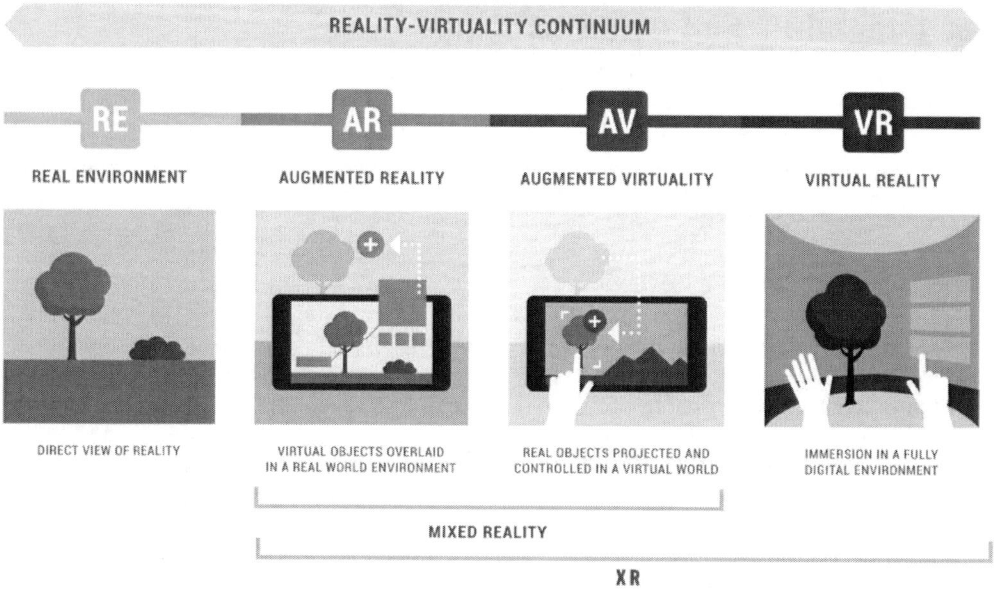

Figure 1. Reality Virtuality continuum illustrated (https://creatxr.com/the-virtuality-spectrum-understanding-ar-mr-vr-and-xr/).

The first category is the one that everyone is familiar with, the real environment. This comprises everything that people interact with on a day-to-day basis, including the physics and non-visible characteristics of the natural world. While at first glance, it is easy to dismiss the real environment as intuitive, there are a lot of complexities that are important to understand before delving into the other three categories of the reality-virtuality continuum. For instance, when the average person picks up an object, let's say an apple, they don't think about the weight of the apple or how the light dances off the waxy surface of its skin but these are important characteristics that distinguish it from a similar sized object like an orange. While this may sound trivial, these are very important characteristics to keep in mind when trying to replicate the experience of picking up an apple in a virtual environment. Another visual reference that people take for granted is the occlusion of a distant object by another object that is close by. For instance, when a window is open, it is easy to see outside. However, if the blinds are closed, the viewer can only see the parts of the environment not covered by the slats of the blinds. In other words, the slats occlude the environment behind them, removing the outside from view. Another trivial fact of day-to-day life that must be considered when creating a virtual world.

The second category of the reality-virtuality continuum is Augmented Reality or AR. AR allows the user to place digital objects into the real environment via a viewing device like a head mounted display (HMD) or a mobile phone. Augmented reality is a powerful tool for helping people to contextualize external data into an environment. For instance, when backing up a car, the driver may look down at their backup camera to find two lines that move back-and-forth with the steering wheel as they turn it. These lines help them to understand the boundaries of the vehicle and where it will be positioned given their input. This is a simple example of augmented reality in practice.

Figure 2. Augmented reality back-up camera illustration https://search.brave.com/images?q=dynamic+backup+camera+lines&source=web.

The third category of the reality-virtuality continuum is Augmented Virtuality or AV. AV is the inverse of augmented reality; instead of placing a virtual object in the physical world, AV allows you to blend traits from real world objects with virtual ones. A good example of AV is streaming a video of physical hand posture into a virtual space. This helps the user to manipulate objects in the digital world intuitively without controllers. Of all the categories within the continuum, Augmented Virtuality may be the least well known and applied but it is still a potent technology with compelling use cases.

The final category of the reality-virtuality continuum is Virtual Reality or VR. Virtual reality is a fully synthetic environment in which the developer controls all aspects of what the viewer experiences. Common applications of VR include gaming and training. Of the categories within the reality-virtuality continuum, VR is the most prevalent with an estimated 171 million VR users globally (Statica).

The History of Virtual Reality

The value of immersive experience has been recognized by human beings since the dawn of artistic expression. An example of an early immersive experience is the panoramic painting of Edinburgh by Robert Barker. Barker, credited as the inventor of the "panorama," painted landscapes that allowed the viewer to see more than the traditional field of view. This allowed the audience to experience the painting's landscape as if they were standing in the scene that Barker painted.

Later, in 1838, an English scientist Charles Wheatstone demonstrated the phenomenon known as Stereoscopy, where the brain interprets two distinct photos shown separately to the left and right eye, as one three dimensional object. This allowed the user to experience photos with "depth" adding to the authenticity of the experience. He later patented the "View-Master" stereoscope used for "Virtual Tourism."

Figure 3. Painting by Robert Barker, named London (https://www.britannica.com/art/panorama-visual-arts).

As time went on, demand for more immersive experiences led to the amalgamation of art and technology in the form of simulators. One of the earliest technology driven innovations was Morton Heilig's "Sensorama" machine in 1957, which allowed a person to experience sights, motion, sound, and even smells of the films that he created. The 10 minute motorbike ride through NY showed different images to each eye creating a stereoscopic effect. Additionally, the sensorama vibrated and blew air at the participant to simulate the thrill of an actual motorbike ride. People enjoyed this experience but it wasn't practical at scale due to the cost and complexity of the hardware.

In 1960, Heilig would attempt to create a more compact immersive technology called the Telesphere mask. This device was mounted on the wearer's head and allowed them to see stereoscopic scenes without the need of a large installation like the sensorama.

Fast forward to 1968, Ivan Sutherland and Bob Sproull created the first Augmented and Virtual reality headset named the "Sword of Damocles" that allowed the wearer to view basic computer generated graphics. Unfortunately, the "Sword of Damocles" suffered from the same issue that the "Sensorama" had, it was bulky and unable to be easily translated to a consumer device.

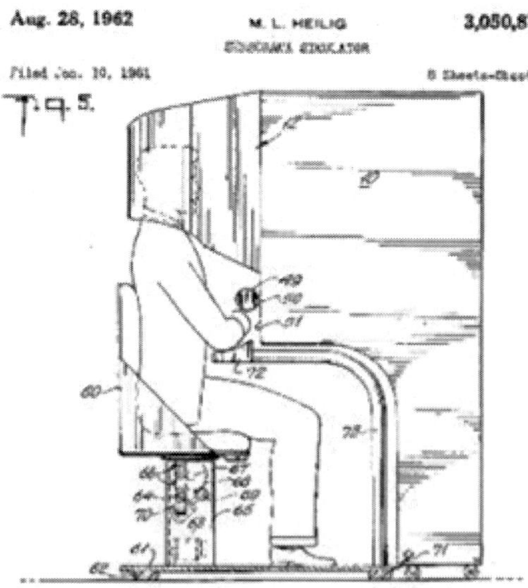

Figure 4. The Sensorama, patent #3050870.

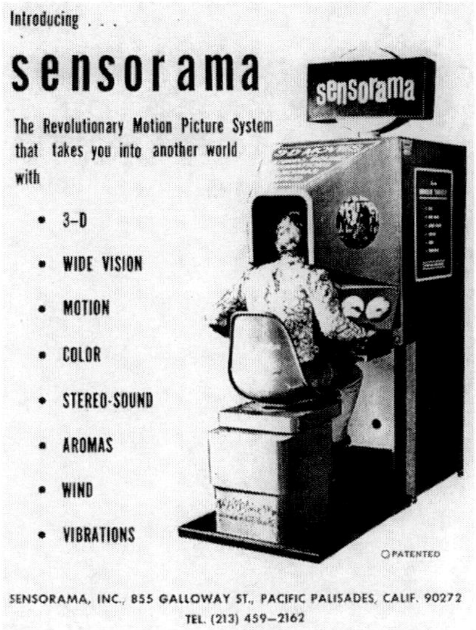

Figure 5. Sensorama Ad.

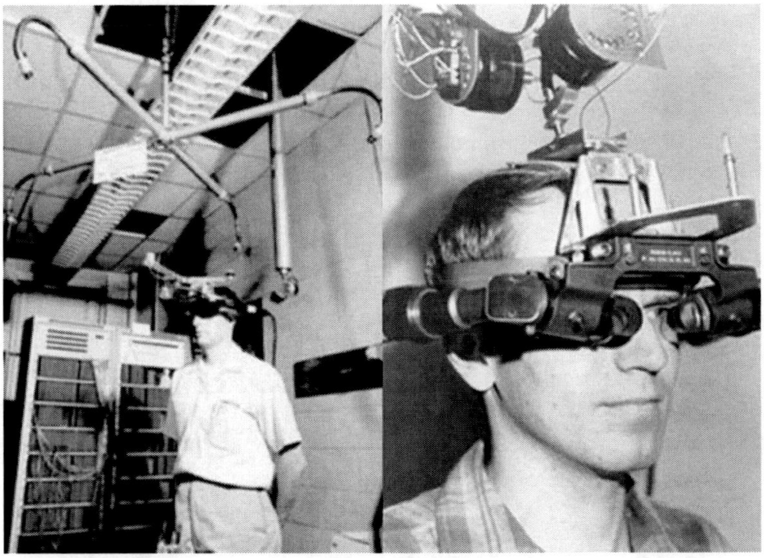

Figure 6. Ivan Sutherland's first VR Head Mounted Display, The Sword of Damocles (http://etsanggarp.blogspot.com/2016/03/).

Twenty years later, in 1986, Tom Furness led the "super cockpit" project for the Air Force that immersed the pilots in computer generated cockpit simulators. The "super cockpit" was "portrayed in a way that takes advantage of the human's natural perceptual mechanisms" (https://www.britannica.com/technology/virtual-reality/Education-and-training#ref884322) and was designed to help pilots to gain a more cohesive understanding of the man-machine symbiosis involved in flying an aircraft.

One year later, the term Virtual Reality was coined by Jaron Lanier of visual programming lab (VPL), providing researchers a ubiquitous term to describe the field. VPL later went on to be the first company to sell Virtual reality goggles called the "EyePhone." The "EyePhone," sold with an optional pair of haptics gloves, marked a major milestone in the haptics space

Over the next decade, several Virtual Reality landmarks would be achieved, including NASA training astronauts (Project VIEW), the creation of Virtuality Group arcade, the Sega VR for Genesis, the Sega VR-1, Nintendo Virtual Boy, and PTSD treatment (Georgia tech & Emory University).

In 2012, Palmer Lucky breathed life back into the consumer VR market by starting a kickstarter for a device he called the Rift. Lucky's Kickstarter raised 2.5 million dollars, ten times the goal of his campaign, eventually leading to the creation of Oculus. Later, in 2014 Oculus would be acquired by facebook who would oversee the introduction of the oculus go, quest, and quest 2.

The History of Augmented Reality

The 1968 "Sword of Damocles" is notable because it was both an early VR headset as well as one of the first demonstrable AR headsets. The first objects that were able to be visualized were wireframe geometric shapes like a cube. The headset was suspended from the ceiling and the wearer was required to spin around on a platform in order to use it.

The term Augmented reality was coined by Tom Caudell, a researcher at Boeing Computer Services, in 1990. Tom Caudell & David Mizell were asked to create a solution to the challenges workers faced with installing wiring harnesses in airplanes. They proposed a head mounted display that would superimpose the cables over the plane instead of relying on written instructions or pictures.

In 1998, Augmented Reality got its debut on the big screen with the introduction of the "Yellow Line" on NFL games. This yellow line allows the audience to understand the location of the first down and is an important marker for fans watching football at home today.

Figure 7. Cube visualization in "The Sword of Damocles" (https://courses.cs.washington.edu/courses/cse490h1/19wi/exhibit/ar-tech.html).

As the use cases for AR continued to grow, it necessitated tools for incoming developers to use. In 2000, Hirokazu Kato came up with an open source library called AR toolkit that helped to standardize methods of creating AR. This library utilizes video tracking to localize virtual graphics over the user's display.

In 2014 Google released its first augmented reality headset called Google Glass. This was the first attempt by a large organization to create a mainstream AR "accessory." This device, while being light and relatively comfortable for users, did not take spatial cues from its environment, limiting its use cases to a hands free display.

In 2016 Microsoft debuted the Microsoft Hololens, an augmented reality headset that uses LIDAR to spatially map the environment. This headset marked one of the first production devices that allow the user to introduce content that would interact with the real world. For instance, if the user placed a virtual object behind a wall in their space, the wall would occlude the virtual model, much like they do in real life. This is an important achievement because the introduction of a headset that understands its environment, allows for deeper integration and believability of AR content.

One of the newest hardware advances in the AR space is the introduction of the LIDAR sensors into mobile devices. For instance, the 2020 iPhone 12 Pro Max features a LIDAR sensor that allows the user to create 3d renderings of the real world without any additional tools. This means that the future of AR is in the pockets of the 5+ billion smartphone users globally.

Human Factors in XR

An important, yet often forgotten and/or misunderstood part of the XR ecosystem is the human factors component. While most would consider themselves an expert in being human, many don't really understand the capabilities and limitations of the sensory and cognitive systems that are leveraged to absorb information and understand the world. The purpose of this section is to provide an introduction to some of the Human Factors theory, mechanisms, and processes through which people perceive the world around them. Some suggestions are also provided for how to leverage this knowledge to better support the human element when creating or improving business processes and leveraging emerging technology.

Information Overload

With the advent of the Internet, access to information has grown exponentially, presenting a whole new problem of how to access only what is needed, where, when, and in the right context. Here are some interesting facts about the abundance of information, at our fingertips:

- More information produced in the last 30 years than previous 5000 years
- All printed knowledge doubles every four or five years
- Weekday edition of the New York Times contains more info than the average person would encounter in a lifetime in seventeenth-century England.
- >4000 books are published around the world every day.

- The average office worker uses 250 pounds of copy paper a year, twice the amount consumed 10 years ago.
- Today, English language approx. half a million words
  - 5x Shakespearian times
  - Average American recognizes approximately 20,000 words
  - Evening news is limited to around 700 words
  - One Sunday Edition of the New York Times
    - 1600+ pages
    - >12pounds
    - >10M words
    - 18hrs/day@500wpm = 18 days to read

Given the deluge of information that exponentially grows over time, it is important that tools be created to better manage the vast information resources available. Additionally, a foundational understanding of how people intake information is required so that it's possible to design new information solutions that are optimal for human use.

Verbal and Non-Verbal Communications

The intercommunication between people is a complex mix of visual, verbal, tactile, and auditory channels. As humans, the simultaneous use of these channels provide a rich contextual conversation that can be difficult to mediate via digital means. There are two types of these communications channels in human to human conversations: verbal and non-verbal cues. Verbal communications are an important communications channel that includes pitch, rhythm, volume, and tone in addition to the actual words spoken. In fact, the relative importance of the words that are used is minor in comparison to how the words are said. Research shows that only 7% of the message in a conversation is communicated in the words that are used. Approximately 38% of the communications take place in how the words are said and 55% comes from facial expressions, body language, and other non-verbal cues. Unfortunately, technology that is used everyday to communicate is not well equipped to handle the nuance of human discussion. Text based communications is arguably the predominant form of mobile communications today, and is notoriously lacking as a communications channel, which is why it is often easy to misunderstand the intended meaning of a typed text. For instance, the reason that a message, written in all caps, seems like the author is yelling, is because the receiver is missing the other contextual information of the conversation. As a result of this "cues filtered out" condition, there is a tendency to ascribe meaning to the capitalization of the letters that may well not be intended. The invention of emojis is an example of a relatively modern invention intended to allow us to add some level of contextual richness back into the text messages we exchange.

Sight

The tools that the human body is equipped with are miraculous in their ability to understand nuance. For instance, the human eye only weighing an ounce, can distinguish up to 900 shades of grey (https://www.ncbi.nlm.nih.gov/pmc/articles/PMC3043920/#:~:text=Even%20for%20high%20brightness%20and,around%20900%20shades%20of%20gray), up to 10 million different colors (Judd and Wyszecki, 1975) and take in more information at a glance than the largest telescope known to man. With that said, it does have an optimal environment in which it can operate. Studies show that people read anywhere from 10-30% slower from a computer screen than from written text and that black text on a white background is preferable to white text on a black background (http://citeseerx.ist.psu.edu/ viewdoc/download?doi=10.1.1.12.2890&rep=rep1&type=pdf). Understanding facts like these is important when designing a human machine interface that relies on human visual systems.

The light sensing cells in a human eye consist of rods and cones. Rods are for light and dark sensing while cones perceive colors relative to a mix of red, green, and blue. While the eye is well suited for distinguishing color, it can be tricked into seeing things that are not there. For instance, if you look at the flag below for 20 seconds and then look at the white margin on the page, the black, green, and tan colors on the flag should appear as red, white and blue. This is due to cone fatigue from the presented colors.

Figure 8. Afterimage optical illusion illustration (https://faculty.washington.edu/chudler/afterflag.html).

Understanding the fundamentals of how eyes perceive color is helpful when creating XR training because a poorly designed user interface (UI) can be detrimental and uncomfortable for the user. Color reverberation (Greenberg, 1998; Hill, 2001) is a phenomena where “the edges of two directly adjacent colors appear to merge, blur and glow, giving the illusion of motion” (https://webdesign.tutsplus.com/articles/why-you-should-avoid-vibrating-color-combinations--cms-25621). If at all possible, these color combinations should be avoided.

Another consideration that must be made when designing using color is the accommodation of color anomalies. Color anomalies encompass the absence or abundance of the color cones in the eye. In the case of color blindness, it is reported that “about 8% of men and 0.5% of women worldwide have some sort of color vision deficiency, though the numbers tend to shift depending on a population’s heritage.”

Some Vibrating Color Combinations

red/green	red on green	green on red
blue/orange	blue on orange	orange on blue
green/magenta	green on magenta	magenta on green
yellow/cyan	yellow on cyan	cyan on yellow
blue/magenta	magenta on blue	blue on magenta
orange/yellow	yellow on orange	orange on yellow (not so bad)
blue/green	green on blue	blue on green

Figure 9. Vibrating color illustration (https://www.pfw.edu/training/accessibility/examples/color.html).

"There are seven official diagnoses of color blindness: four different types of color blindness fall in the red-green category, two are in the blue-yellow spectrum and one version describes a type of vision completely lacking in color (https://www.allaboutvision.com/conditions/color-blindness/types-of-color-blindness/). The names for the various forms of color blindness are:

- Protanopia (aka red-blind) – Individuals have no red cones.
- Protanomaly (aka red-weak) – Individuals have red cones and can usually see some shades of red.
- Deuteranopia (aka green-blind) – Individuals have no green cones.
- Deuteranomaly (aka green-weak) – Individuals have green cones and can usually see some shades of green.
- Tritanopia (aka blue-blind) – Individuals have no blue cones.
- Tritanomaly (aka blue-weak) – Individuals have blue cones and can usually see some shades of blue.
- Monochromia- Individuals see no color at all

Designers should be mindful that approximately 10% of their audience is probably colorblind and accommodate this through easy to read signage and displays in XR. *Example: supplement with shape, text or pattern changes. Also avoid Red, Green and Black combinations or else important information might go unnoticed and unheeded.*

Sound/Auditory

In addition to sight, the body is also equipped with an auditory system that can distinguish between hundreds of thousands of different sounds. These sounds can have different pitches, tones, directionality, patterns, etc. all of which provide additional layers of meaning in an environment. For example the Doppler effect occurs when something speeds by and is indicated by a change in pitch and sound intensity, which is naturally interpreted as approaching and departing. Conversely, the absence of sound can also be highly informative regarding the operations and situational awareness (Parents can tell you that when children get quiet, it is often time to beware). Hearing is an important part of every experience and contributes to the retention of the environment that is presented. Additionally, it provides an auxillary channel through which information can be conveyed, enriching the experience overall.

There are several human factors theories, studies and phenomena that can help to better understand how people sense and process sound. Some examples are as follows:

Dichotic Listening (Cherry, 1953, Moray, 1959) dichotic listening tests, first used by Donald Broadbent in the 1950s, investigated human attentional capabilities by presenting different stimuli to the right and/or left ears of subjects (Hugdahl, 2015). His findings in this space resulted in the development of his Filter Model of Attention, and provided valuable insights into how we attend and process information presented auditorily. This can help us to better design the auditory components of User interfaces. In summary, he found:

When subjects are simultaneously presented two auditory signals, they can only attend to one at a time. In practice, it is best to present only one auditory signal, or attempt to minimize multiple simultaneous auditory signals as we can only attend to one at a time. Use time, location, and pitch separation of signals that may coincide to minimize cognitive interference.

Doreen Kimura (1961) used Dichotic Listening tests in the early 1960s to investigate lateral asymmetry of auditory processing in the brain, and found that healthy subjects have a right ear superiority for verbal stimuli, and left ear superiority for melodic (music) perception. This revealed the general predominance in humans for left hemisphere speech perception and right hemisphere melodic perception.

In practice, it may be more effective to present speech information to the right auditory channel and melodic related content to the left.

Shadowing tasks (Cherry, 1953), where participants are asked to repeat the content of a message they are listening to resulted in poor recall of the shadowed message. This indicated that processing required to shadow occurred in working memory and is not preserved in long term stores.

In practice, having a subject repeat a message while shadowing will result in poor retention of the verbal content.

Gated Response

Gated or filtered response refers to the neural process of filtering out redundant or irrelevant stimuli from the environment and protects our limited cognitive resources from overload (Jones et al., 2016). It is considered primarily automatic, however is also involved with deliberate attentional processing in the brain, suggesting they are not modally independent. An example of how humans utilize gated responses is called "the Cocktail party effect." This effect describes the phenomenon of being in a noisy environment, such as a cocktail party, engaged in conversation with someone, when you hear some salient words, like a name, or sound that impinge on your attention, while otherwise engaged.

This translates to use in XR when the developer uses specific, unique sounding names or commands in order to improve attention and retention of material in a simulation.

Touch

Touch is the most tangible of the senses and a good crutch when the other senses are unable to be used (Bronkhorst, 2000). In the body, there are 3 different touch receptors embedded in the skin: thermoreceptors, mechanoreceptors, and chemoreceptors. Thermoreceptors respond to change in temperature, chemoreceptors respond to chemicals applied externally or released from within the skin, and mechanoreceptors respond to mechanical inputs like vibration. In XR development, mechanoreceptors are most pertinent as this is what haptics generally act on. Interestingly, despite the body relying heavily on mechanoreceptors, it commonly will misperceive mechanical inputs. A good example of this is the two point discrimination test. This test is performed by blindfolding the participant, touching two distinct points on their skin and then asking them to discern if there are two stimuli or one. Then, the points are moved closer together until the participant cannot tell the difference between 2 stimuli vs 1 stimuli. This gives a minimum threshold for the haptic resolution needed to authentically replicate the feeling of touch.

Cognitive Psychology Thinking Computational Model

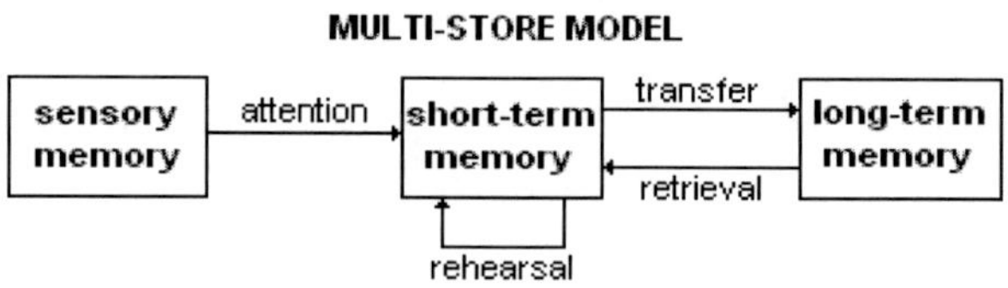

Figure 10. Atkinson and Shiffrin (1968) multi-store model of memory diagram.

Sensory Memory

Sensory memory stores are the shortest term memory stores. They act as a sort of buffer that enables us to retain sensory inputs long enough to process these impressions after the stimuli have ended. While these impressions are accurate, they are fleeting in nature. Each of our senses has a corresponding sensory memory store and they vary in duration from a few hundred milliseconds to several seconds in length, depending on the sensory system. The naming convention for the systems are as follows:

- Sight- Iconic
- Sound- Echoic
- Touch-Haptic
- Smell/taste- Olfactory

For the purposes of this chapter, only iconic, echoic, and haptic modalities with considerations of kinesthetic memory will be discussed in relevance to XR technologies.

Iconic Memory

When a person looks at the world their brain takes a snapshot of what is seen (https://human-memory.net/sensory-memory/). This snapshot enters into the brain in the form of a sensory memory known as an iconic memory store. As discovered by George Sperling in 1960, iconic memory is notoriously fleeting, lasting less than a second in young adults. This memory store can be observed by looking intensely at something and then closing the eyes. The brief after image that is seen is an example of iconic memory.

Echoic Memory

Echoic memory is an important tool to use when teaching in an XR training. Echoic memory is the mental echo experienced after an auditory stimulus is heard and can last 2-3 seconds in the working memory. This type of memory acts on the phonological loop and is a crutch for other, more fleeting types of memories. An example of this is when a person tries to recall a phone number. Generally, they will read the phone number off of a page and repeat it to themselves until they get it written down. Unbeknown to them, they are using echoic memory, in their working memory to store the information of the phone number. Something that is important to note about echoic memory is that it gets corrupted by other sounds in the environment. In the situation of a person trying to remember a phone number, if another person calls their name, breaking their focus and distracting their echoic memory, they will forget what they were repeating. During the development of XR, it is important not to interrupt memory tasks with extraneous noises, causing the user to forget what they are focused on.

Haptic Memory

"Haptic memory is the form of sensory memory specific to touch stimuli. Haptic memory is used regularly when assessing the necessary forces for gripping and interacting with familiar objects" (Johansson et al., 1993). Bliss, Crane, Mansfield, and Townsend (1967) were some of the first scientists to study haptic memory, finding that haptic memory stores are similar to iconic memory stores in duration and capacity. The parietal lobe, part of the brain responsible for processing sensory information, includes the somatosensory cortex which helps the body to interpret touch and filter extraneous signals. In XR, haptics are important in enhancing the immersion a person feels. Aligning visual, auditory and haptics feedback helps with continuity and creates a higher fidelity experience for the user.

Split Attention Effect

Split Attention Effects occur when the same modality is being used for various types of information with a given display (Moreno and Mayer, 2000). Split Visual Attention, for instance, could use a combination of text and images, causing the user to split their attention between the two types of materials to understand the information being presented.

An Auditory split attention effect can happen when an audible sound is played and another unrelated sound occurs (someone tries to talk to you while you're counting change). This can also occur across channels, for example when unrelated sounds occur while the user is focused on visual material (e.g., When looking for a house number and you turn down the stereo in your car). This is less intuitive as it would seem that loud sounds shouldn't affect your ability to see, however the split attention between both stimulus channels results in increased cognitive load and degraded processing capacity.

Split attentional effects can also be exacerbated when there is poor "Spatial Contiguity." Spatial Contiguity is a term referring to distance or time separating two pieces of information. Split attention effects where information elements are separated by excessive time or distance, within or across channels, inhibit information absorption by the user.

When designing XR user interfaces, it is important to not overwhelm the user by submitting them to multiple forms of unrelated stimulus at the same time. That is, design the experience such that the information conveyed to the user focuses all stimulus, on the same topic such that all information elements are cohesive and reinforce each other exhibiting Spatial Contiguity. Example, a verbal description is given while the visual item being described is being presented simultaneously.

Making displays as intuitive as possible can go a long way in helping enhance usability relative to interface designs. Integrating information pertaining to specific features of a graphic, into the graphic, provides important contextual cues that clarify the data and its relationship to said features. This concept, known as Temporal Continuity, facilitates user's understanding of diagrams by reducing the working memory load, thereby preserving greater cognitive bandwidth for the task being performed. In practice this means labeling content locally to increase clarity. Further enhancements can be realized by providing narration instead of on screen text.

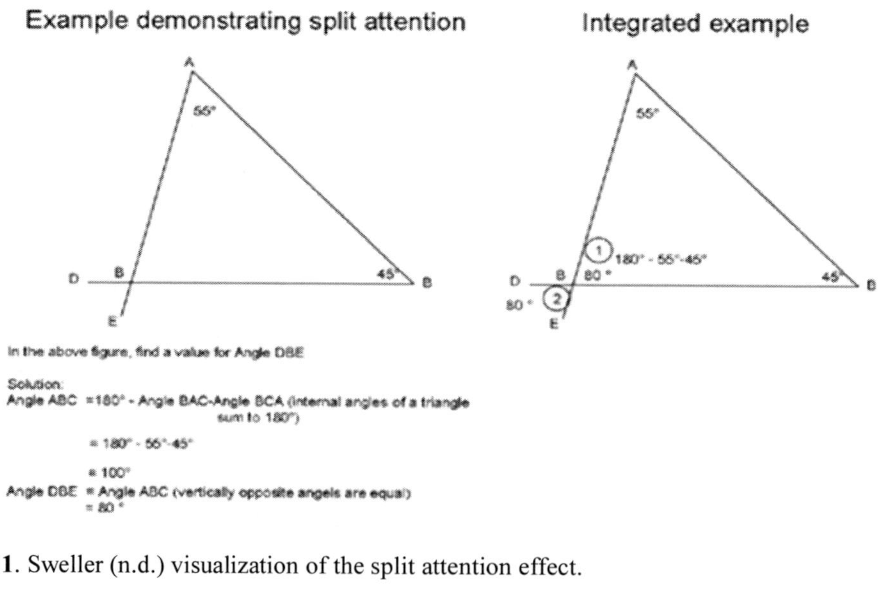

Figure 11. Sweller (n.d.) visualization of the split attention effect.

Automated Behaviors

One of the most powerful capabilities humans have is the ability to automate a complex series of behaviors and actions to perform increasingly complex tasks. Over time, and with practice, these tasks or collections of synchronized and complex movements are perfected into unified flowing patterns that achieve some meaningful end. As amazing and useful as this automation capability is, it can also, conversely be problematic when we attempt to alter these well learned behaviors to accomplish an alternative task. There are a multitude of different sub processes that underpin these automated behaviors. These sub processes are important to understand in order to best leverage human capabilities, and in some circumstances protect people from their own preconceived habits.

The Stroop Effect (1935) is a phenomenon that reveals much about how we process information through interference between automated reading ability and color reporting (https://www.verywellmind.com/what-is-the-stroop-effect-2795832). The Stroop Test consists of creating a list of color words with each word written in a color other than what the word represents. For example the word “green” would be written in any other color but green, same for all other color words in the list. The subject is first asked to scan the list, reading as quickly as possible each color word in order. The second time, the same subject is asked to scan through the list and report the color of each word, as opposed to what the word says. There are sizable discrepancies in how fast an individual can read the word (very fast) vs the color of the word (more slow). which highlights a number of psychological principles, one of which is Automaticity.

It is theorized that because we are so efficient at the process of reading, that we’ve automated this process to such a degree, that reading has become compulsive, making it nearly impossible for us to see a word and not automatically read it.

This explains the very rapid performance of the first trial of the experiment when the subject is asked to read the list of color words. When the subject is asked in the second trial to report the color of the word, instead of reading the word itself, the behavior of reading is so automated and compulsive that it interferes with our ability to attend to only the color of the word, forcing us to cognitively suppress the reading response and redirect our attention to reporting on the color instead. This redirection or task switching instantiates a time penalty for each color word being reported on, as the user first starts to read, then suppresses reading, redirects their attention to the word color, interprets the color then reports on it. This causes the reporting of the word’s color to take much more time than simply reading the list of words.

Figure 12. Stroop effect demonstration.

Kinesthetics

Kinesthesis, often referred to as "Muscle Memory," is not a memory store at all, but rather the process of muscle proprioceptors sending positional information from the muscles to the brain (Edwards, 2011). Kinesthesis involves the learning of movements that are commonly performed by an individual through numerous repetitions and requires the individual to have a sense of their body and bodily changes throughout a range of motion expressed while performing a specific, intentional motor skill. The receptors in the muscles will transfer positional, directional, rate of speed, force, and other descriptive characteristic information of the movement to the brain, informing it about what the body is doing. Eventually the brain will learn these ques well enough to enable the individual to replicate the movements, even when the senses are inhibited (person not watching what they are doing). This is because the brain has formed and recorded patterned connections among the body part locations, speed, position, muscle tension, etc. enabling them to be accurately reproduced time and time again with exacting precision. The greater frequency and duration the repetitive motions are practiced, the better the motions are learned and less effort and attention required to perform the motor skills, resulting in movements becoming nearly instinctual.

The automation of tasks, enabled by the process of Kinesthesis, is one of the very powerful capabilities we have that enable us to perform complex tasks accurately, reliably, and repeatedly. This capability provides us the mechanism to build complicated procedures out of a series of incremental, granular muscle movements in a "LEGO-esque" way. This process is why we can very quickly master an initially complex sequence of movements and turn them into a comprehensive activity such as driving an automobile using a stick shift, riding a unicycle, or performing gymnastic acrobatics. While this is incredibly useful, this natural tendency to automate can pose potential safety hazards for us when these automated motor patterns are expressed in the wrong environmental or situational context. These automated behavior patterns are generally easy to extinguish in a reasonable amount of time. However, it requires focused attention to extinguish the automated behavior, while the new actions are performed.

Some examples that highlight where contextual mismatch has historically caused problems include:

- Switching from stick shift to automatic transmission vehicle after having driven it for several months. Often immediately following the switch, a user may find themselves stepping towards a non-existent clutch pedal and/or reaching for a phantom gear shift knob.
- Attempting to Drive to a Dr's appointment first thing in the morning, at the same time you usually drive to work. Often individuals can end up driving to work (automated process guides you to your standard destination) instead of going to the Dr's office as intended.
- Traveling overseas to another country where they drive on the opposite side of the street, such as traveling from the US to England. If you're not paying close attention, you may end up driving on the wrong side of the lane into oncoming traffic.

In practice, this means that developers must be aware of automated behaviors and how new actions can positively or negatively affect performance in a similar task. The individual will need to be actively focused in order to extinguish the formerly automated task.

Chunking

Chunking refers to a phenomenon where the brain takes a collection of disparate pieces of information and groups them together into a relevant whole. The postulated reason the brain does this is that it creates less cognitive friction to group similar types of information together rather than processing it as a whole. George Miller (1956) suggested that chunking is the reason that people break phone numbers down into segments. For instance, instead of trying to remember the number 3428767877, a person might break it down into 34 287 678 77. Effectively, this reduces the number of distinct characters that must be remembered from 10 characters, to 4 groups. Chunking can also be observed when a person is asked to remember a date. For most, it is much easier to recall the date June, 14th 1947 than it is to remember 06141947. Miller found that the average person has enough working memory to recall 7 +/- 2 chunks. In practice, this means that developers must create a framework to organize information into larger logical chunks so that it is easier for the user to comprehend.

Serial Position Effects

Serial position effect is the tendency for people to recall the beginning and end of a series most accurately. The recency effect pertains to the information most recently heard, at the end of the series, while the primacy effect refers to the information that was read first. Coined by Hermann Ebbinghaus (1913), the recency effect is most commonly used by marketers who start their advertisements with an attention grabbing statement and end the ad with closing comments that are designed to be memorable. In XR, this applies when trying to teach the user a skill. The designer should begin with important information that the user needs to remember (primacy), spend the “down time” in the middle having the user practice the skill, and finish with a learning lesson. This method optimizes the user’s retention of the material and utilizes the least valuable time to build automated responses like muscle memory.

XR Integration and Extensibility

The power of XR technologies lie in their ability to contextualize information for its user. Thus, it is important to understand the ecosystem of tools surrounding XR that enhance its user experience and practicality. Part of the ecosystem focuses on user experience while others center around the data feeds in and out of the devices.

User Experience

In immersive experiences, embodiment is the "sense that emerges when [a body]'s properties are processed as if they were the properties of one's own biological body" (Kilteni et al., 2012). This sense of embodiment makes the experience believable for the user and increases engagement. An example of an XR add on that contributes to a sense of embodiment is the use of haptics. In XR There are many ways that haptics can manifest in an immersive experience such as force feedback and vibrotactile feedback. Force feedback means that the haptic device the wearer is using exerts external force on the body. In the case of haptics gloves, force feedback is exerted by an exoskeleton that supports the hand. While the type of haptic required for each experience is different, some experiments have found "significant superiority of force feedback over no haptic feedback regarding embodiment" (https://www.frontiersin.org/articles/10.3389/frvir.2020.573167/full).

Figure 13. Haptx haptic gloves in use (https://siliconangle.com/2019/12/19/vr-glove-maker-haptx-raises-12m-announces-partnership-advanced-input-systems/).

However, the challenge that designers face with current force feedback systems is they are hardware intensive. In other words, the devices required to have a force feedback simulator are comparably expensive and can be cumbersome to set up.

Vibrotactile feedback is another form of haptics that is more widely used because it is natively built into most controllers. In contrast to force feedback, which is more ecological, vibrotactile feedback is symbolic https://www.frontiersin.org/articles/10.3389/frvir.2020.573167/full). An example that differentiates vibrotactile from force feedback can be illustrated by picking up an object in VR. In the case of force feedback, when the user's virtual hand comes in contact with the object, the exoskeleton restricts movement, physically replicating the bounds of the digital object. Whereas, in a vibrotactile experience, the controller will vibrate when the virtual hand makes contact, but the user is free to pass their hands and fingers through the object. Ultimately, the type of haptics used in an experience must be weighed by the scalability vs experiential benefits.

Another experiential detail that must be contemplated when designing an XR experience is audio. In this space, hardware is not the constraint as most devices use traditional audio devices like headphones or speakers placed near the ears. Instead, it is the sound design that is most important. One example of how audio can be used to improve a sim is by tying sounds to an object so that it draws the users attention to the direction of the sound. This is commonly done with avatars when they are talking directly to a user so that they know which way to look. If the audio isn't tied to an avatar or location, the voice is omnipresent. While omnipresence can be confusing if the intention is to get the user to look at a particular place, it can also be helpful for giving general guidance that doesn't pertain to the environment. Thus sound and the location that it is coming from should be carefully constructed as a tool to guide the user's attention.

Data Integration

Data is another essential part of crafting a valuable XR experience and it can be broken down into two different camps, incoming and outgoing. Incoming data is information that is imported into an experience from the outside world and is important because it forms the foundation on which the contextual part of an experience is built. In VR, imported data can come in many different forms. For instance, in a flight simulator, the creator may want to replicate weather effects in real time or show real flight information to the user in the virtual cockpit. This allows the training to be unique every time and improves the accuracy of the experience.

In the case of Augmented Reality, data can be fed in and overlaid onto the user's physical environment to provide additional guidance or context. For example, an operator working at a manufacturing facility is tasked with doing a pump swap in the field. In order to execute this task, the operator needs instructions on how to perform the task, inventory information about the parts they need, the location of the pump, and safety training for their task. Without AR, the operator is required to look up each piece of information individually and then understand how it applies uniquely to their situation. With AR, each of those pieces of information can be introduced in a cohesive way. Instead of searching discreetly for each piece of data, the AR experience integrates it directly into the experience. The AR operator experience looks something like this: the operator receives a work order for a pump swap to their AR headset. It notifies them that there are 2 pumps left in stock so they request that the parts be delivered to the work location. They then navigate to the work location using a visual overlay, like google maps. Once at the work location, they launch the work instructions for the task which shows them the safety considerations for the job first and then walks them through the procedure in a hands free manner. In this example, the augmented reality solution necessitates the integration of an inventory management system, geospatial location data, an EHS guidance tool, and a procedure database. The integration of these data points, to provide all information required to accomplish a given work statement at the point of use and time of need, is an example of a concept called information kitting.

Outgoing data is equally important but used differently in the XR space. Instead of adding context to the virtual world, outgoing data ties the action in the virtual world to analytics that can be used to improve performance, drive automation, etc. For instance, performance data relating to how a trainee acted in a training sim helps the trainer to understand where they have opportunities for growth. This data can be analyzed asynchronously, meaning that the trainee can perform a training without the need for the trainer to physically be with them, opening up the availability of training in the workforce and automating training records. Another example of outgoing data being used in XR is monitoring of task execution in manufacturing. In this example, an operator is asked to isolate the energy going to a machine using the approved lock out tag out procedure. Using AR, not only can they determine the location & steps necessary to complete the action, incoming data, but they can also use outgoing data to record the timing of the steps and who completed them for compliance records. XR allows for more granular user data to be generated around proficiency of the user.

Information Kitting

Information exists in disparate systems, repositories and formats, often requiring excess time and effort to find all necessary information needed to accomplish a given task. This increases the cost of doing business.

In addition to the time wasted gathering information, the disruption in mental flow compounds the problem, due to human limitations in ability to multitask when completing mentally demanding work. Information kitting reduces the time and cognitive load of a task by collecting, sorting and assembling information into a format optimized for consumption. This improves the efficiency of information work by streamlining the business processes required to perform a given task. We also simultaneously increase quality by ensuring adherence to best business practices, reducing reliance on tribal knowledge, and cognitive burden to perform business tasks.

Combining Information Kitting and XR

By leveraging XR capabilities in conjunction with the benefits of information kitting, we can not only collect, optimize and organize the information in the process to be efficient, we can also directly place the work instructions and reference artifacts local to the work being done as a digital overlay onto the physical world.

By embedding the information needed to do the job into the environment itself, the information can simply be referenced in context, instead of processed, remembered, recalled, projected and then used, thereby dramatically reducing cognitive workload.

XR and Remote Assist

One very useful feature that further enhances the visualization capabilities of XR is the ability to provide remote assistance to an end user from one or more experts remotely located. Remote assist enables two-way, real-time communications among multiple parties, providing access for a user to expertise they are lacking or other perspectives on a given situation or work statement. The XR user is able to call others into an impromptu video call where the XR user can see the folks called through their webcams, and they appear in one or more window panes, similar to a video conference. The remote callers can see what the XR user is looking at from a first person point of view, via the cameras on his head mounted display. This configuration allows the whole team to see what the XR user is seeing and use that first person view video feed to survey the situation or work being done. Additionally, much like a party line, each of the individuals on the call can converse back and forth, asking and answering questions as they discuss the situation. The remote individuals as well as the XR use can also use an annotation feature of Hololens to draw annotations, diagrams, symbols, etc. on top of the real world to demarcate specific features of interest to guide the discussion.

Figure 14. The use of AR enabled work instructions enables easier, better and faster understanding by demonstrating the task required while overcoming language barriers.

Figure 15. By providing kitted information, containing all that is needed at the point of use and time of need, AR enables mechanics and shop floor personnel to focus their attention on the task at hand. This dramatically reduces frustration, distractions, and time wasted searching, compiling and organizing information.

Skillset

When considering Augmented and Virtual reality for use in engineering, a basic foundation of a few different skill sets should be considered. A good place to start is to get experience with a game engine such as Unity or Unreal engine. These game engines enable physics, lighting and other important visual characteristics of the experience to be tailored to the end users' needs.

Another key skill to hone is 3d model creation and rendering. In order to use a gaming engine to create an immersive experience, one must first acquire the models needed within the world that's being created. Some models, like a landscape, can be generated within the game engine using prefabricated 3d models and textures. However, if the desired product is unique, custom 3d models will have to be created. Generally, in an industrial application, a mix of prefab and custom models will be used so an understanding of how to create both is helpful. Additionally, the sizing of models must be taken into account when designing a solution. A high resolution model with a large polycount is great for viewing individually, but when added into an XR scene, it could cause performance issues (rendering issues, artifacting, lag).

Model optimization is a skillset that enables the models used in XR to be right sized for the application. One example of an optimization process used to reduce model size is decimation. Decimation works by removing nodes on the mesh of a model, effectively reducing its resolution. A delicate balance between model resolution and file size ensures that the application is believable/helpful but also can be run with no performance issues.

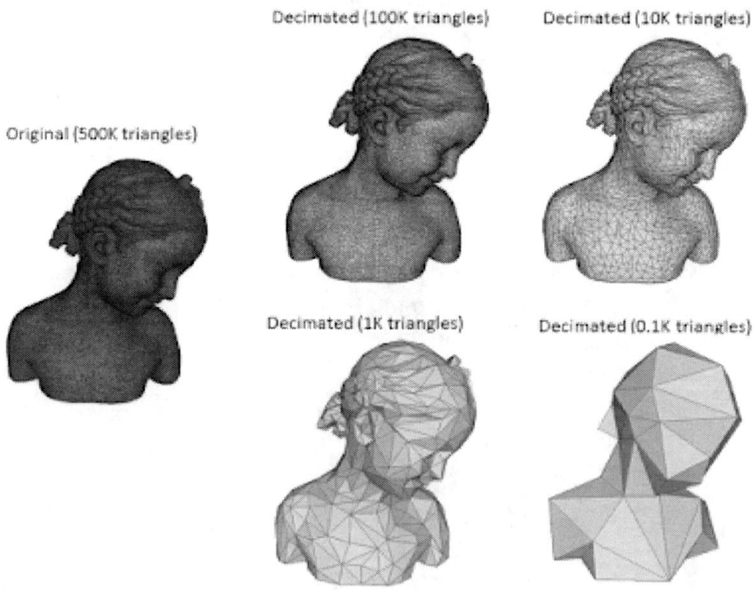

Figure 16. Decimation resolution visualized (http://kmamou.blogspot.com/2015/06/a-simple-c-class-for-3d-mesh-decimation.html).

Polycount - The Number of Vertices Found within the Mesh of a 3d Model. Higher Polycount Generally Means Higher Resolution

Beyond the experiential benefits of XR, another advantage XR provides is being able to record data relating to the actions of the user in the app. For instance, if a company wants to use VR to train employees, it is important to understand how the employee is performing within the simulator. In this case, there are several options for recording the data but the predominant standard, currently, is called xAPI. xAPI allows the developer to use actor, verbs, and objects statements to record what the user is doing. For instance, in the VR training example, the company could record a statement that like:


```
“actor”: {
“name”: “John Doe,”
“mbox”: “mailto:Johndoe@example.com,”
},
“verb”: {
“id”: “http://adlnet.gov/exapi/verbs/initialized,”
“display”: {
“en_US”: “initialized”
}
},
“object”: {
“id”: “http://examplexAPI.com/verbs/initialized,”
“definition”: {
“name”: {“en-US”: “Example xAPI”
    “type:” “https://xAPI.org/xapi/example
```


In this case, the data shows that John Doe initialized the activity “Example xAPI.” While this may seem basic, the actor, verbs, and objects structure allows for flexibility in the activity, user, and environment that can be reported. Ultimately, xAPI is a very important XR supporting technology because it allows the value of the tool to extend beyond the experience it creates.

Conclusion

XR technologies are currently going through an explosive renaissance, much like the internet did in the late 1990s, and the landscape will continue to change as new tools emerge that allow the user to garner experiences that more closely reflect and augment their real environment. As a result, the incredible rate of technological advancement can be daunting to a technologist focused on delivering XR solutions in the context of engineering. Thus, it is important to remember that XR is just another tool in the proverbial toolbelt, and applications should be tailored to the user in a way that provides value beyond what any other technology can deliver. Not because it is flashy and exciting. In order to realize the value of XR in industry, a concerted effort must be made to interrogate the business needs, as well as study the human doing the task, ensuring that the experience is optimized for both business value and human performance.

When done effectively, this considers all of the sensory channels, including: sight, auditory, olfactory, and haptic modalities of the environment. XR tools, like any other tool, must be optimized for the task at hand and molded to the end user's capabilities and needs. Using the tactics and techniques laid out in this chapter, along with a basic understanding of how cognition and sensory channels work, allow XR practitioners to effectively design tools to help their end user, ultimately delivering business value.

References

Afterimages. UW Faculty Web Server. (n.d.). Retrieved December 22, 2021, from https://faculty.washington.edu/chudler/afterflag.html.

All About Vision. (2021, November 8). *What are the different types of color blindness?* All About Vision. Retrieved December 22, 2021, from https://www.allaboutvision.com/conditions/color-blindness/types-of-color-blindness/.

Atkinson, R. C., Shiffrin, R. M. (1968). "Chapter: Human memory: A proposed system and its control processes." In Spence, K. W., Spence, J. T. (eds.). *The psychology of learning and motivation*. 2. New York: Academic Press. pp. 89–195.

Bliss, J. C., Crane, H. D., Mansfield, P. K., & Townsend, J. T. (1967). Information Available in Brief Tactile Presentations. NASA CR-623. *NASA contractor report. NASA CR. United States. National Aeronautics and Space Administration*, 57-86.

Britannica, T. *Editors of Encyclopaedia* (2018, March 22). panorama. *Encyclopedia Britannica*. https://www. britannica.com/art/panorama-visual-arts.

Bronkhorst, Adelbert W. (2000). "The Cocktail Party Phenomenon: A Review on Speech Intelligibility in Multiple-Talker Conditions." *Acta Acustica United with Acustica*. 86: 117–128. Retrieved 2020-11-16.

Cherry E. C. (1953). "Some experiments on the recognition of speech, with one and two ears." *Journal of the Acoustical Society of America*. 25 (5): 975–979. Bibcode:1953ASAJ...25..975C. doi:10.1121/1. 1907229. hdl:11858/00-001M-0000-002A-F750-3.

Cherry, K. (2020, May 25). *How the Stroop effect works*. Verywell Mind. Retrieved December 22, 2021, from https://www.verywellmind.com/what-is-the-stroop-effect-2795832.

Color and contrast. Information Technology Services Training - Color and Contrast - Examples for Accessibility - Purdue University Fort Wayne. (n.d.). Retrieved December 22, 2021, from https://www. pfw.edu/training/accessibility/examples/color.html.

Ebbinghaus, Hermann (1913). *On memory: A contribution to experimental psychology*. New York: Teachers College.

Edwards, William (2011). *Motor learning and control : from theory to practice*. Belmont, CA: Wadsworth Cengage Learning. ISBN 9780495010807. OCLC 695606314.

Elaine. T. (1970, January 1). *Garp Virtual reality*. March 2016. Retrieved December 22, 2021, from http://etsanggarp.blogspot.com/2016/03/.

Encyclopædia Britannica, inc. (n.d.). *Education and training*. Encyclopædia Britannica. Retrieved December 22, 2021, from https://www.britannica.com/technology/virtual-reality/Education-and-training# ref884322.

Hugdahl, Kenneth (2015), "Dichotic Listening and Language: Overview," *International Encyclopedia of the Social & Behavioral Sciences*, Elsevier, pp. 357–367, doi:10.1016/b978-0-08-097086-8.54030-6, ISBN 978-0-08-097087-5.

Introduction to psychology. Lumen. (n.d.). Retrieved December 22, 2021, from https://courses.lumenlearning.com/wmopen-psychology/chapter/reading-touch-and-pain/.

Johansson; Gordon; Wrestling; Cole (1993-06-15). "Memory Representations Underlying Motor Commands Used During Manipulation of Common and Novel Objects." *Journal of Neurophysiology*. 69 (6): 1789–1796. doi:10.1152/jn.1993.69.6.1789. PMID 8350123.

Jones, L. A., Hills, P. J., Dick, K. M., Jones, S. P., & Bright, P. (2016). Cognitive mechanisms associated with auditory sensory gating. *Brain and cognition*, *102*, 33–45. https://doi.org/10.1016/j.bandc. 2015.12.005.

Judd DB, Wyszecki G (1975). *Color in Business, Science and Industry*. Wiley Series in Pure and Applied Optics (third ed.). New York: Wiley-Interscience. p. 388. ISBN 978-0-471-45212-6.

Kimpe, T., & Tuytschaever, T. (2007, December). *Increasing the number of gray shades in medical display systems--how much is enough? Journal of digital imaging*. Retrieved December 22, 2021, from https://www.ncbi.nlm.nih.gov/pmc/articles/PMC3043920/#:~:text=Even%20for%20high%20brightness%20and,around%20900%20shades%20of%20gray.

Kimura D (1961). "Cerebral dominance and the perception of verbal stimuli." *Canadian Journal of Psychology*. 15 (3): 166–171. doi:10.1037/h0083219.

Kmamou. (1970, January 1). *A simple C++ class for 3D mesh decimation*. A Simple C++ Class for 3D Mesh Decimation. Retrieved December 22, 2021, from http://kmamou.blogspot.com/2015/06/a-simple-c-class-for-3d-mesh-decimation.html.

Konstantina Kilteni, Raphaela Groten, Mel Slater; The Sense of Embodiment in Virtual Reality. *Presence: Teleoperators and Virtual Environments* 2012; 21 (4): 373–387. doi: https://doi.org/10.1162/ PRES_a_00124.

Kurniawan, Sri & Zaphiris, Panayiotis. (2001). *Reading Online or on Paper: Which is Faster?*.

Liu, J. (n.d.). *Evolution of AR technology*. History Exhibit. Retrieved December 22, 2021, from https://courses.cs.washington.edu/courses/cse490h1/19wi/exhibit/ar-tech.html.

Miller, George A. (1956). "The magical number seven, plus or minus two: some limits on our capacity for processing information." *Psychological Review*. 63 (2): 81–97. doi:10.1037/h0043158. hdl:11858/00-001M-0000-002C-4646-B. ISSN 1939-1471. PMID 13310704.

Moreno, Roxana and Mayer, Richard E. (2000). "A coherence effect in multimedia learning: The case for minimizing irrelevant sounds in the design of multimedia instructional messages." *Journal of Educational Psychology*. 92 (1): 117–125. doi:10.1037/0022-0663.92.1.117. ISSN 0022-0663.

Richard, G., Pietrzak, T., Argelaguet, F., Lécuyer, A., & Casiez, G. (1AD, January 1). *Studying the role of haptic feedback on virtual embodiment in a drawing task*. Frontiers. Retrieved December 22, 2021, from https://www.frontiersin.org/articles/10.3389/frvir.2020.573167/full.

Richard, G., Pietrzak, T., Argelaguet, F., Lécuyer, A., & Casiez, G. (1AD, January 1). *Studying the role of haptic feedback on virtual embodiment in a drawing task*. Frontiers. Retrieved December 22, 2021, from https://www.frontiersin.org/articles/10.3389/frvir.2020.573167/full.

Schiff, E. (2016, January 25). *Why you should avoid vibrating color combinations*. Web Design Envato Tuts+. Retrieved December 22, 2021, from https://webdesign.tutsplus.com/articles/why-you-should-avoid-vibrating-color-combinations--cms-25621.

Sensory memory: Types of memory facts & information. The Human Memory. (2020, November 25). Retrieved December 22, 2021, from https://human-memory.net/sensory-memory/.

Sweller (n.d.). *Visualisation and instructional design. School of Education University of New South Wales, Sydney*, NSW 2052. (pp. 1501-1510). [1].

The virtuality spectrum - understanding AR, MR, VR and XR. CreatXR. (2021, January 10). Retrieved December 22, 2021, from https://creatxr.com/the-virtuality-spectrum-understanding-ar-mr-vr-and-xr/.

Vailshery, L. S. (2021, January 22). *Active virtual reality users forecast worldwide 2014-2018*. Statista. Retrieved December 22, 2021, from https://www.statista.com/statistics/426469/active-virtual-reality-users-worldwide/.

VR glove maker haptx raises $12M and announces partnership with Advanced Input Systems. SiliconANGLE. (2019, December 18). Retrieved December 22, 2021, from https://siliconangle.com/2019/12/19/vr-glove-maker-haptx-raises-12m-announces-partnership-advanced-input-systems/.

Chapter 8

Continuing Engineering Education Practices and Preferences

Charles E. Baukal, Jr.*

Koch Engineered Solutions Institute℠, John Zink Hamworthy Combustion, Tulsa, Oklahoma, USA

Abstract

This chapter presents the results of a survey of 65 working engineers in the energy sector on their formal continuing engineering education (CEE) practices and preferences. The purpose of the survey was to determine what type of continuing education is currently being done in a particular industry and what engineers would prefer if they had their choice. Demographic data such as gender, age range, and ethnicity were collected. Current CEE practices such as number and type of courses taken each year were surveyed. Participants were also asked for their CEE preferences such as which types of courses (technical, management, EH&S = environment, health, & safety, legal, other) they would prefer more of and what was their preferred delivery format (face to face, hybrid, online). This information is important to those developing curricula and designing and delivering continuing professional development for engineers.

Keywords: continuing engineering education, preferences, practices

Introduction

The need for CEE has been well-documented (Baukal, 2010, 2017). Continuing education is critical for working engineers because of the breadth of processes and equipment they design and use and because of rapid changes in technology (Baukal, 2012). For example, plant engineers take courses to learn how to design, operate, and maintain different types of equipment specific to their operations (Valencia et al., 2008). Chapter 4 discusses methods for teaching plant operators and engineers how to operate process heaters.

Calls continue to be made for improving engineering education. The U.S. National Academy of Engineering established a Committee on Engineering Education to answer the question, *What will or should engineering be like in 2020*? (National Academy of Engineering, 2004). The Phase 2 report from that committee titled *Educating the Engineer of 2020* (National Academy of Engineering, 2005) called for the reinvention of engineering education.

---

* Corresponding Author's Email: charles.baukal@kes.global.

While much research has been done on educating engineering *university students*, very little is available in the literature on the continuing education of *working* engineers. This is not surprising as engineering disciplines have received relatively little attention from learning sciences researchers (Johri and Olds, 2011).

There are many types of CEE ranging from informal to formal. This chapter does not consider informal training such as reading journals and trade magazines, attending conferences, and mentoring (on-the-job training). It does consider different types of formal training, some leading to credits and some not. It considers not only technical training, but also management, EH&S (environmental, health, and safety), legal, and other types.

No previous research was found that described the continuing professional development practices and preferences of working engineers. This information is especially important to those developing curricula for this type of training. It is also important for those designing and delivering this type of training. This chapter reports the results of a survey of working engineers to determine their CEE preferences and practices (Baukal, 2018).

Methodology

A survey (shown in the Appendix) was given to students attending Koch Engineered Solutions Institute^{SM} (KESI; formerly the John Zink Institute^{SM}) continuing professional development courses in Tulsa, Oklahoma in September through November 2017 and to a handful of participants attending a seminar given by the author at a conference in New Orleans in September 2017. Nearly all of the participants worked in the oil and gas business, primarily in the downstream sector. The results reported here only include those 65 survey participants who identified their position as "engineer" or listed an engineering degree as their highest degree. The survey was anonymous and voluntary. No incentives were given to the participants to take the survey. Some of the data collected on learning style and verbal-visual preferences are not reported here and are part of an ongoing study of learning preferences for working engineers (Baukal, 2017).

Demographics

The survey consisted of 20% females and 80% males. The age group distribution is shown in Figure 1. Approximately two-thirds of the participants were under 35 years old. This is as expected since most students attending KESI are newer to the industry which is why they are taking the training.

The ethnicity distribution shown in Figure 2 shows the participants were a little over half Caucasian/White with a significant number reporting Asian ethnicity.

Figure 3 shows that the native country of most of the participants was the United States although there were a significant number of countries from outside the U.S. including Brazil, Canada, Chile, China, Dominic Republic, Egypt, India, Italy, Korea, Malaysia, Nigeria, Pakistan, Saudi Arabia, South Korea, Sweden, and the UAE.

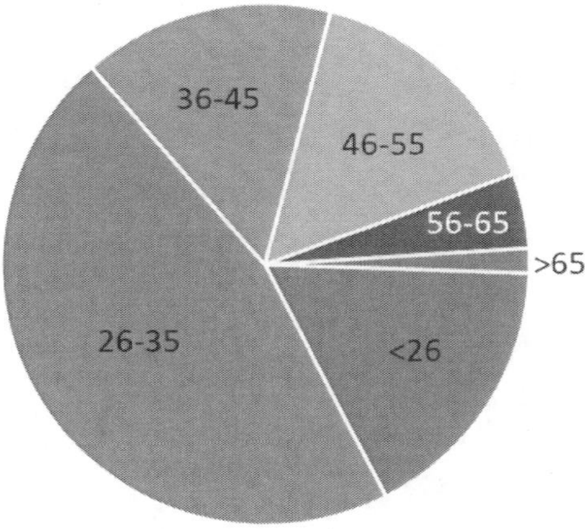

Figure 1. Age distribution for the survey participants.

Figure 4 shows the native language of about two-thirds of the participants was English. Other native languages included Arabic, Chinese, Hindi, Italian, Korean, Mandarin, Portuguese, Spanish, Swedish, Tamil, Urdu, and Yoruba. The large number of non-native English speakers can present a challenge when the training courses are in English. However, most of the participants attending KESI courses have strong English-speaking skills. It is assumed that those with weaker English skills would not attend a course strictly in English. That is not always the case when courses are held at the plants where employees may be forced to attend classes with their colleagues by their supervisors.

According to Figure 5, most of the participants had lived in the U.S. for more than 20 years although there was a large number who did not live in the U.S.

Figure 7 shows nearly half of the participants were chemical engineers with a large number of mechanical engineers as well. That is as expected because most of the students worked in the chemical/petrochemical industry. Other majors included civil/environmental, general engineering, engineering management, industrial technology, and nuclear engineering.

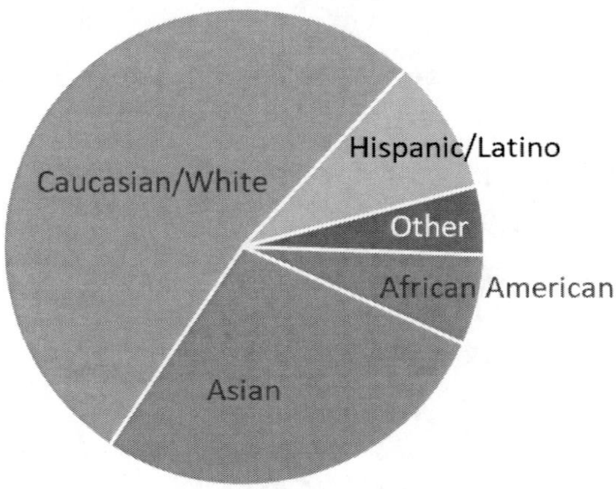

Figure 2. Ethnicity distribution for the survey participants.

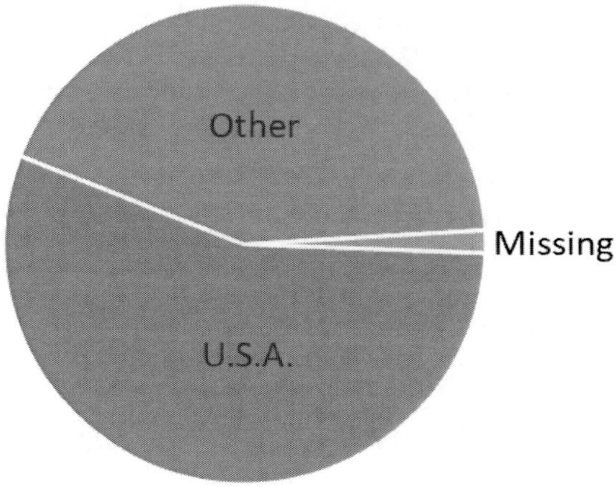

Figure 3. Native country of the survey participants.

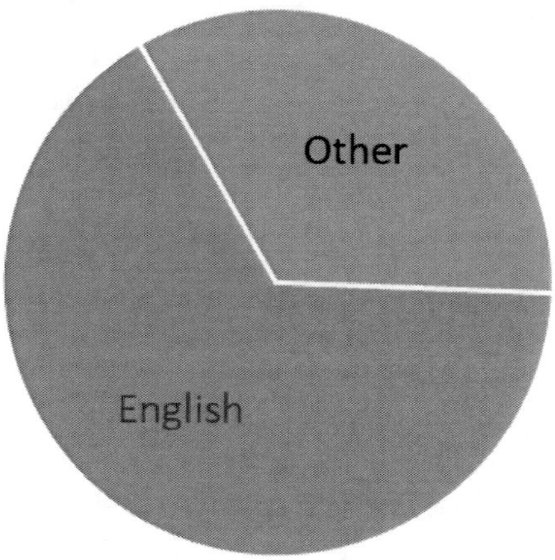

Figure 4. Native language of the survey participants.

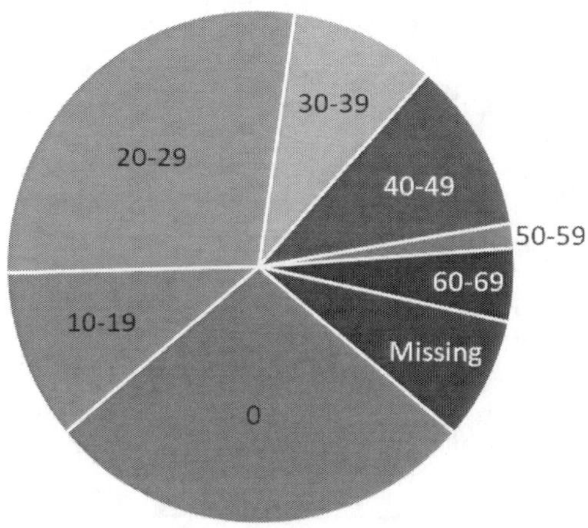

Figure 5. Participants' years living in the U.S.

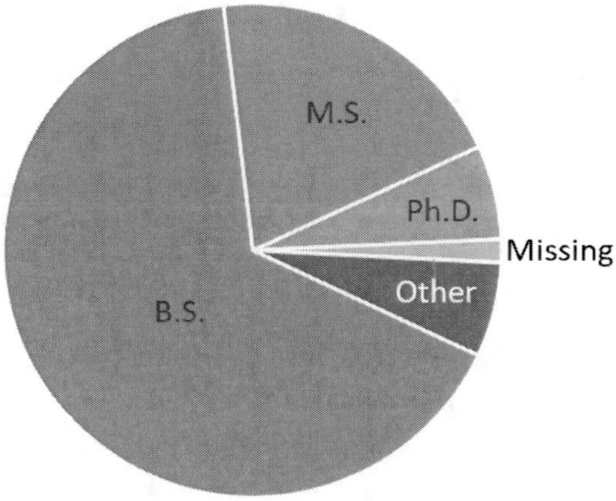

Figure 6. Participants' highest degree.

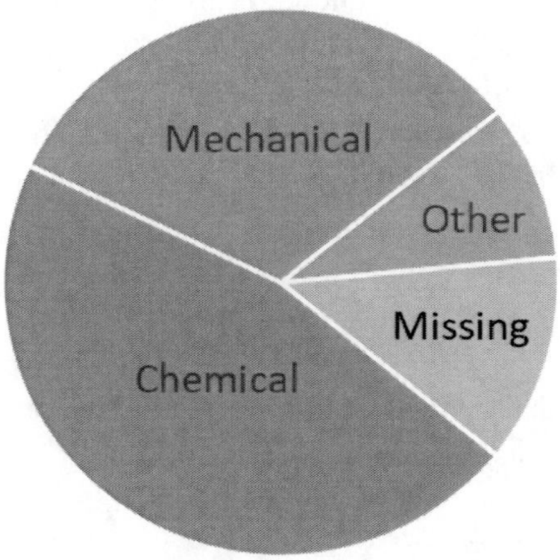

Figure 7. Participants' degree major.

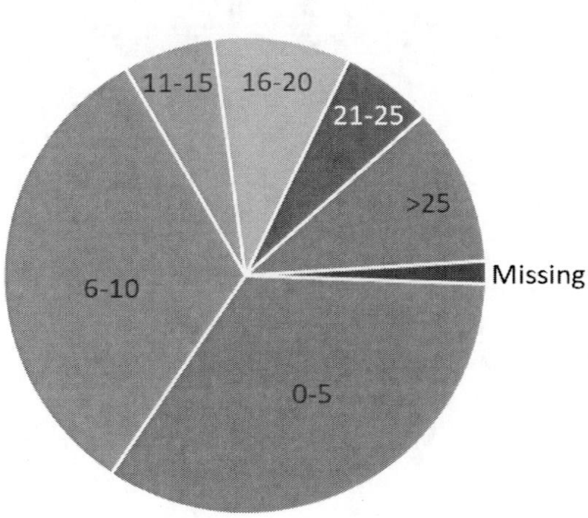

Figure 8. Participants' total work experience.

Most survey participants had a B.S. degree in engineering as shown in Figure 6 with a significant number having some type of graduate degree.

Approximately 23% of the participants had a Professional Engineering (PE) license. According to Figure 8, most of the participants had less than 10 years of total work experience. As previously mentioned, that is expected for students needing training courses because they are relatively new in the industry.

Continuing Engineering Education Practices

Most participants took up to 5 courses of all types each year as shown in Figure 9. Relatively few took more than 5 courses, some as many as 25 or more.

Figure 10 shows what fraction of the survey participants took various types of courses. Over 90% took technical courses while less than 10% took other course types not listed. Far fewer of them took management, EH&S, and legal courses.

Figure 11 shows the fraction of the training taken that was mandatory. It shows approximately a linear relationship for the fraction of participants compared to what fraction of their training was mandatory. A little over 20% of the participants did not take any mandatory training.

Figure 12 shows the fraction of training taken by the participants that was required by others besides their employer. An example would be safety training required by clients to enter their facilities. The graph shows that 60% of the participants were not required to take any training by non-employers.

Figure 13 shows what fraction of participants' training was taught by their employers. For less than 10% of the participants, their employer did not teach any of their training. For about 8% of the participants, their employer taught all of their training. For the rest it was approximately a linear relationship.

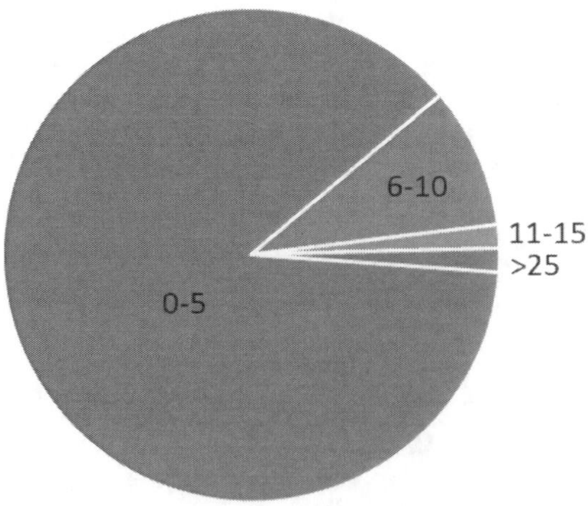

Figure 9. Number of training courses of all types taken per year by the participants.

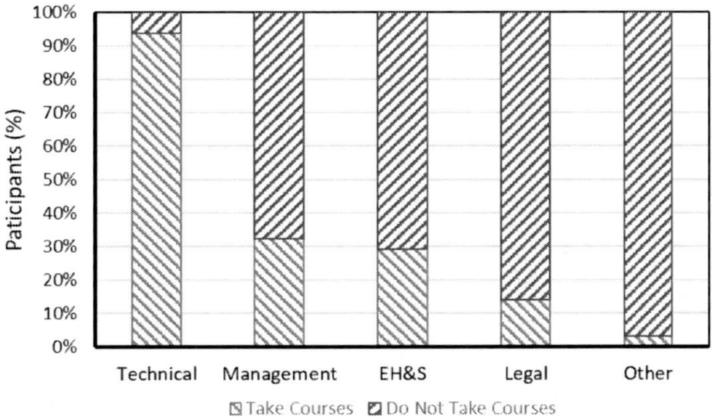

Figure 10. Types of courses taken (or not taken) by the participants.

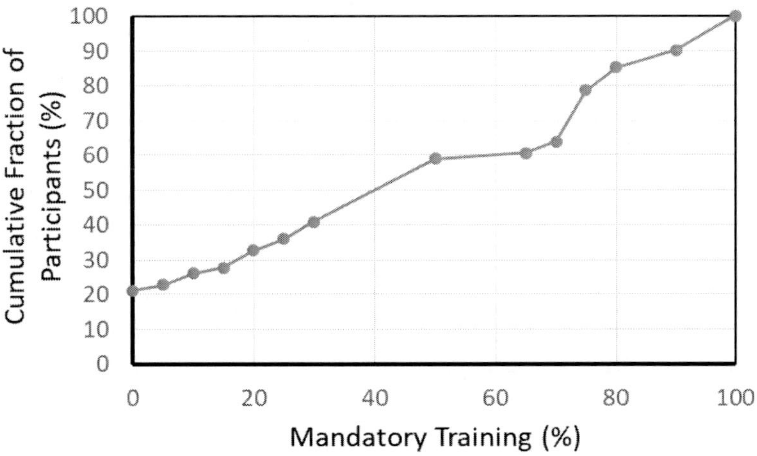

Figure 11. Fraction of the training taken that was mandatory vs. cumulative fraction of the participants.

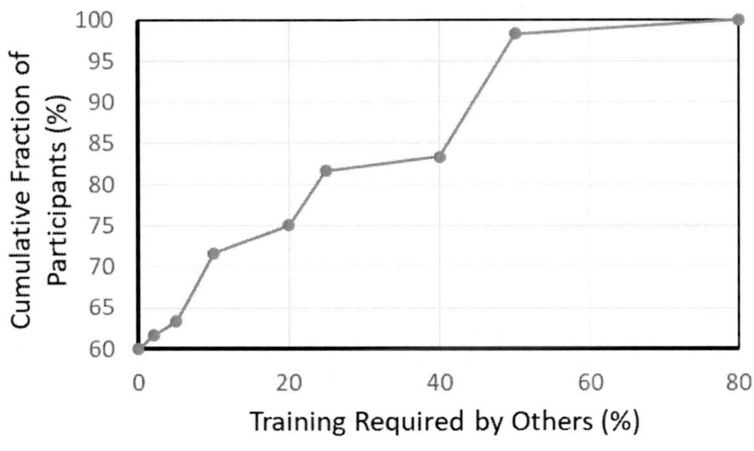

Figure 12. Fraction of training required by others besides participants' employer.

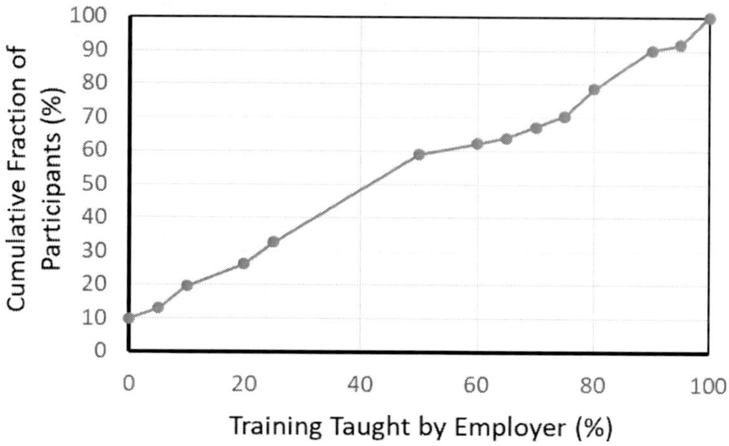

Figure 13. Fraction of participants' training taught by their employer.

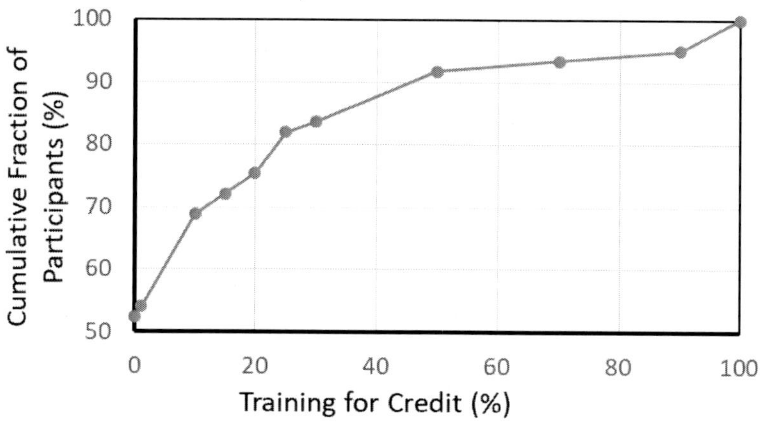

Figure 14. Fraction of training taken by participants for credit.

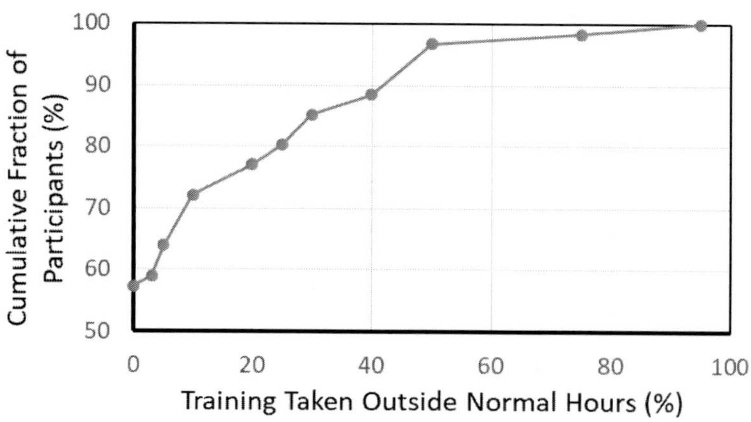

Figure 15. Training taken by participants outside normal working hours.

Figure 14 shows how much of the participants' training was for credit. For a little over half of them, none of their training was for credit. For approximately 5% of the participants, all of their training was for credit.

Figure 15 shows less than half of the participants were required to take training outside of normal working hours with about 10% having to take half or more of their training outside normal working hours.

Continuing Engineering Education Preferences

Figure 16 shows which types of courses participants would like to take more of. A large majority would like to take more technical courses and about half would like to take more management courses. Relatively few would like to take more legal, EH&S, or other courses. The two "other" courses identified were financial and self-improvement.

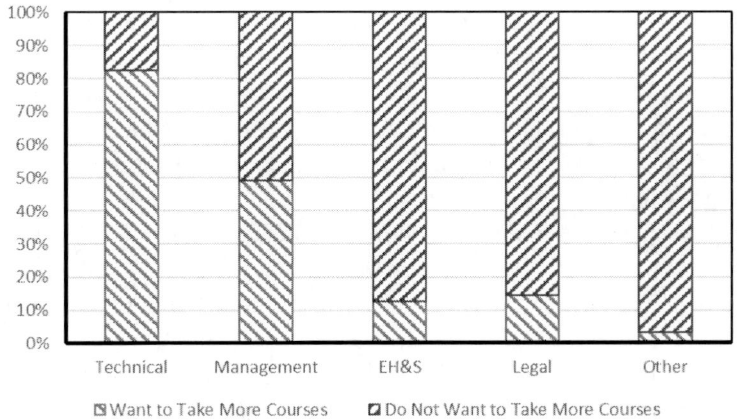

Figure 16. What types of training participants would like to take more of.

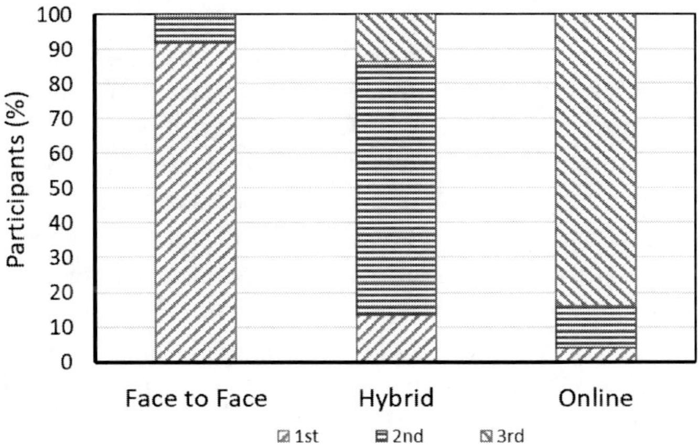

Figure 17. Participants' preferences for training course format.

Figure 17 shows participants' preferences for how they prefer their training to be delivered. Hybrid is the combination of face-to-face and online training. Over 90% preferred face-to-face training as their first choice, 84% selected hybrid as their second choice, and 73% chose online as their third choice. Some of the participants only listed one choice (almost always face-to-face).

Conclusion and Recommendations

While the results reported here are based on a relatively small sample, some trends in the data are evident. One somewhat surprising result is that the majority of those sampled took 5 or fewer courses per year. Many larger companies typically have more than that number just for their EH&S and legal training courses required to be taken by their engineers. The vast majority of courses taken by the working engineers were technical as might be expected. The legal courses are likely tailored to the specific company and probably need to be developed and taught internally. Depending on the content, the technical and management courses could be taught by either internal or external subject matter experts. While the survey did not ask which courses were mandatory, it is expected these would be mostly EH&S and legal courses. Most participants were not required to take any training courses by non-employers such as clients. The amount of training taught by the employer varied widely. Most of the training taken was not for credit. A significant amount of training was taken outside of normal working hours.

The preferences for more training were very similar to current practices where the vast majority of participants would like to take more technical classes. Nearly half would also have liked to take more management classes. Less than 15% would like to take more EH&S, legal, or other classes. The overwhelming majority preferred their classes to be delivered face-to-face, their second choice as hybrid, and online as their last choice. Since this survey data was collected, it would be interesting to see if participants still rated online as their last choice, given all of the online training that has been done due to COVID-19.

Clearly much more research is needed in this area. A larger and more diverse sample is required to make generalizable conclusions. For example, the present survey consisted mostly of those under 35 years old, so some more older participants are recommended. More participants from engineering disciplines besides chemical and mechanical engineering are needed to see if there is any dependence on discipline. The participants in this survey were from the energy industry so it would be important to consider other industries to see if there are any similarities and differences. It would be useful to understand the size of the organizations where the participants work to see if that impacts how many courses are required to be taken by working engineers, especially non-technical courses such as EH&S and legal. It would be interesting to know if the number of courses taught by the employer varied by employer size where it might be expected larger companies have more resources for such training. It would be interesting to know if the training taken for credit was primarily of any particular type (e.g., technical). It would be useful to know what type of training is taken outside of normal working hours and how it is delivered. It might be expected a significant portion of that is online but there is currently no data to validate that assumption.

References

Baukal, C. E. (2010). Continuing engineering education through distance learning. *European Journal of Engineering Education*, 35(2), 225-233.

Baukal, C. E. (2012). Strategy for delivering continuing engineering education. *International Journal of Training and Development*, 16(4), 284-299.

Baukal, C. E. (2017). *Learner Preferences for Continuing Engineering Education.* Mauritius: Lambert.

Baukal, C. (2018). Continuing Engineering Education Practices and Preferences. *2018 American Society for Engineering Annual Conference and Exhibition*, Salt Lake City, UT.

Valencia, R., D. Link, C. Baukal, and J. McGuire (2008). Consider classroom training for plant operators. *Hydrocarbon Processing*, 87(11), 55-59.

National Academy of Engineering (2004). *The engineer of 2020: Visions of engineering in the new century.* Washington, DC: National Academies Press.

National Academy of Engineering (2005). *Educating the engineer of 2020: Adapting engineering education to the new century.* Washington, DC: The National Academies Press. Retrieved from http://www.nae.edu/Publications/Reports/25677.aspx, 2005.

Johri, A. and B. M. Olds (2011). Situated engineering learning: Bridging engineering education research and the learning sciences. *Journal of Engineering Education,* 100(1), 151-185.

Appendix: Verbal-Visual & Learning Strategy Preferences Survey

Question	Circle Your Choice/Fill in the Blank
Gender	1 = Female 2 = Male 3 = Other
Age	1 = <25 2 = 26-35 3 = 36-45 4 = 46-55 5 = 56-65 6 = >65
Ethnicity	1 = African American 2 = Asian 3 = Caucasian/White 4 = Hispanic/Latino 5 = Other ______________________________
Native Country	1 = U.S.A. 2 = Other ______________________________
Native Language	1 = English 2 = Other______________________________
Length of time lived in the U.S.	_______ years
Highest earned degree and major (e.g., high school diploma, B.S. in Mechanical Engineering, etc.)	______________________________________
Do you have a professional engineering license?	1 = Yes 2 = No

Verbal-Visual & Learning Strategy Preferences Survey (Continued)

Question	Circle Your Choice/Fill in the Blank
Primary Position/Function/Role	1 = Engineer 2 = Operator/Technician 3 = Management/Administration 4 = Teacher/Educator/Professor 5 = Full-Time Student 6 = Other ______________________
Employer	1 = Industry 2 = Academia/Education 3 = Government 4 = Non-Profit 5 = Other ______________________
Question	Circle Your Choice/Fill in the Blank
Management Level	1 = Non-supervisory (do not supervise anyone) 2 = Middle Management (supervise at least 1 person) 3 = Senior Management (VP, President, etc.)
Total work experience (including teaching if applicable)	1 = 0-5 2 = 6-10 3 = 11-15 4 = 16-20 5 = 21-25 6 = >25

Verbal-Visual Preference

In a learning situation, sometimes information is presented verbally (e.g., with printed or spoken words) and sometimes information is presented visually (e.g., with labeled illustrations, graphs, or narrated animations). Please place a check mark indicating your learning preference.

○	○	○	○	○	○	○
Strongly more verbal than visual	Moderately more verbal than visual	Slightly more verbal than visual	Equally verbal and visual	Slightly more visual than verbal	Moderately more visual than verbal	Strongly more visual than verbal

Learning Strategy Preference

Please read the ATLAS assessment (blue sheet) and then mark the result here (Place ONE check mark next to your learning strategy preference):

○ Navigator	○ Problem Solver	○ Engager

The description of your learning strategy group in the ATLAS "Groups of Learners (page 2 of the ATLAS instrument) is reasonably accurate in describing you as a learner.

○	○	○	○	○	○	○
Strongly agree	Moderately agree	Slightly agree	Neither agree nor disagree	Slightly disagree	Moderately disagree	Strongly disagree

Continuing Education

Question	Circle Your Choice/Fill in the Blank
Average number of training courses you take each year (circle one)	1 = 0-5 2 = 6-10 3 = 11-15 4 = 16-20 5 = 21-25 6 = >25
Typical types of training courses you take each year (circle all that apply)	1 = technical 2 = management 3 = environmental, health, & safety 4 = legal/compliance 5 = other: ____________________
Approximately what percentage of the training courses you take each year are mandatory (where the rest are at your option)	________%
Approximately what percentage of the training courses you take each year are required by someone other you're your employer (e.g., customers, licensing boards, government, etc.)?	________%
Approximately what percentage of the training courses you take each year are taught by your employer (where the rest are taught by those outside your organization)?	________%
Approximately what percentage of the training courses you take each year are for credit (e.g., college credit, Continuing Education Units or CEUs, Professional Development Hours or PDHs, etc.)	________%
Approximately what percentage of the training courses you take each year are taken outside of normal working hours (e.g., nights, weekends, holidays, etc.)	________%
If you could take more training each year, what type would it be (circle all that apply)?	1 = technical 2 = management 3 = environmental, health, & safety 4 = legal/compliance 5 = other: ____________________
Rank your preferences for the following methods of instruction for training courses (1 = most preferred, 2 = 2nd most preferred, 3 = least preferred)?	___ face-to-face (classroom) ___ online (computer-based training) ___ hybrid (classroom + online)

Chapter 9

Live-Streamed Curricula in a Virtual-Demand World

Regina M. Godfrey*

Koch Engineered Solutions Institute℠, Tulsa, Oklahoma, USA

Abstract

Virtual learning is not just a trend; it's become the norm. Due to a multitude of societal, environmental and cultural changes, many virtual platforms have surfaced as the preferred delivery method for curriculum distribution and continued education opportunities. Entering this new realm has created some initial challenges, but benefits far outweighed the challenges. In this chapter, you will learn about approaches, discoveries, and outcomes related to delivering educational content virtually in a new age.

Keywords: virtual learning, virtual platforms, e-learning, distance learning, curriculum distribution

Introduction

The world of education is ever-changing and never more so than now, as demands for digital offerings in continued learning are increasing. What used to be a hope and vision for technological enhancement has become a virtual expectation. Covid-19 has accelerated some of those advancements.

Beyond those demands, the digital arena has grown by leaps and bounds over the past two decades while a multitude of platforms struggle to keep up with the competition for fear of becoming obsolete in their features and deliveries. Migrating from traditional learning in the classroom to the digital arena is no easy feat and takes planning, preparation, practice, and comparative analysis to determine the best methodologies in relation to your particular preferences, expectations, and, ultimately, your goals.

A virtual classroom or setting is an online distance learning environment in which instructors and learners interact via technical tools within a given software such as Zoom or Microsoft TEAMS. This allows schools, companies and other institutions to host singular classes, grouped courses, global webinars and, essentially, any remote curriculum instruction while also maintaining a functional classroom environment. Live-streaming is the next tier of that virtual learning which promotes real-time learning in a collaborative setting. Stepping up from virtual learning to actual live-streaming requires additional equipment, manpower, an online video education platform, and a reliable internet connection. Beyond those tangibles,

* Corresponding Author's Email: gina.godfrey@kes.global.

you'll need IT support and buy-in and will likely need legal approval from your institution or company for the proposed content delivery.

Each year, the Association for Talent Development (ATD) measures the delivery methods organizations use to gauge and meet training needs. Data clearly demonstrates the measurable and significant increase in virtual e-learning platforms over the past several years. According to the 2020 ATD *State of the Industry* report, traditional classroom curriculum-sharing has decreased from 53% in 2018 to 40% in 2019, supporting the notion that technology-based learning methodologies accounted for more than half of all learning hours in 2019. In their estimation, 70 percent of organizations are now employing virtual classroom setups (Christopher, 2015).

Ultimately, the virtual learning environment (VLE) must be one that promotes flexibility, reaches participants globally, sustains energy while delivering the curriculum and training, and promotes community not limited by time or place. But before getting started, it's important to understand some of the developments and nuances along the evolutionary path of education.

Evolution of Learning

One might conjure the image of hieroglyphics on a cave wall when thinking of the earliest forms of communication, educational instruction, and learning. From there, it has evolved over the ages to smaller, intimate settings such as the earlypioneer classrooms to formalized learning in the 17th century where larger congregations assembled and, finally, to the next tier of classroom/auditorium presentations and panel discussions experienced on a grander scale today.

Along the way, we've discovered that the setting is not the only factor. In fact, the learners themselves are categorically diverse and respond differently to the delivery of curricula whether they be auditory learners, visual learners, kinesthetic learners or tactile learners (see Chapter 4). The way we acquire knowledge is directly aligned with how we interpret data. Although learners may have a preference for a particular delivery method, most people are, in fact, a multi-blend of all four learning types and retain information best when a combination of delivery methods are implemented in mainstream education and knowledge share. This "combination," if you will, has also evolved with the introduction of newer technologies and curricula delivery methods such as video, interactive modules, podcasts, and live-stream forums like instructor-led courses, webinars, and conferences.

Almost every human being reacts to visual stimuli and responds to color, graphic depictions, illustrations, and other visual interpretations when it comes to knowledge transfer and general information. It's safe to assume that a preferred delivery method would then appeal to visual learners and rentention of information is directly correlated to visual representation. It is for this reason that many learning management systems (LMS) defer to the graphically-aesthetic portrayals whereby click-to-learn methods are implemented and utilized. As such, virtual and online learning have never been more critical than today, providing flexibility to both educators and learners.

"The coronavirus (COVID-19) outbreak and rapid spread came as a shock to countries around the world, infecting and killing thousands of people around the globe which led to the World Health Organization (WHO) to declare it a pandemic" (Sintema, 2020, p. 1). Because of the COVID-19 pandemic, the widespread effects created an instant global health disaster with far-reaching ramifications within economic and social sectors. Worldly influences such as the pandemic and mandatory quarantines shifted the educational delivery methodologies and essentially forced digital distance learning during the international crisis. This created a sink-or-swim scenario whereby institutions and companies alike evolved to the realm of digital platforms and distance learning, resulting in changed attitudes and approaches to open the doors to the future of education. Further, these associated quarantine mandates created an ideal window for e-learning and virtual studies. This grew beyond the extent of classrooms all over the globe, and manifested itself in companies with required work-from-home policies, and even household living rooms via virtual family gatherings. What was traditionally promoted in face-to-face exchanges has become hosted in virtual encounters via computer screens and chat rooms.

Traditional Methodologies

When one thinks of traditional pedagogical learning, they envision a teacher at the head of the room, directing students through recitation techniques and frequent memorization exercises. This was the method in which most were accustomed to as early-aged learners. The next tier in the learning stages progresses beyond this recursive process to more analytical approaches whereby the learner's brain is challenged to problem-solve and implement critical thinking.

Any learning experience consists of content and dialogue (engagement). Whether in-person or virtual, the instructor's role is to ensure a proper balance of well-founded, appropritately-defined, and targeted content while also fostering meaningful dialogue in the process. Any forum for learning has traditionally involved tactics designed to stimulate and engage the learner while delivering the content for maximum retention. So it seems this becomes more prevalant and necessary in the digital age for remote curricula delivery. Engagement must be ratcheted up to much higher levels to sustain energy and foster two-way participation during the learning process. Many tactics have become more readily used to enhance and ensure that engagement is ever-present in virtual applications. Tactics such as gamification (Kim et al., 2018) and interactive polls not only entertain and solicit dedicated participation and feedback, but also allow the instructor to gauge interest and learning retention along the way. And just as a traditional classroom instructor might give pre-course assignments, virtual platforms also allow for pre-engagement activities to motivate participation, gauge progress, and measure end-result goals and projected outcomes.

The launch of the Internet in the 1990s created an instant-information world. Over the course of the past three decades, learners are now well-positioned to embrace new technologies and are in a sense conditioned to utilize virtual platforms with ease and confidence. Due to the more recent innovative and extensive far-reaching tools such as smart devices, search engines, and social media channels, learners have an even greater access to information at their fingertips. A virtual gateway to infinite knowledge is just one click away! Today, it's common to see students on school buses downloading research materials and attending remote labs for course credits. Long gone are the days of hard copy encyclopedias and trips to the library for research. Other adult professionals might spend their morning commute listening to a remote podcast or attending a virtual webinar on a given subject matter. Given all of the instant possibilities at one's fingertips, learners have come to expect more of their instructors and the methodologies implemented to deliver content, curricula, and general training overall. Suffice it to say that as technology has evolved, so has the learner. This means the instructor and the content needs to change as well to keep up with technology and the learner.

Societal, Environmental and Cultural Impacts

Due to the migration of traditional classroom settings to virtual distant learning, the arena today consists of remote employees coming from diverse cultures and settings. Similarly, the training methods for these learners has completely morphed in the process creating a higher demand for e-learning, online training, webinars, virtual classroom courses, video simulations, and live-streams. The positive takeaway is that learners can adapt to the environment according to their immediate needs and learn "on the go", so to speak. Learners today often want and need bite-sized training available anytime and anywhere.

While we know that higher socioeconomic status and intellect positively correlate with more advanced motor-skill development, one's culture and environmental influences directly impact the rate and level of that development. Self-motivation also plays a crucial role in this development. Suffice it to say that behavioral influences dictate the level of investment in the learning process from one learner to the next. Many assumptions are made in virtual deliveries of content since much may be hidden from the instructor, such as body language and affirmations. Thus, the instructor must endeavor to define a happy medium to appeal to a broader global audience on the whole and become a virtual chameleon with blended deliveries and creative approaches.

Environmental policies and geographical settings also play a role when it comes to virtual delivery and its resulting impact. Subject areas such as world views, security, and privacy management must be considered in the decision to migrate from traditional in-person classroom and training to e-learning and virtual offerings. Cross-cultural apsects of instruction are critical and make it relevant for participants of any culture to assuage and adapt to this relatively-newer remote learning arena. However, virtual learning is crossing all barriers to become the preferred metholodology in reaching larger, more widely spread and diverse audiences who are expected to continue their learning objectives regardless of environmental or societal limitations. Metaphorically speaking, it becomes a "survival of the fittest" approach when faced with the need to transform from traditional methodologies to mainstream digital and virtual learning modalities.

There are many influences which affect the decision to go virtual and/or live-stream and also many benefits that present themselves to qualify the decision to host virtual deliveries. Some of these benefits include, but are not limited to, making education accessible regardless of natural disasters, travel bans or locations that are far from the host institution, campus or company headquarters. Other key benefits are the lack of distractions from other learners in virtual settings andthe time and savings for all parties involved as virtual deliveries conserve energy and manpower. Yes, the instructor(s) must still exert energy toward preparation on the front end, but a wider lens shows that the benefits are more far-reaching both in audience size and impressions.

There is also an undisputed reduction in cost in virtual exchanges as no travel is required for the instructor/host or for the learner. In virtual deliveres, a secure internet connection and network is all that is needed to transfer knowledge.

Evaluation of Existing Resources

So, how does one begin to ascertain what virtual methodology works best within their institution or organization to deliver curriculum to prospective learners? The first discovery must involve a deep-plunge evaluation of resources at hand (extending beyond manpower to technical tools) in relation to the level of demand, defined timeframe, and overall expectation toward a mutually-beneficial result.

The spread of the COVID-19 virus has forcibly created an arena for heightened and necessary e-learning demands. While in-person, traditional classroom settings have measurably diminished in that timeframe, the need to evaluate options and move swiftly to implement alterior distant teaching methodologies has become the norm and expectation in almost every school, company, and household. No one is beyond the reach or impact of the global pandemic ramifications.

Key questions will inevitably arise in this evaluation stage and should be addressed in advance of any decision making. Which virtual learning platform is best for the institution or organization? What key benefits are absolute and which are less critical? What departments are involved within the institution or organization beyond information technology (IT)? Who becomes the tip-of-the-spear leader in the proposed virtual deployment? Who will fill crucial roles in delivering the live-stream virtual content? What budget is afforded for this new delivery method? What parameters will be the measuring stick to determine success?

Remember that there is no one-size-fits-all method. The first task in the evaluative process is to define the purpose of the training program, followed closely by identifying the intended audiences and demographics, evaluating seasonal influences and target windows to launch most successfully, and ascertain talent needs in relation to budgetary resources and desired outcomes. Identifying the team who will define and implement the virtual strategy is a must. These stakeholders are usually the early adopters of the overall virtual implementation framework. Change management techniques may be needed to gain their initial support, buy-in, and partnership throughout the launch (Huggett, 2017).

The increased demand for group learning portals has allowed learning management systems (LMS) and virtual platforms to flourish. But an even greater demand has emerged for live-streamed virtual delivieries. The bottom line remains that learners require their instructors to move the needle and transmit the information in a timely and relevant fashion. Self-paced curriculum channels and online learning systems aren't for everyone. Many humans by nature require consistent motivation to remain engaged, participatory, and stimulated. Modern virtual approaches blend the two content deliveries whereby learners participate in independent work and pre-engagement activities such as online tests to gauge their knowledge and retention capabilities. This hybrid approach fosters greater creativity and curiosity between the teacher and the taught.

Comparative Analysis of Options

Once existing resources have been defined and a gameplan has been determined, it's time to delve into research of the available virtual platform options in order to compare key benefits for the greatest return. While there are many self-paced, online virtual learning platforms to deliver curricula, exploring options for live-streaming entails more thorough research and experimentation to identify which platform will have the greatest mutual benefit and high-quality, real-time representation. While the learner is the first and foremost priority, it's critical to select a virtual software that meets the instructor's needs, the facility's IT capabilities, the virtual classroom setup, and the sustainable delivery protocols. But it's preliminarily important to define the various deliveries in advance before plunging ahead. In Table 1, Christopher (2015) illustrates the differences between in-person, online and virtual e-learning curricula deliveries. Many factors come in to play when deciphering the best delivery methods for a given situation or specific set of needs.

But given the above and considering traditional pedogogy offerings, virtual software platforms have become increasingly prominent, all but necessary, and quite common over the past few years. While the list of virtual platforms is ever-growing, some of the more recognizable and highly-utlized software options at the time of this writing include: Zoom, Microsoft Teams, Webex, Adobe Connect, Blackboard, Schoolology, Canvas, and Tovuti, for example. Recently, and with measured frequency, most of these platforms are now relatively affordable options as demand has predicated the need in 2020-2022, due to quarantines around the globe.

Table 1. Virtual classroom training compared to other types of training (Christopher, 2015, p. 5)

Component	Traditional Face-to-Face Training	Self-Paced e-learning	Virtual Classroom Training
Information Flow	Two-way information flow between facilitator and learners	One-way information flow between technology and learner	Two-way information flow between facilitator and learners
Facilitator	Physically present with participants	Not applicable	Physically separated from participants
Location	Physical classroom or meeting room	Accessed via a computer or mobile device	Accessed via a computer and sometimes a mobile device
Materials	Handed out during class	Downloaded from course site	Emailed or mailed to participants in advance or downloaded from virtual classroom
Typical Exercises	Facilitated discussion; small group work	Drag and drop exercises; online simulations	Facilitated discussion; small group work in breakout rooms and polls
Duration	Varies from an hour to multiple days	Modules broken into small chunks of time, such as 30 minutes	Typically chunked into 60-90-minute sessions

Self-paced, online courses have filled a large void; however, the essential instructor-to-student interaction can be lost in the translation. It's crucial to have that "in the moment" experience to deliver information and curricula for the most-assured retention. Live-streaming instruction rises to the top of the priority list and becomes the foremost option in comparative analysis. One must weigh the risks and establish an effective virtual classroom that makes use of multiple communications channels within the software such as screen sharing, multiple co-hosts and presenters, interactive chat rooms, polls, question and answer boards, conferencing and/or break-out room options for small group discussions, and high-quality audio-video projection. It's recently become crucial for virtual enhancements to feature heigtened flexible display settings and custom accessibility. To do so, consider bandwidth within your infrastructure to ensure that it's strong enough to handle multiple users. Leverage the instruction data when it comes to blended learning programs and stay flexible and willing to collaborate on continual enhancements.

Know your demographics, establish learning requirements, check your pricing structures (if you charge for curricula and training), evaluate content in relation to software compatability, and absolutely seek the most reliable reporting data analytics. Explore the costs of hardware, software, and technology infrastruture. When considering a live-stream virtual delivery approach, plan for technical support and consider investing in a contracted external videographer and/or producer to set up a maximum audio-visual production, as well as support and troubleshoot any IT issues or deliverability snags during the live-stream exchange. Configure and prepare condensed content to allow for breaks within the program. Delivering content in more manageable time blocks makes it easier on students. This is most critical when you don't have a pulse on the participants' body language or engagement. Remember that the learner's attention span could be diminished in this digital, distant delivery and distraction can ensue quite easily when not in a traditional face-to-face content exchange. Therefore, assigning pre-defined breaks, rewarding participants for on-time virtual attendance, polling for feedback, and enlisting gamification strategies have all proven to be essential methodologies to sustain energy in trainings.

Establish quality instructional standards and consider investing in an external coach for virtual presentation prowess. Not everyone is born to present, especially in virtual settings, but presentation skills can be taught and implemented with expert coaching and role-playing scenarios in advance of the virtual content delivery window. Remember that instructing face-to-face allows for higher engagement due to affirmation of learners and active verbal participation. Without those cues, many presenters and instructors can fall short of expectations and, therefore, must be trained themselves for virtual delivery. Because participants and/or learners are mostly muted in these virtual learning platforms, this can feel awkward for the presenters and create some uneasiness for some presenters and make it difficult to thrive in this environment. Presentation skills coaches and workshops are readily available and are very affordable in relation to the return on investment. Even if your content is superior and the curricula is rock solid, it's absolutely essential that instructors become famliar with virtual e-learning settings in advance in order to present the best quality outcomes. Working with a coach to hone critical virtual presentation skills is a must. Rehearsing is that much greater of a mandate in this scenario as well.

In any form of virtual and/or distance e-learning, it's critically important to implement quality assurance metrics such as score-driven pre- and post-tests for gauging progress, polls and exercises, and post-learning session evaluation/critique forms to assess learning outcomes and instructor quality. Direct feedback from those tools should be implemented in virtual settings. Further, it's important to consider that individuals in almost every institution and organization seek continuing education units (CEUs) and professional development hours (PDHs) along their career path. Establishing a partnership with an accreditation-authorized organization such as the International Association for Continuing Education and Training (IACET), for example, strengthens your curriculum offering and enhances your continuous improvement strategies. It also indicates to the learners that your organization delivers high quality content.

Accredited partnerships give you a competitive advantage in relation to the learner who will be more driven to participate and realize the value of their investment in your virtual program more fully. Many of these accredited partners mandate specific and monitored criteria for the student to earn the pre-determined CEUs such as in-classroom or virtual attendance for a dedicated percentage of time, participation levels and engagment goals, minimum passing scores on post-exams, and established reporting processes. These requirements further validate your offering and will garner a stronger consideration by the learner as they assess which virtual course or presentation to attend. Once a course or virtual presentation is complete, fulfilling the PDHs or CEUs is as simple as recording the student's participation hours in the course, as well as their pre- and post-test scores, then generating a course certificate of completion with the accredited partner's logo for authenticity and validation. Keep in mind that rigorous record-keeping is a crucial step when working with any accredited partner. Ultimately, one of the greatest benefits of distance learning is that students and participants can continue their education while they are remote or unable to travel and still earn valuable PDHs and CEUs to fortify their career learning path. Those credits also symbolize to current and potential future employers that high quality training has been completed.

Now that you have evaluated your existing resources and defined those which you will need to establish, how do you determine which virtual platform is best for your institution or organization? Through experimentation and evaluation, we have discovered some key takeaways for various current digital platforms as follows:

- Zoom – Early adopters set the tone for widespread global usage of Zoom. These broad applications created initial security issues, but this platform has now become an innovative and secure video environment for reliable translation. With proven delivery, exceptional reporting tools, continual IT and infrastructure enhancements, supportive customer service, design for maximum audience headcount, and archival data storage, it has become a preferential and trusted choice for many distant learning and instructional forums.
- Webex – While this is mostly known as the intellectual video-conferencing solution for today's workforce, this platform's features include audience-enabled animated emojis, gesture recognition, and user-friendly polling.
- Adobe Connect – Hinging on the Adobe-established name, this platform offers versatility in its features such as storyboarding, but currently remains an expensive option for users even on a bulk basis.
- Microsoft Teams – This platform has a trusted appeal, once again hinging on the Microsoft brand name, but is currently more limited in widespread usage due to corporate firewalls precluding availability.

Virtual Platform Implementation and Launch

Virtual platfoms allow the administrators to control time and pace when it comes to knowledge share. There must be preliminary and subsequent continual communications to define two-way expectations in order to prepare the learner for the launch into this new virtual arena. As for engagement, virtual live-streamed courses and curricula must incorporate video to intrigue and captivate the learner with both visual and verbal appeal. Beyond video and repetition, it's imperative that several mock demonstrations and rehearsals have been conducted via the virtual portal to establish comfortability among the instructors and ease of use for all production crew and facilitators of the platform.

The initial step before launching is to establish a workspace that is ideal for delivery and learner comfort. If the space allows, consider a "TED Talk" presentation setup for content sharing, whereby the instructor is optimally-positioned in front of an oversized monitor backdrop. This design promotes a professional, quality look and maximizes the virtual user experience (Figure 1). For even greater instructor prowess and delivery, consider positioning a smaller monitor in front of the presenter -- hidden from virtual view -- to allow the presenter to reference subject matter or PowerPoint presentation slides, and also serve as a focal point for consistency among all subsequent presenters and instructors, as well as a guide to controlled pace and delivery (Figure 2). The last thing you want to do is feature a talking head behind a podium with lackluster delivery techniques. Ensure that lighting is ideal and consider additional lighting, a soundboard for audio integrity, and an additional overhead microphone (Figure 3). While it's crucial that instructors understand the criticality of preparing and practicing for delivery in virtual platforms, it's just as important that you provide the audio tools to ensure maximum production quality.

Figure 1. Example of virtual, live-stream classroom (back of room).

Figure 2. Example of virtual, live-stream classroom (front of room).

Beyond transitioning a traditional instruction method to a virtual setting, there are other areas which must be vetted in advance. You'll want to test all virtual technologies and set routines and parameters which should be consistently adhered to, pre-determine assigned roles, and establish clearly-defined goals and expectations of the instructors toward the end-result. Create extensive, yet user-friendly and mobile-friendly survey tools to measure engagement and assess the user experience.

Remember that multi-media-oriented teaching methods equal greater enrichment and retention. For this reason, you'll want to utilize all the tools possible within a given virtual platform. Consider blackboards within the technology to promote that feeling of in-classroom engagement and participation. Screen share among various presenters to set the tone for collaboration and change the voice and delivery style. Remember that virtual modules allow for custom content from one presentation or course to the next. Digital outlines, schedules, manuals, exams, and other components of the learning process can be revised easily and instantly in a virtual setting. Simply said, if it's not printed, it can be revised on the spot and updated in real-time. This highly-personalized capability is perhaps one of the greatest benefits of virtual instruction. You'll want to carefully and critically assess those first virtual experiences in order to implement necessary enhancements for subsequent deliveries. Each instuctor-student virtual encounter, then, serves as a springboard for the next.

Figure 3. Example of virtual, live-stream classroom setting and technology set up.

Items needed for presenters on screen in a live-stream setting will likely include a streaming video camera (webcam), external microphones or lapel mics (you'll need several for seamless instructor transitions), lighting equipment, tripod, dual monitors as aforementioned, strong Wi-Fi signal, and ultimately a reliable and privacy-protected online video platform such as Zoom. Explore all of the viable features within the virtual platform and assess which options will fortify the user experience. Something as simple as a timer feature can ensure on-time returns from breaks, timed polling or quizzing, countdowns, and other versatile methods to retain momentum.

If virtual delivery is newer to your organization or institute, you will have to consider that traditional curricula, which might be spread out more extensively in the classroom, will now need to be condensed to manageable sessions of time in order to retain audience attention span and engagement. It's more effective to offer segmented curricula and allow for greater breaks in between segments. This should be communicated with the learner in advance. The presenter will be required to exude a greater deal of energy and emotion to garner that same engagement and keep the learners interested, but also entertained. This is new territory for most instructional presenters and is further justification to invest in a coaching program for virtual presentation skills. You'll need savvy presenters who can react in the moment when any virtual latency or glitches arise such as blank screens, audio disruptions, lagging issues, or anything that could distract or frustrate the learner. Something to note is the level of audience forgiveness when technical or instructional glitches occur. Virtual platforms allow a more flexible schedule to address those mishaps in order to pause for resolutions or improvise with unplanned breaks in the delivery. Based on our experience, the audience acclimates and adjusts accordingly.

Ensure that you have proper coverage and have pre-assigned monitors for the Chat feature and/or question and answer boards during the livestream delivery. Emphasize the criticality of this role to those who are assigned and ensure that they are comfortable with the virtual platform and can rapidly respond to learner inquiries in a live setting. You will find that learners feed off one another in virtual settings just as they do in person. If one student posts a question, others will follow suit. For that reason, it's suggested that multiple monitors are prepared and ready to respond in a timely fashion to ensure that audience engagement is encouraged and rewarded. Here is where a responder can award a virtual prize, for example. Remember that if you enlist a prize or reward, your institute or organization will be responsible for the fulfillment of that recognition in its tangible form. If budget allows, this prize for participation approach can be extremely fun and motivating. It can also entice the learner to attend future virtual events hosted by your institution or organization. While this motive isn't completely equitable to the instant gratification of rewarding a prize in person, it does provide brief moments of levity to reignite the learner attention span and desire to participate more purposefully.

Factoring in all of the above, one can surmise the many benefits of transitioning to a live-stream virtual setting including the ability to broadcast live anywhere, any time, across all time zones. Additionally, the high-level recording and editing features within these virtual software platforms allow instructors to re-host the segments at a later timeframe, edit segments for categorized custom presentations, and also archive segments for continual reference in the future. These recordings also have multi-purpose benefits including the ability to use the segments as a teaching tool for future instructors or edit and offer smaller segmented curricula which can be marketed by specific subject areas for a potential revenue stream. Because virtual platforms allow you to enlist multiple presenters, screen sharing is limitless and not sequestered to one blackboard per instructional segment. This collaborative arena promotes a more active knowledge share and experience for the learner especially when pre-engagement activities are infused to tee up the knowledge transfer experience and establish transitional segues. More skills are employed by the instructors resulting in more adopted techniques among the learners in a given virtual setting. One of the greatest lump benefits of virtual e-learning comes down to savings. Because the curriculum and content delivery can be presented and/or taken anywhere, time and money are both saved, not to mention the elimination of paper exchange and the benefit of unlimited spacial storage in this virtual arena. Time savings is also a factor, as instructors follow a pre-set schedule of content delivery, allowances for breaks, and the removal of interruption from muted students who are virtually typing their questions or comments in Chat windows versus traditional classroom settings.

Conclusion

While face-to-face instruction is still believed to be the accepted tenet of education, a shift in paradigm has occurred mainly due to the spread of COVID-19 and has accelerated virtual collaboration exponentially. Institutions, organizations and companies found themselves faced with the higher demand to teach for engagement at a moment's notice in order to sustain and thrive in the training realm. This crisis teaching approach birthed a new view and opened doors to viable virtual platform options that could be implemented quickly and globally with assured reliability. In most instances, the virtual learning and online learning exploded during quarantine mandates, but shifted back to traditional in-person classroom settings, and then shifted back to virtual and online when surges in the virus infection rate rose again. This hybrid instructional back and forth methodology created more diverse and flexible learners who will likely experience this same on again, off again environment as wordly influences could dictate future protocols.

Another seemingly wave of the future involves the expansion of e-learning into augmented reality-based courses (see Chapter 7) which rely more heavily on the tactile methodologies. Here is where one imagines a mechanic, an engineer or a surgeon honing their talents through augmented practicum. This technology has been used for many years to train airline pilots in flight simulators. While virtual reality has become a more keen and user-friendly approach, augmented reality stretches the learner's mind more robustly. The emergent phenomena of virtual reality, augmented reality, and mixed 3D reality is affecting all environments and is measurably impacting the ways people communicate with technology and the modalities in which they collaborate with each other. Institutions, organizations, and companies are embracing these emerging technologies and implementing them at a more rapid pace than ever before (Becnel, 2019). Technology in this digital age continues to evolve toward these higher-level platforms and learners will essentially come to expect that content will evolve along with the delivery platform as well. The COVID-19 pandemic's unprecedented impact was the controlling force which rapidly launched virtual demands to be implemented immediately in all learning situations and scenarios. This wasn't solely manifested in classrooms, but in companies across the globe and in households of varying cultural backgrounds.

Virtual learning platforms such as the aforementioned Zoom, Microsoft Teams, and Adobe Connect will continue to morph and reassess user demands and technical advances to more succinctly cater to the evolving and ever-changing user experience. Areas such as reporting structures, storage repositories, interactive expansion, and cross-functional capabilities remain at the top of the list for each platform in order to not only sustain but also remain competitive in leading-edge characteristics.

Going forward, a heavier emphasis will be placed on teaching for engagement in all facets of content delivery among virtual deliveries. Technological advancements will be unprecedented in this busy arena and global learners will be challenged in more diverse and critical-thinking strategies. Likely, hybrid approaches and blended learning tactics and principles will define the ever-evolving virtual learning landscape.

In summary, the emergence and global effects of the COVID-19 pandemic and its rapid spread dictated the immediate need for increased virtual distance learning options, ultimately opening the door for many institutions and organizations to embrace this new approach on a global level. This renowned growth pattern has resulted in the migratory structure of workplace teams to a more geographically-dispersed, yet collaboratively-connected, setting which will become the global norm. This newly-defined collaboration methodology will be essential to the future competitiveness of every organization and will drive the further development of the virtual classroom for synchronous learning to occur well into the future. Forward-thinking organizations will keep pace and incorporate new technological developments that promise to further enhance the learning experience for their employees and students. These are both exciting and challenging times for organizations delivering continuing engineering education.

References

Association for Talent Development (ATD) Research. (2020). *2020 State of the Industry*. American Society for Training & Development.

Becnel, K. (2019). *Emerging Technologies in Virtual Learning Environments*. Hershey, PA: IGI Global.

Christopher, D. (2015). *The Successful Virtual Classroom: How to Design and Facilitate Interactive and Engaging Live Online Learning*. New York City, NY: AMACOM.

Huggett, C. (2017). *Virtual Training Tools and Templates: An Action Guide to Live Online Learning*. Alexandria, VA: Association for Talent Development (ATD).

Kim, S, Song, K., Lockee, B. and Burton, J. (2018). Gamification in Learning and Education: Enjoy Learning Like Gaming. Cham, Switzerland: Springer Nature.

Sintema, E.J. (2020). E-Learning and Smart Revision Portal for Zambian Primary and Secondary School Learners: A Digitalized Virtual Classroom in the COVID-19 Era and Beyond. *Aquademia*, Volume 4 (Issue 2), pp. 1-2.

Chapter 10

Continuing Engineering Education in the Process Industries

Myra Fanning* **and Charles E. Baukal, Jr.**

Koch Engineered Solutions InstituteSM, John Zink Hamworthy Combustion, Tulsa, Oklahoma, USA

Abstract

This chapter discusses the practical design and implementation of a continuing engineering education program concerning industrial combustion equipment. While it focuses on the process industries, most of the principles are universal. It discusses the participants (students, instructors, and training coordinators), learning styles, professional training organizations, subject matter categories, training locations (institution, on-site, hub, online), other considerations such as instructor training and accreditation, two case studies, and concludes with some recommendations. Many specific details are covered such as instructor training, room layout, and even instructor apparel.

Keywords: process industries, safety, training

Introduction

Those working in the combustion industry are very aware of the day-to-day concerns of working with combustible materials and dangerous equipment inside a refinery. U.S. employers and employees depend on the Occupational Safety and Health Administration (OSHA) to provide and enforce specific rules and regulations in the workplace to assure a safe working environment is maintained. OSHA and lawyers agree that employers must provide up-to-date *combustion training* regularly, providing the employee the opportunity to learn when to STOP, THINK and ASK before they ACT. Training should also provide instructions on proper installation, operation and maintenance for engineers and operators to follow so that all employees return home safely each day.

Trial lawyers associated with Arnold & Itkin Association (arnolditkin.com) have stated, while almost every workplace is at risk for a chemical explosion, workers at a chemical or industrial plant are exposed to greater risks. By the creation of gases, the build-up of heat, and the reaction, a *chemical plant* can become the source of serious and debilitating explosions. These can result in serious *plant explosion injuries*, such as third-degree burns, and even severe *property damage* that can affect the local community for years to come.

OSHA's general standards for employers include, but are not limited to:

* Corresponding Author's Email: myra.fanning@kes.global.

1. Create a safe work environment free of hazards following all safety and health requirements
2. Regularly check workplace conditions
3. Equip employees with the proper equipment and safety tools
4. Maintain all equipment and tools / replace as needed
5. Warn workers of all hazards on site
6. Train workers in health and safety procedures, frequent review of these standards
7. Notify OSHA within 8 hours of an accident including three or more injured workers or a fatality

In this chapter, we want to focus on #6 *Train* workers in health and safety procedures, frequent review of these standards. Dale S. Beach defines training as “the organized procedure by which people learn knowledge and/or skill for a definite purpose” (Bartleby Research, 2021). Training refers to the teaching and learning activities carried on for the primary purpose of helping members of an organization acquire and apply the knowledge, skills, abilities, and attitudes needed by a particular job and organization.

Trainers may identify training in the workplace as continuing professional development (CPD). The Institute of Continuing Professional Development (ICPD 2010) defines continuing professional development as ‘the systematic maintenance and improvement of knowledge, skills and competence, and the enhancement of learning, undertaken by an individual throughout his or her working life’. Padfield and Schaufelberger (1998, p. 8) defined continuing education as including “all those processes that contribute to the advancement of an individual’s knowledge, skill, understanding, competence, and general professional and personal development” (Padfield & Schaufelberger, 1998). Continuing education, continuing professional development, and professional development are all related terms that generally concern learning outside the normal workplace (Dirks, 2011).

Most trainers are aware of a few common types of training conducted in the workplace: induction training, refresher training, on the job training, vestibule training and training for promotions. Some training sessions may be formal, informal, mandatory, optional, for self-development or a train-the-trainer session. As a trainer, it is important to conduct a learning assessment before training begins. The summary of the assessment will help the trainer determine which type of training should take place.

Most companies acknowledge the importance of training for their employees. This chapter will provide tools and tips trainers have implemented successfully, including those at John Zink Co and Koch Engineered Solutions Institute (KESI), formally known as the John Zink Institute.

Participants

There are various types of participants who attend combustion training. The most important are the students themselves who are typically engineers or operators. The instructors or trainers are responsible for delivering the course content. They may come from a variety of organizations such as the employee’s company, universities, or suppliers. Another type of participant considered in this section is the training coordinator, training leader or facilitator. The roles of each may differ depending on the type of training offered or the training demands of the company. In general, each role is responsible for making the arrangements for the students’ organization or assuring the training session maintains proper timing and engagement. These three types of participants are briefly discussed next.

Students

The students are the most important participants in any training program. KESI often addresses students as VIPs, acknowledging them to be the most important group in the room. If they do not learn the material covered in a course, then the training has been a failure, despite the best efforts of the instructors and the training coordinators. At most plants, the two most important constituencies for training are engineers and operators. While they may work for the same organization, they generally have significantly different needs which is why they have been separated here. Sometimes it may be preferred to have separate classes specifically designed to meet the needs of these two types of students. There may also be some other types of students such as college students, college professors, support staff, equipment suppliers, government regulators, and any others involved in industrial combustion equipment. The goal should be to receive a comment stating, "the best training I've ever had" at least once from the student course evaluations for every class.

Engineers

Engineers generally have a college degree in some field of engineering (e.g., mechanical, chemical, electrical, civil, petroleum), although there are some "engineers" who do not have a degree or do not have a degree in engineering. Engineers with degrees in engineering have taken many math and science classes and generally have a strong background in the fundamentals related to those subjects. Those with degrees in mechanical and chemical engineering will also have taken courses in subjects such as fluid flow, heat transfer, and chemistry, which are important topics when discussing industrial combustion.

However, unless an engineer has an advanced degree related to combustion, they generally will have had few, if any, classes in combustion. Even if they have taken a college course in combustion, it is not likely to have covered industrial combustion. Therefore, engineers normally need training in the specific types of combustion equipment used by their organization. Depending on how long it has been since an engineer has taken relevant courses (e.g., heat transfer) and how frequently they have used a particular subject in their job, they may or may not need a refresher in some of the basics. Even if they have not used a subject for some time, they will have different training needs compared to, say, an operator, for example, who has probably never had any formal training in that subject.

Another important consideration for training engineers is their responsibility within a given organization which usually varies significantly compared to operators. Engineers are generally more concerned about overall performance such as thermal efficiency and pollution emissions, compared to operators who are generally more interested in other things such as reliability and operability. Both engineers and operators will be interested in subjects such as safety because of its importance for preventing personnel from getting injured and because both types of participants will likely have had little if any formal training in combustion safety.

In general, training for engineers may be more theoretical compared to training for operators which may be more practical. This is sometimes a difficult balance to achieve when both types of participants are in the same class. Engineers may have decision and purchasing rights, where most operators may not. Therefore, the engineer's focus and understanding of a particular piece of equipment may vary from an operator. One way to handle this is with prerequisite online training where students can go at whatever pace they prefer so that they can begin class with a minimum expected level of prerequisite knowledge.

Operators

Operators are responsible for the safe and reliable use of the equipment. They generally have different training needs compared to engineers. For example, operators need to ensure there are no unplanned equipment problems that could cause an unexpected shutdown. That means operators need to be very knowledgeable about detecting problems and fixing them before they lead to much larger problems. Operators monitor the equipment on a daily basis, where engineers may see the same equipment on a much less frequent basis. Operators need to be knowledgeable about preventative maintenance. They may do some of this maintenance or they may communicate with a maintenance group to handle these issues. Whether they actually do the maintenance or not, they need to be knowledgeable about what needs to be done and when to make sure the equipment is available for continued operation.

Operators usually have little, if any, formal training in subjects such as heat transfer, fluid flow, and chemistry. If these subjects will be covered in a training course on industrial combustion equipment, it should be assumed this is the first time these students will have taken those topics and that it will not be refresher training. In general, operators will be less interested in theory and more interested in the practical implementation of what they are being taught. Although operators may not have decision or purchasing rights, they can provide valuable input to the purchasing discussion after becoming aware of equipment that they may not have previously seen or used. This may lead to a valuable discussion on upgrading equipment or modifying daily processes and safety procedures seeing as they are the ones working on the equipment daily.

Others

There are a number of other potential types of students that may take a course in industrial combustion. One type is college students, especially those studying and doing research in the area of combustion. Baukal et al. (2010, 2011) describe industrial combustion training for university students, delivered by a combustion equipment supplier. Part of this training included the students visiting the supplier to see combustion equipment in actual operation. College professors involved in combustion research may take industrial combustion training to better understand how equipment is used in industry to more precisely tailor their research to the needs in industry.

Another potential type of other student includes suppliers who work with industrial equipment, but do not actually use that equipment. For example, these suppliers might supply fans, blowers, flame arrestors, refractory materials, flame detectors, ignition systems, control valves, pollution detection equipment, and other related equipment. Suppliers may be taking a combustion course to learn more about how their products are being used so they can be further improved.

Government regulators may take an industrial combustion training course to better understand how the equipment is being used so proper regulations can be written and implemented. It is certainly more difficult to write good regulations without a proper understanding of the context. Support staff such as administrative assistants, company buyers and other office personnel may attend training so that they may better understand phone calls from customers explaining their type of equipment, understand conversations during departmental meetings, and to better meet the needs of the customer.

Instructors

The instructors are responsible for delivering the course content to the students. They may come from a variety of sources and often have a variety of backgrounds. Two key requirements for the instructors are that they should be very knowledgeable of the subject matter (usually referred to as subject matter experts or SMEs) and they should be skilled presenters. It is recommended that all instructors complete a presentation skills training session themselves before presenting. Some of the typical sources of instructors are discussed next.

Employer

Employers may provide instructors to train their fellow employees. This could be in a variety of formats. In on-the-job training, a more experienced employee mentors a less experienced employee in specific policies and procedures. This usually takes place over an extended period of time. The trainee or apprentice usually "passes the course" when the experienced employee is satisfied the trainee has acquired the necessary knowledge and skills to do the job. The curriculum is usually very fluid and varies depending on the preferences of the mentor and the capabilities of the mentee. The length of the training may also vary considerably depending on both the mentor and apprentice, as well as on the context. For example, if the unit where the mentor and apprentice work is particularly busy, the training may take more time because the mentor may have less available time for training.

Another type of trainer for an employer is someone dedicated either full-time or part-time to training. These trainers may have a training background, or they may have been subject-matter-experts with many years of experience who are now teaching. This type of trainer often teaches in a traditional classroom to multiple students. The courses may range in length from less than a day to many months depending on the scope of training required. For example, a refresher course will typically be fairly short, while training of new employees may take much longer depending on the position. Some companies train new employees as part of a cohort where a group goes through training at the same time. There may be multiple trainers in that scenario, but usually one person oversees the entire program who is sometimes referred to as the training coordinator. This position is discussed in more detail below.

Supplier

In the past, it was fairly common for organizations to have resident experts in a wide range of technologies. For example, most refineries had experts in process heaters, pumps, valves, reactors, and other common equipment in the plant. To save money, many of the specialists in the plant were replaced by a smaller group of generalists who had less specific knowledge about any given equipment but more knowledge broadly of the equipment in the plant. To complement the generalists, larger organizations often had specialists as a corporate function who would visit plants to help as needed. Today, many plants do not have the same level of expertise on site that they have had in the past, so they often rely on the equipment suppliers for help including training. Since the suppliers designed and built the equipment, they have specialized knowledge that is often needed by end users. The suppliers then may be available to train employees in the plant. These external trainers are sometimes referred to as external providers (see Chapter 11). However, the suppliers may not have intimate knowledge of the process where the equipment is used, so sometimes suppliers may partner with the plant to jointly deliver training which is sometimes referred to as a hybrid provider (see Chapter 11).

The potential problem with suppliers providing training is that the information may be slanted toward their equipment. If the plant only uses their equipment, this may not be a problem, but if the plant uses equipment from multiple suppliers, using one supplier to provide training is potentially problematic as the single supplier may not be able or willing to speak about competitors' equipment. If the training is somewhat generic, this may not be an issue, but if detailed and specific knowledge and skills are needed, using a single supplier to train may not be the optimal solution.

There are many potential benefits to using a supplier to provide training. They are probably more knowledgeable about their equipment than anyone at the end user's site. They typically know about issues industry-wide and are not limited to issues only at a given site or for a given company. They often have training materials specifically developed for their equipment and they usually have experienced trainers who update and deliver the content on a regular basis.

University

Companies may partner with universities to teach certain types of content. University instructors are another type of external provider (see Chapter 11). Typically, this type of instructor will focus more on the theory side of the process or equipment. However, in some cases, training provided by a university professor can be very targeted when that professor has an established relationship with the organization, possibly through joint research programs. Many engineering professors do consulting in addition to their normal university duties. They are generally very experienced presenters, although their audience is normally college students who are usually much younger and less experienced than personnel working in industry. It is important the professor training plant employees is also a skilled trainer of adults (Darkenwald and Merriam, 1982; Praeger et al., 1998; Merriam et al., 2007). According to the theory of andragogy, adults should be taught differently than both children and even young adults attending college right after high school (Knowles et al., 2005).

For any of the above instructors who are not employees at the company they are conducting training, the training coordinator may want to review basic company policies and principles with the external instructor before presenting to avoid inappropriate dialogue.

Training Coordinators

Many organizations have one or more persons dedicated to the administration of the training program. At plant sites, these persons are often called training coordinators. In many cases, these coordinators were operators before becoming coordinators. Besides the administration of the training program, these people also typically do some training. They organize the training to ensure employees receive the proper training which is usually determined in conjunction with supervisors and plant management. Training coordinators make sure adequate facilities are available to conduct the training and work with management to schedule the training. For example, during a plant maintenance turnaround, it may be advantageous to train some who are not directly involved with the turnaround, but it would not be wise to schedule training for others who are intimately involved in the turnaround.

Plant coordinators try to keep up-to-date on the latest information and technology regarding training. They may also train other trainers in presentation skills. They work with plant management to determine when new types of training may be needed. For example, after an incident, it may be recommended that certain types of job positions receive certain types of training to avoid future incidents. Another example is when new equipment is installed in the plant and personnel need to be trained on installation, operation, troubleshooting, and maintaining that equipment. Plant training coordinators serve as the hub for all functions related to training in the plant. Plant training coordinators may also be a mediator between the training supplier and their purchasing department as they prepare to submit a required purchase order and service agreement, including terms and conditions to assure the training has been approved and payment will be made.

Learning Styles

There is an important aspect of training that is often forgotten: *make learning and training fun*. It may have taken years for some to actually believe that training can be fun, but it has since become an expectation and has been proven to be effective, if the activities are completed with a purpose. Not all learners see through the same mental lens. There are many paths into the learner's mind. If a trainer only utilizes one path of learning, the results are going to be less successful. A variety of delivery techniques benefits all types of learners because it keeps learning interesting and engaging. There is a golden rule that trainers know very well: people retain 10% of what they read, 20% of what they hear, 30% of what they see, 50% of what they hear and see, 70% of what they say, and 90% of what they say and do (Kornikau and McElroy, 1975, p. 370). "One must *learn by doing* the thing, for though you think you know it—you have no certainty until you try." *Sophocles, 5th century B.C.*

Several techniques may be used to strengthen retention and understanding. A trainer cannot expect the participants to walk out of the classroom and apply the knowledge without allowing time for practice. There are three (3) types of learners in the room: *Prisoners* (they made me come, I know it all … I could be teaching this class…), *Vacationers* (Whew! 8 hrs away from that phone, I'm in the AC, which beach do I hit first, in a nice hotel, no kids, mind is somewhere else), and *Learners/Explorers* (call them the sponges. WOW! They may come in as L/Es but the goal is to keep them there).

After much study and researching, trainers learn that there are other thoughts and opinions regarding adult learning. Malcolm Knowles, known as the father of adult learning theory shares in this book *The Adult Learner: A Neglected Species*, published in 1973 six assumptions to adult learning. Some think that the Knowles assumptions overlap themselves, so much so that Ann Herrmann-Nehdi developed the *Neurolinguistic Programming* (NLP) theory which narrows it down to three (3) types of learning styles, making it easier for trainers to understand. These include: visual, auditory, and kinesthetic learning styles. To summarize each: visual learners prefer pictures, graphs, illustrations. "Show me," is their motto. Auditory learners pay close attention to every word, voice and sound present in the classroom. "Tell me" is their motto. Kinesthetic learners are "hands on" learners. They enjoy class activities. Their motto is "Let me do it." Identifying which participant prefers which particular learning style may be difficult at first, but experienced instructors generally can do so before the first lunch break. It is important to silently identify the learning styles and types of learners early and make every attempt to engage them, praise them, challenge them and eventually have them all leave as satisfied participants (Biech, 2005).

Dave Meier (2000), founder of The *Center of Accelerated Learning* in 1980, takes this concept to another level. Accelerated Learning (AL) has only one goal: to get results. There is a difference between AL and fun-games, gimmicky, "creative" approaches that do not really bring results or have a purpose. They are used more so to take up time or even waste time. The theory behind AL is to blend both fun and serious lecture in a way that enhances learning and produce positive outcomes using the following principles:

1. Total learner involvement enhances learning
2. Learning is not the passive storage of information but the active creation of knowledge.
3. Collaboration among learners greatly enhances learning.
4. Activity-centered learning events are often superior to presentation-centered ones.
5. Activity-centered learning events can be designed in a fraction of the time it takes to design presentation-centered ones.

Examples of activities that may be incorporated into a training session are:

- Quiz cards – serve as a review of the previous discussions (see Figure 1)
- Molecule modules – class activity to share connection between various chemicals (see Figure 2)
- Gamify activities
- VR Training
- Puzzles

It is important to continue to research the internet, attend conferences/workshops, and events to learn of new activities that may be used in training sessions to enhance learning. Many participants did not sign up for the course to hear a lecture the entire session. Sivasailam "Thiagi" Thiagarajan, reminds trainers that: lecturers ignore the fact that real learning requires active participation, not passive listening; Lecturers tend to focus on what they want to tell the audience rather than on what members of the audience want to know and Lecturers do not receive useful feedback.

Gamification is a way for companies and training coordinators to improve learning outcomes by incorporating easy-to-use web and mobile based learning platforms into their training sessions. Gamification is about *transforming the classroom environment and regular activities into a game.* Most humans enjoy a little competition and a fun way to compete. Gamification provides just that in a fun, quick and beneficial way. The "prisoners" often forget they're learning and begin laughing and enjoying the training session. Due to the results the "sponges" receive, they're self-confidence is off the charts. They soak up all the information provided and are not afraid to let others know it as they answer the questions correctly. KESI students often request more games or quizzes in future courses. They appreciate having them available as a refresher of the course information.

Figure 1. Quiz card example.

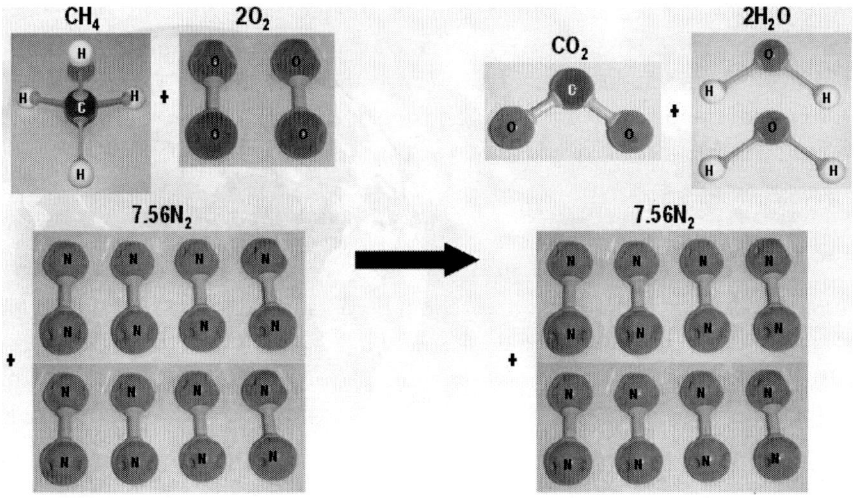

Figure 2. Molecule modules example.

It is important to note that gamification may provide the opportunity to:

- Introduce the content with text, images, audio or multi-media
- Know what the students know in real time, and plan for remediation
- Extend learning with Summary Screens at the end of a question

No matter where your training takes place, such as: classroom, webinars or e-learning using gamify activities will enhance the learner's training experience. BravoZone, provided by C3 SoftWorks is one of the software platforms often used by many companies and training coordinators. They provide various templates for trainers to customize training materials into a game. The games have "catchy" titles and music that draw the interest of the students. Some example games include: Catapult, Moonshot, Billionaire and Spinoff. There are many gamification systems software platforms available. Some platforms offer the ability to issue quizzes to use to check understanding or as a refresher. One often used at KESI is UMU (www.umu.com). An UMU quiz may be presented at the conclusion of a topic.

In Chapter 9, due to the Covid-19 pandemic new technologies that exploded in 2020 and 2021 were discussed. Most employees were forced to work from home. Many employees considered it a great opportunity for employers to develop new skills to enhance their position or advancement. Finding ways for employees to continue training was crucial. How fortunate it was for Zoom and Microsoft TEAMS to be in operation and available to provide a platform for employees to communicate with other team members, participate in webinars and for training organizations to offer training thru e-learning. Many trainers quickly learned to incorporate "fun" into their webinars, hoping to maintain interest and engagement while sitting still, glaring at a computer or laptop for an hour.

In November 2021, Anders Gronstedt (2021) wrote:

> Most industry experts agree that we're at the end of Web 2.0 and mobile apps. The twin forces of augmented and virtual reality and work-from-home habits are ushering in a new era of Web 3.0 (or Web 3D). The metaverse is upon us, and it has profoundly affected the learning industry. Accenture, for example, paid $18 million to equip 60,000 employees with their own consumer Quest headsets for training and collaboration (and probably some play because work and play is converging).

The real-time, spatial, 3D, immersive, and embodied internet can be deployed on high-quality VR headsets that cost less than an airline flight. The workplace is moving to workspace, and VR training is evolving from one-off, single-user simulations on shared headsets to ongoing, programmatic social learning on personal headsets. (See Figure 5)

Forward-thinking companies are creating shared VR learning spaces where students and instructors are interacting with each other as lifelike human avatars. Navigating hyper-realistic performance simulators via VR headsets, phones, iPads, and PCs, they learn with and from each other in hands-on simulations. A few recent client projects illustrate how these early-stage building blocks of the metaverse are already transforming learning:

- *Pfizer, Novartis, and Bristol Myers Squibb* harness VR learning to practice crucial life-saving skills in our safe, virtual reality pharma labs. Instructors can demonstrate how to transfer liquid and hand the pipette to a student standing to their right, who can grab it and rehearse the same task, just like in real life.
- HVAC giant *Daikin* offers the ultimate customer experience of X-ray vision. Customers don the Quest headset and peer inside a giant chiller to watch how water and refrigerant work together to keep building temperatures comfortable. A sales rep can walk around the digital twin with customers and point out features and benefits.
- *Alarm.com* installers use their hands in VR to route cables and connect wires and test whether they've secured two doors with a card reader. A ticking time-bomb is counting

down time. If the installation task is not completed in time, the place explodes. If they get it done sooner, they get to add their names to a leaderboard.

Although cost is often a strong factor when purchasing software and equipment for training, Gronstedt warns us about purchasing the least expensive equipment. "This is not the time to shoehorn VR technology into existing learning programs. At best, your users and decision makers will wonder why you didn't just do it on a PC screen. At worst, they will get nauseous. Instead, partner with a professional developer and invest in the potential of real-time game engines and experiential learning theories. A professionally developed simulation in a game engine might be more expensive up front, but it's a malleable and scalable investment that can be amortized over many years and numerous applications. The same VR sim can overlay the real world in augmented reality (AR) or be featured in "pancake mode" on mobile and PC screens."

Training Organizations

Association for Talent Development (ATD)

Most careers provide a professional organization that offers support and resources; the training profession is no exception. The Association for Talent Development (ATD) is a professional membership organization supporting those who develop the knowledge and skills of employees in organizations around the world. This association was previously known as ASTD. The ATD mission is "Empower professionals to develop talent in the workplace." The ATD vision is "Create a world that works better.

ASTD began in 1942 during a training committee meeting of the American Petroleum Institute (API) held in New Orleans, Louisiana. In 1943, 15 men came together in Baton Rouge, Louisiana to conduct their first *American Society of Training Directors* meeting. Their goal was to create an organization devoted to raising the standards for the industrial training profession, and to further the training professional's education and development. In 1964, the association changed its name to the *American Society for Training & Development*. In 2000, the organization chose to refer to itself just by the letters ASTD to recognize that the profession looks far beyond training in its quest to create a better working environment for all involved. In 2014, the organization changed their name once again to the *Association of Talen Development* (ATD*)* to reflect the global nature and broad scope of its members' work. ATD has an international as well as US membership base (more than 120 countries; 100 U.S. chapters; 18 international strategic partners and global networks). The association's members work in every industry and sector. Some data for ATD are shown in Figure 3.

ATD hosts many conferences in the U.S. and internationally. These include: International Conference & Exposition, ATD TechKnowledge Conference and EXPO, Core 4, TalentNext, SELL, Learn from the BEST, ATD-Yale Foundations of Management Excellence, LearnNow workshops, and summits in China, Japan, Korea, and Mexico. Additionally, each fall ATD hosts the ATD Chapter Leaders Conference (ALC) in the Washington DC area. ATD also has awards programs that recognize organizations and individuals for their excellence in talent development.

Figure 3. ATD 2019 by the numbers.

The ATD International Conference & Exposition (ATD ICE) conference is the world's largest conference, attended by professionals dedicated to workplace learning and development. It is annually held in prominent cities; places you don't mind spending a week in for training. Over 130 local chapters in the U.S. and over 30 international partners from around the world attend these conferences. Attendees can walk away with numerous training ideas and training materials. The networking opportunity that is available at these conferences is priceless. Vendors of most of the products, promo items and software training departments utilize are also present in the exhibition hall during the conference. This is great opportunity to try new products and software before making a purchase or to gather additional information first-hand to share with your company. It is recommended to bring along an empty piece of luggage to carry the numerous giveaways you will receive. Some attendees use the giveaways as prizes in upcoming training sessions.

ATD ICE conference attendees look forward to hearing (and seeing in-person) world-wide known keynote speakers such as: Dr, Ronan Tynan, Dr. Kelly McGonigal and Captains Mark and Scott Kelly (2017), President Barack Obama, Marcus Buckingham and Comedian Connie Codesta (2018), Oprah Winfrey, Seth Godin and Eric Whitacre (2019), Patrick Lencioni, Mel Robbins and Angela Duckworth (2021), 2020 ATD ICE was cancelled due to the pandemic. To get a better idea of the magnitude, below are the numbers from ATD ICE held in Washington, D.C. in May 2019. This is also a great time to have one-on-one conversations with your training SMEs or listen to them in small and informal lunch, dinner settings and book signings. Imagine spending time with mentors in the industry such as, Bob Pike, Becky Pluth, Jim Smith, Ken Blanchard, Saardek Love or Sivasailam "Thiagi" Thiagarajan.

Trainers are often interested in advancing their career by achieving a training certification. ATD is a trusted organization that provides several certifications, depending on your professional level and interest. Other organizations offering training certification are: Langevin Learning Services (Specialty Certification, Master Trainer, Training and Development Diploma), Training Industry, Inc. (Certified Professional in Training Management and Certified Training Practitioner), The Training Associates (The Training Associates Certification), The Institute for Performance and Learning (Certified Training and Development Professional), and ICAgile (HR and L&D certification). In this chapter, we will focus on ATD certification.

The ATD Certification Institute (ATD CI) is an independent organization created by ATD to take the lead and set industry standards for the talent development profession.

ATD CI Vision: To provide world-class credentials for talent development professionals.
ATD CI Mission: To elevate the talent development professions by:

- Creating talent development profession-wide standards
- Providing an independent assessment of those standards through credentialing
- Supporting career paths through testing
- Ensuring recognition through certification and continued professional development through recertification.

ATD CI offers two credentials that are based on the Talent Development Capability Mobility: the Associate Professional in Talent Development_(APTD) and the Certified Professional in Talent Development (CPTD). Some may be aware of a past certification entitled Certified Professional in Learning and Performance (CPLP). APTD and CPTD has replaced that certification.

APTD certification is for talent development professionals who are in the early part of their careers or whose professional roles and aspirations are focused on the foundational areas of talent development. CPTD is a professional certification for TD professionals with at least five years of experience. It is broad and measures a professional's knowledge and skill application across the breadth of talent development capabilities. CPTD certification focuses on knowledge and skills shown in the Talent Development Capability Model shown in Figure 4. CPTD certification is more intense. After mastering 22 capabilities within the model, the CPTD recipient should be able to add value to the company with their enhanced understanding of training development strategies and solutions.

The Talent Development Capability Model was introduced in January 2020. The model is used as a professional blueprint of standards trainers need to know to be more effective in the workplace. The capabilities are organized in three domains: Building Personal Capability, Developing Professional Capability, and Impacting Organizational Capability (see Figure 4). ATD's research shows that successful talent development professionals need knowledge and skills from all three of these areas to be most effective.

Trainers often ask, "why should I get certified?" Here are a few comments from APTD and CPTD recipients:

- Allows me to speak the "language" when meeting with high level clients
- Validates my knowledge of training and experience
- Expanded job opportunities
- Provided credibility

Why Should I Get Certified? (td.org)

Figure 4. ATD Talent Development Capability Model.

This model may also be used for personal evaluation:

- Assess current knowledge and skills against what is needed to succeed in the future
- Set performance objectives
- Identify a personal development plan based on your gaps
- Use it to determine possible career path
- Pursue capability model-based credential – APTD or CPTD

As a trainer, it is beneficial to become involved in a local ATD chapter or ATD Nationals. The opportunity to become a member of both is recommended, known as a *Power Member*. Members have access to a bank of presenters, speakers, and consultants to share with their Human Resource department when the need arises for the training of employees. *Power Members* receive discounts on ATD products or conference registrations.

Trainers and vendors from various companies often share learning tools and ideas used within their training department at the monthly meetings. Networking with other trainers allows the opportunity to share ideas and best practices which makes provision for growth and enhancement to one's training career. As a reminder, there are a variety of trainers in varies fields within the US and Internationally. Imagine the different training techniques, tools and ideas that can be shared.

As an example, one local ATD chapter brought in a vendor who provides VR headsets to training departments within the fire department industry (see Figure 5). Having the opportunity to experiment with the headsets and equipment was so valuable. This activity challenged the user to put out a "mock" fire using a fire extinguisher that reflected off a screen nearby. Immediate feedback was provided, sharing various levels of your performance. The National Fire Protection Association (NFPA) understand the value of providing up-to-date training and hands on participation.

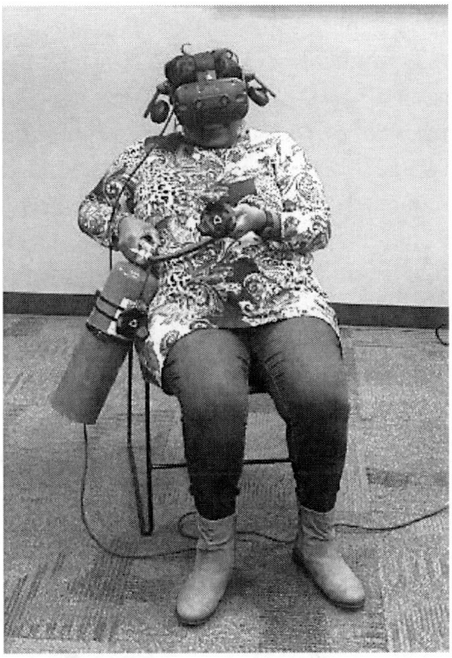

Figure 5. NEOK ATD meeting in Tulsa, OK.

Subject Matter Categories

There are many ways that subject matter could be categorized. It could be based on the type of process or equipment. It could be based on the unit in the plant. It could be by job or task function. Here, subject matter has been broadly categorized as fundamentals and operations, which encompasses some of the other methods. For example, operations would cover specific equipment in the plant such as valves, compressors, piping, and reactor vessels. It would also cover specific job functions such as outside operators who are responsible for the operation of specific equipment. Often the most powerful training includes aspects of both fundamentals and operations. Both categories are briefly discussed next.

Fundamentals

Fundamentals refers to the basic theory behind a process or piece of equipment. One philosophy for training is that personnel should be taught what they need to do, but not necessarily why they are doing it. While that philosophy may be more efficient in terms of the amount of time required for training, it unnecessarily narrows the amount of knowledge a student has about the process or equipment. If an issue arises that has not been covered in the training, the student may not be properly equipped to think through how to handle that issue as they have not been given an adequate foundation that includes the theory behind the process or equipment. This does not mean everyone needs to know how to design every process in the plant from scratch, which would not be possible nor is it necessary. It does mean that those involved should at least have a working understanding of some of the basics.

An example can illustrate this philosophy. Most plant operators know that the weather can have a significant impact on equipment performance, but they may not understand why or be able to predict what will happen based on weather forecasts. One of the pollution emissions from a process heater, nitric oxide or NO, usually decreases when the air humidity increases (Baukal and Bussman 2009a,b). This is simply because the added water in the air reduces the flame temperature, which reduces NO. By learning one of the fundamental principles of NO formation, which is that anything that increases temperature normally increases NO, assuming all other variables are held constant. By understanding that simple principle, plant operators would also be able to understand what happens when the air temperature increases or decreases, which happens frequently during the normal atmospheric temperature fluctuations due to the changes in weather from day-to-day and season-to-season. If the plant is operating close to their allowable pollution limits, relatively simple adjustments can be made to compensate for things like the weather which are beyond the control of the plant. This can help the plant avoid costly penalties for going over their permitted limits. Some basic understanding of the fundamentals can help operators make better decisions as it is virtually impossible to cover all possible scenarios in a training course.

Operations

Training focused on operations may include topics such as installation, operations, troubleshooting, and maintenance. This general type of subject matter covers the practical aspects of equipment and processes. This nuts-and-bolts training often has a hands-on component. Where training on the fundamentals is usually more knowledge-based, training on operations is usually more skills-based. Operations-type training may include examples of the right way to do things and the wrong way to do things, including the consequences of doing things the wrong way.

One of the challenges with operations-type training is that it may be difficult to practice skills on operating equipment. It is rarely possible to simulate abnormal, even emergency, situations in an operating plant, both because of the potential safety and financial ramifications. In those conditions, other techniques may need to be employed during the training, so employees are adequately prepared to handle non-routine conditions. Simulators are routinely used to demonstrate the proper procedures for handling a wide range of conditions without adversely affecting plant production (Baukal and Bussman, 2009c). Actual equipment not installed in the plant, scale models, cutaways, photographs, drawings, and videos are other methods that may be used to train personnel in plant operations.

Training Locations

Institution

There are many institutions, colleges and universities that offer combustion, process technology and instructional training courses. The Koch Engineered Solutions Institute (KESI), John Zink Company's educational organization, has offered combustion training since 1950 to engineers and operators from around the world who are looking to optimize their plant or maximize their career.

The North American Process Technology Alliance (NAPTA) provides a plethora of industry and educational institutions who offer technical programs for all ages. Many of these institutions offer scholarships, certification and accreditation. The Koch Engineered Solutions Institute is one of the few to offer accreditation and certification provided by International Association for Continuing Education Training (IACET).

On-Site

Field service personnel have the responsibility to visit various refineries around the world and often they discover the need for training. JZI became aware of the need and began offering on-site training courses: basically, taking the classroom to the refinery. "Stay there. JZI will bring the classroom to the workplace."

There are some advantages to teaching a class at the plant location. When a company has chosen to offer on-site training, they are able to have many of their operators and engineers in the facility training room at one time, all discussing issues that pertain particularly to their own equipment. Customization is another advantage to on-site training. A survey may be distributed to the requestor in advance, asking for specific information regarding their training session. This survey basically asks for the desired outcome of the training session; what problems currently exist at their location, how many participants, how many shifts, how many days are allowed to cover the training session. A basic agenda may be given, but the on-site coordinator and the instructor may see the need for additional topics to cover. Subject to economic conditions, many facilities may have limited resources. Therefore, allowing 2 or 3 of their operators to be away from the refinery at the same time can be critical to their daily operation. Conducting an on-site training allows several operators and engineers to attend the same 1-day course together, receiving the same information regarding fundamentals, troubleshooting and maintenance of combustion. They can discuss current issues and perhaps resolve them as a group. If there is a need for various night shift workers to attend a training session, arrangements can be made to offer the training to them during their shift hours. Another advantage is the ability to take the classroom to the field. It is recommended that this be done in small groups to allow all the participants the ability to hear the instructor. At this time, problem areas can be emphasized visually to all attendees and then a discussion can take place on location or back in the classroom. This can be noted as a "hands-on" experience for the participants.

There are also some potential disadvantages for on-site courses. Often times it is difficult for a company to send their operators and engineers to a 2-day or 3-day course offered offsite. The reason may be limited resources, budget constraints or conflicts in the company schedule.

Funding several operators and engineers to attend training offsite can be costly. They must consider airfare, lodging and meals. The non-productive days for traveling can also cost the company, especially if it means traveling on a Saturday or Sunday. For hourly employees, the company would be responsible for paying overtime to the employee if applicable. When a company decides to conduct on-site training rather than sending their employees offsite, generally only two or three SMEs are assigned to conduct the training. Although the outcomes are generally positive, they do not have the advantage of hearing other SMEs in the field of combustion. When operators or engineers attend offsite courses, several instructors may be available to share their expertise and answer questions. Many times, the space available for on-site training is limited. On-site training space can range from a small trailer located on the plant premises to a break room in the middle of several offices. Not all plants are equipped with an actual training room which may have a projector, large projector screen, adequate table and writing space or chairs designed for long periods of sitting. The noise level surrounding the available training space may also be high and distracting. All of these may become a disadvantage for the participants as well as the instructors and can create a negative learning experience.

When on-site training sessions are held at the customer's facility, there is a possibility that participants may get called away to assist co-workers or to take a phone call. Often, participants do not see the urgency to return to the classroom after a short break or lunch break on time. This causes a disruption in the classroom as they return to their seats, after the session has started.

As noted in the previous section, often the attendees include supervisors of one or two groups of those present. This may cause participants to hesitate to answer questions honestly, speak of current issues that they are experiencing because they know it may be of a different opinion than that of their supervisor, who is also present in the classroom. Whereas, when they attend an offsite course, they generally attend without their supervisor and are more open during discussions about issues at their facility.

Overall, on-site training at a customer's facility may not have all the advantages of attending offsite training courses, but it can be a benefit to the company. Providing a learning opportunity for operators and engineers to learn proper operating skills, expand their safety knowledge of their equipment and learn how to troubleshoot issues they have on a daily basis is crucial. It becomes a win-win for all parties involved.

Hub

In addition to offering on-site courses, companies may select to offer a Hub training session in a particular city or state. "Hub" refers to a centralized location, similar to how airlines use hub cities to efficiently move people from one location to another. In this case, the instructors and students are all brought to a convenient location, typically in a city where multiple plants are located.

Some companies may identify only a small group of employees who need training or a refresher course, but not enough to justify the expenses to ask for on-site training at their facility or send their employees to an offsite course. Field service personnel are then challenged to find other refineries in a particular area that also have small groups of employees needing training. If other regionalized interest is found, a centralized location is identified, and an announcement and notification is sent to all refineries within a certain mile radius that a supplier will offer a burner or flare operations course on a particular date.

There are some specific advantages of hub classes. Companies are able to send their employees to a location within driving distance which lowers travel and lodging expenses. This training is typically a 1-day course, so employees only miss one day of work at the plant. Hub training also allows great networking to take place. Generally, the participants are all in the same line of work so they have the opportunity to share operating experiences and discuss ideas and solutions.

Online

Since the late 1990s, online training, sometimes referred to as e-learning (electronic learning), computer-based training (CBT) or Web-based training (WBT) has made its way into the learning environment. All of these forms are growing in popularity and are available in many companies. JZ currently uses web-based safety training to supplement in-house training. Modules are available on nearly a hundred different topics. The Test Center personnel view a new safety topic on a weekly basis. Each module has an instructional section that includes text, graphics, audio, and sometimes video, followed by a quiz to demonstrate that the key concepts have been grasped. Training records for each employee are automatically updated after the quiz has been successfully completed. Another version of this form of training is where a live class is conducted over the Internet including email interaction between participants in a type of on-line chatroom. A related type is where the class lectures can be viewed either over a television or over the Internet with email used to ask the instructor questions. This method, though not live, is ongoing over the course of the semester. More and more universities are offering some form of web-based training.

JZI has gone a step further and offers an online course entitled, "Process Burner Theory." This is an online course designed to provide basic fundamentals of combustion, fluid flow and heat transfer as they pertain to burners used in process heaters and furnaces. The course consists of 16 modules, each of which takes approximately 15 minutes to complete via the Internet. There is a quiz given after each module to check for understanding and achievement. This is a great tool for those operators who really need a foundation in the combustion field but are not able to travel to attend a JZI Tulsa training course.

Worldwide organizations and conferences had to make a transition in 2020-2021 due to the pandemic. Some were forced to cancel their events or reschedule for 2021 or 2022. ATD ICE and many other conferences began offering their sessions virtually. Later in 2020, some felt it was safe to offer in-person sessions as an option. KESI was not exempt from this dilemma. Fortunately, KESI was able to transition rather quickly. They were able to hire a professional videographer, prepare their instructors on how to present virtually, reduce presentations to fit into smaller timeframes, provide the proper setting and successfully offer their Fall 2020 courses as e-learning courses.

PTech Organizations

A recent API study analyzed all the factors that affect a plant's efficiency. According to the study, human error had greatest impact on asset utilization. An American Chemical Society study concluded that when plants hire qualified, trained workers, companies documented an average savings of $16,000 for every new hire made. It is also possible for companies to see an increase in plant asset utilization from one to four percent. The question then becomes, "Where should a company search for qualified plant workers?"

The Process Technology (PTech) program provides classroom instruction and practical experience to prepare students for employment in a variety of process related jobs (www.naptaonline.org). The material prepares individuals to monitor, operate, and maintain equipment used in the processing of raw material into marketable chemical/petrochemical products. This includes instruction in materials handling, extraction, distillation, evaporation, drying, absorption, heat transfer, and reaction processes. It also includes chemical and fire safety, codes and standards, and general plant day-to-day operations. Students involved in PTECH programs participate in a curriculum that emphasizes safe and efficient work practices, basic occupational skills and employability skills. The content is organized into competency-based courses of instruction which evaluate specific occupational competencies. Each student is required to successfully complete the competency exam before receiving credit for participating in the program. The process technology program generally requires a two-year educational commitment and most of that time is hands-on experience in a lab similar to one at an actual processing plant.

Refineries have discovered that hiring PTech graduates may decrease employee selection costs up to 80 or, even 90 percent, reduce two-year new employee turnover up to 50 percent, drop job-training costs up to 40 percent and result in up to 37 percent fewer safety-related incidents. Process technology positions are not only available in oil refineries, power plants or chemical plants, but they exist in pulp and paper manufacturing, water treatment plants, power generation plants, food and beverage processing and pharmaceutical plants.

After gaining employment, PTech graduates soon learn to approach each workday with curiosity and flexibility. Graduates may begin the day by performing basic duties, but due to customer demands, weather changes, unexpected equipment malfunctions and new technology, the day may end with unexpected adventure. PTech employees often work rotating shifts, some only working 12 hour shifts / 14 days per month. These hours are a huge attraction to those looking for a career choice. The pay and benefits are often compared to college graduates who have attended college for four years or more. There is a famous story of a PTech grad who bragged to his neighbor who was an engineer for NASA of his six figure salary. The graduate spoke of several bonuses of $4,000 or more received within one year. In the meantime, the neighbor only recalls a one-time bonus of $100. The story ends with the neighbor informing his wife of a career change to process technology (NAPTA, 2021).

After successfully completing the curriculum, a State Certificate of Achievement or an Associate of Science Degree, PTech graduates soon discover the advantage of entering a workforce where there are more PTech positions than there are qualified employees. Today, not many college graduates have that experience.

The average age of a process technology operator today is fifty, which means as many as half of all process operators will be eligible for retirement within the next five to ten years. This will likely create a situation in the refinery workforce of not having enough trained workers available to replace the retirees. Therefore, there are organizations and corporations who support PTech graduates and work with employers to help meet the demand.

The North American Process Technology Alliance (NAPTA), formerly known as Gulf Coast Process Tech Alliance (GCPTA) provides educational support to those who are seeking a process technology career. NAPTA interacts closely with many colleges that have endorsed the process technology curriculum. These endorsed colleges have an active advisory committee, have completed a comprehensive program audit and have been approved for endorsement. Both are closely involved and supportive of developing standards for PTech programs to assure that qualified graduates are prepared for today's process technology workforce.

Other Considerations

Instructor Training

Instructors must not only have deep subject matter expertise, but also have the ability to convey that knowledge to novices. Simply being an SME is not enough. Instructors must also have proper training in how to effectively deliver training content.

Preparation

Do you immediately panic when asked to present to an audience? Do you immediately check your calendar, hoping that you have a previous engagement? Most of us can recall our first speech or the time our presentation didn't go as planned. There are several steps that should be followed so that most of your presentations and training sessions are successful. Let's begin with preparation.

A well-known mentor and trainer, Bob Pike (2003) is famous for reminding speakers that preparation is essential to achieving an outstanding training outcome. To do that, he recommends the six Ps of an effective presentation: "Proper Preparation and Practice Prevents Poor Performance." Dale Carnegie once said, "*There are always three speeches for one you actually give: the one you practiced, the one you gave and the one you wish you gave.*" A part of that preparation is enhancing a person's presentation skills. A consultant and KESI trainer came up with the following Ps for KESI training: *Plan, Prepare, Proofread, Practice and Present with Passion.*

Often presenters do not take the time to do the majority of any of the Ps and it is evident when it becomes stage time. The audience will be able to tell if you did not prepare or practice and instead decide to just "wing it." This is not recommended, especially now that Goggle is right at our fingertips. A favorite presentation method is what can be called the "NEWS AT TEN METHOD." Tell the audience what you are going to tell them. Tell them! Tell them what you have told them (Thomas, 2021).

Presentation Skills

Great presenters have two qualities: appropriate skills and confidence to use those skills well. Datta Groover (2021) would add listening as another important skill. These skills should include: good eye contact, great posture and appropriate facial expressions and hand gestures (non-verbal aspects). Most exhibit timely pauses and silences and speak with clarity. "*Silence is a language unto itself. Its alphabet may be based on the absence of language and its vocabulary may be incredibly limited. But silence lubricates the ideas, examples and antidotes shared in communication*" (Jakes, 2021). Although silence may sometimes feel awkward and uncomfortable, trainers are taught that silence allows the participants time to reflect on what was just said. It also allows the trainer to regroup or refresh. Perhaps even take a sip of water to avoid a dry mouth, evaluate the feel of the room or even recheck the presentation notes. Good presenters engage the audience by encouraging participation, inviting questions throughout and answering questions well. They are careful in the appropriate use of clichés, diction, slang or vulgarities, gender references, acronyms and jargon. It is imperative that a good speaker knows their audience! Take time to review the list of attendees to determine if the room will include mostly operators, majority engineers or is this a special session for support staff or a variety of different titles.

Utilizing listening skills is another very important tool for a presenter. This may sound strange, and one may ask, "How can one listen and talk at the same time?" Here are questions to ask oneself while speaking:

- Is your audience on board with you?
- Are they following you?
- Are you speaking over their heads, or do you need to take it up a notch?
- Did they fail to laugh at what you thought was funny?
- Is your energy level equal to theirs or is it too high or too low?
- Are you racing ahead of them, speaking too fast?

One cannot stress enough, "know your audience." This also includes being "tuned in" to the audience throughout the presentation. This is achieved by "listening" to the audience for both their verbal and non-verbal cues. One way to answer this question is to incorporate gamification.

As one reflects on the best speakers that they have ever heard, one common element most likely experienced was their *passion* for their topic. "*People may hear your words, but they feel your attitude*" *John Maxwell* (2021).

Although, a presenter may have passion for their subject, the other key presentation elements are not guaranteed to fall into place. In the training industry, a Subject Matter Expert (SME) is often called upon to present a particular topic. The SME will most likely have a passion for their topic because it is in their field of study, but that does not guarantee that they have taken the time to hone their presentation skills. Therefore, it is important to remember that not all SMEs are good presenters. Although the presenter is an expert in their field and very knowledgeable of the topic, the material may not be received well by the participants simply because it was not presented well. The presenter's tone, body language, and mood often affect the audience more than the actual presentation.

Trainers step up to the podium with different presentation styles. It is important to identify one's style and enhance it. Be authentic. Several presenters may be given the same topic, same agenda, and outline, yet each will most likely deliver the content in a different manner. This is great because there are various levels of learners. Audiences may range from high level executives or supervisors or engineers to new operators or employees who have not heard a majority of the content being presented. Whatever a person's style may be, they must remember the key elements of a great presenter. These elements improve the likelihood of the information being received, understood, and placed in practice.

In summary, Craig Strachan (2021), keynote & seminar speaker has prepared what he calls, *The A-Z of Public Speaking*. Perhaps it could be used as a checklist before entering the classroom.

Have a great **attitude** on and off the stage
Body language and gestures enhance the message
Make **connections** with the audience
Delivery and content are key
Speak with **energy** and **enthusiasm**
Facts tell, stories sell
Get to the point
Use **humor** to make connections
Inspire the audience
Use original **Jokes**
Know the audience
Record and **listen** to one's speeches
What is the **message**?
Notice how the audience is responding to the speech
Speak at every **opportunity** (stage time!)
Is **PowerPoint** enhancing or detracting from the speech?
Q cards can be useful – especially for a long presentation
Rehearse the presentation
Simple messages are easy to understand
Stick to the **time** allotted
Does the audience **understand** the message?
Use **visual**, auditory, and kinesthetic phrases – address all the senses
When in doubt, leave it out
Use real-life e**X**amples
You are not the star
Zzzzz – don't put the audience to sleep

Behind the Scenes

Training needs to be assessed to determine how to design and develop the training course. Once this step has been taken, it is time to check in on one's presentation skills. Although it may seem easy and simple to follow, practice is highly recommended. As the saying goes, "it's easier said than done." So, prepare instructors with tools to succeed by using Pike's suggestion notes in the beginning of this section. Pike believes that 80 percent of being a good trainer and getting people involved depends upon adequate preparation. Participants rarely know what went on behind the scenes, but they will notice if the presenter did not take time to prepare. It is recommended to make every effort to be ready for the first participant who enters the training doors; first impressions are very important.

Establishing an environment conducive to learning is a critical aspect of starting a training session with positive momentum. Facilitators may spend hours the day before (and sometimes late hours) arranging the tables and chairs, arranging the manuals and student packets for participants to gather as they enter, assuring that all have a writing utensil, notepad or even a highlighter to note “ah ha!” information received. It’s always nice to provide a tote for participants to put their classroom materials in. Placing candy dishes with mints and chocolates and a box of Kleenex adds a special hospitality touch. Preparing name tags and/or name tents are useful tools not only for the participants to network, but it also allows the instructor to identify the student by name and not, “you in the green shirt.” It is recommended that the name is placed on BOTH sides of the name tent. This is convenient for those speaking from the back of the room. Some instructors utilize hands-on materials, such as product equipment, puzzles, handouts, etc. It is important to have those items available and ready for use. Although most of these tasks are the responsibility of the facilitator, it is helpful to the instructor to know in advance that these things are available to the participants in case they want to reference them during their presentation.

The room arrangement may have a significant impact on a presentation. The speaker should review the learning objectives. Consider if the learning objectives require participants to construct a product, work in groups to achieve an outcome, or play a game designed for learning. If so, is the classroom arranged in such a way that allows movement without confusion? Arranging the room in advance to meet the course objectives is fundamental to preparation and gives comfort to the presenter. The presentation will flow with ease as the presenter advances from one concept or activity to the next. Therefore, preparing the classroom just minutes before start time is *not* recommended.

Many trainers and facilitators take the following tasks for granted and fail to check: room temperature, lighting, outside noises, visibility of all visual aids from every angle of the classroom, projector and screens, restrooms, emergency exits, room capacity (fire code), adequate learning space for each participant (9 square feet/person table space), microphones, a full easel pad, markers, extra pens/paper/manuals, back up instructor in case the assigned instructor becomes sick or unavailable. Perhaps there are several others to add to this list. Prepare, prepare, prepare!

Tools

Today, many trainers have become very comfortable using a PowerPoint presentation; however, one must not allow PowerPoint to become a crutch while presenting. PowerPoint is a very effective tool to use. It provides various visual aids for better understanding, especially for those who are visual learners. But, when it is overused, the electronic slides can lose their effectiveness. Using too many graphics and various backgrounds within one presentation may cause confusion. Conversely, displaying only text on slide after slide can be boring and show a lack of preparation/creativity on the part of the presenter. A rule to remember, not only should you know your presentation, but also prepare a cheat sheet of the presentation (printed handout). If possible, use a minimum of 2 slides per page on the cheat sheet so that you’re able to see it from a distance. One never knows when the projector or PC might decide to malfunction.

If an instructor plans to use any form of gamification throughout their topic, it is highly recommended that a group practice the activity at least twice before stage time. Do not hesitate to not only check that it operates properly but also re-check each question and spelling. A duplicate question or question worded incorrectly or confusing could destroy the entire game and final scoring results. The students will not be happy if that happens. Remember, most of us are very competitive.

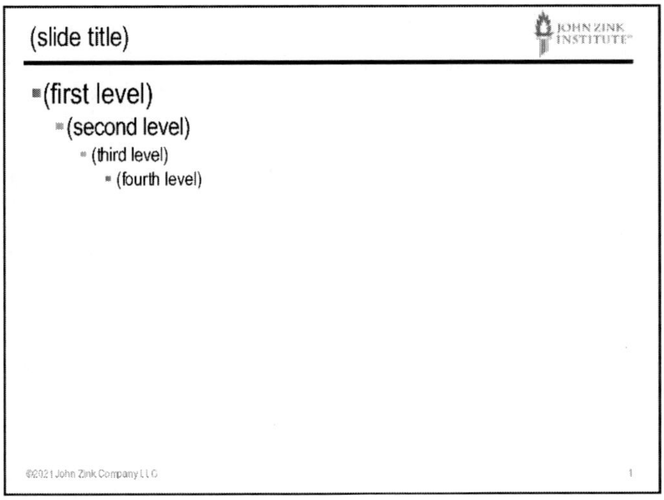

Figure 6. John Zink Institute PowerPoint template slide example.

Many companies recommend that all company sponsored presentations use one basic template (see Figure 6). This template should be very minimal, and not overpower the information noted on each slide. From a legal perspective, each slide should also include course title and copyright information; it may be helpful if a page from the hardcopy version is separated from the entire presentation.

Visuals should support the training and not distract from it. They are most effective when (Biech, 2005):

- They are relevant to the subject.
- They are visible and understandable.
- Page orientation is consistent, using either landscape or portrait.
- Words are large enough to read.
- They are oriented to the listener: "Here are four ideas you will…"
- Color is used appropriately.
- The typeface varies in boldness and size.
- The print is in both upper and lower typeface.
- The typeface enhances the readability (usually a san serif font).
- Bullets delineate each point.
- They enhance the speaker's performance rather than replace it.
- The visual becomes an extension of the speaker.
- They are tied together with a common element, for example, a sketch or graphic.
- They are customized for the group.

Most trainers can relate to the 6x6 rule when preparing their presentations. This simply refers to the number of lines down and words across the visual (slide). Other tips to remember are:

- When presenting a list, design it so the text can be revealed one line at a time.
- To further emphasize the line item, change the color of the newest item or have the previous items fade subtly. (*Try not to overdue this action)*
- Headings should be about 44 pt and body 32 pt, but no less than 24 pt.
- Select one primary transition throughout the content for each module.
- Limit colors to two per slide plus black.
- Fade to black to signal a new module or to pause for discussion or an activity. *(this can be achieved by selecting "ctrl" and letter "b" on the keyboard)*
- Use a subtle background
- Ensure that there is enough contrast between typeface and background; light on dark is preferred by many.
- Use clip art sparingly and don't try to be cute. *(important to obtain permission to use the clip art)*
- When using animation, select one type and use throughout a module or content section.
- Keep sound effects brief, and make sure they add impact *(remember to check the volume level and speakers in advance.)*

The first three slides of your presentation should include: Title page (Figure 7a), Learning Objectives (Figure 7b), and an Outline (Figure 7c) if not previously provided. To conclude, a summary slide is a great option (see Figure 7d).

It is highly recommended and timesaving to prepare flip chart pages in advance if they will be used for class discussion or a class activity. Perhaps make a few notes on pages and then use a sticky note on the side as a reminder as to when to share that information. Often, the instructor is not the greatest artist and may need to trace an object or picture from a professional. That is highly acceptable. Just remember to do that in advance. Plan, plan, plan!

Today there are post note flip charts. What a neat invention! These are great for trainers to utilize because they enable a presenter to remove a paper (point) from the easel and place it on a visible wall for participants to reference at a later time. Perhaps one may want to encourage participants to add comments to a large sheet on the wall by attaching a smaller post note from their student packet if they have another comment relating to that discussion. Wow! A class activity opportunity! Being in a world heavy in technology, there are various interactive whiteboards available. Great tool to use if each of your students have access to a laptop, PC or iPad to use during the training course. Obviously, this requires extended time and practice of the instructor. Unfortunately, many instructors are already stretched thin, and training is not their main job responsibility. This may be a task for the training coordinator, assisting the instructor beforehand or during the presentation.

Do not forget about handouts. Handouts reinforce the key points that a presenter is trying to make and serve as a reference once participants return to their workplace. They may also provide additional information that was not added to the final presentation; need to know vs. nice to know discussion. Be careful of the timing of distributing handouts. If the handouts are distributed too early (before the point of the discussion), the participants may become distracted and not pay attention to the current discussion. For example: During instructor workshops, presenters often wait until the very end of the workshop to distribute handouts. They serve as a summary, not a crutch or sign of laziness and excuse to say, "It's in the handout."

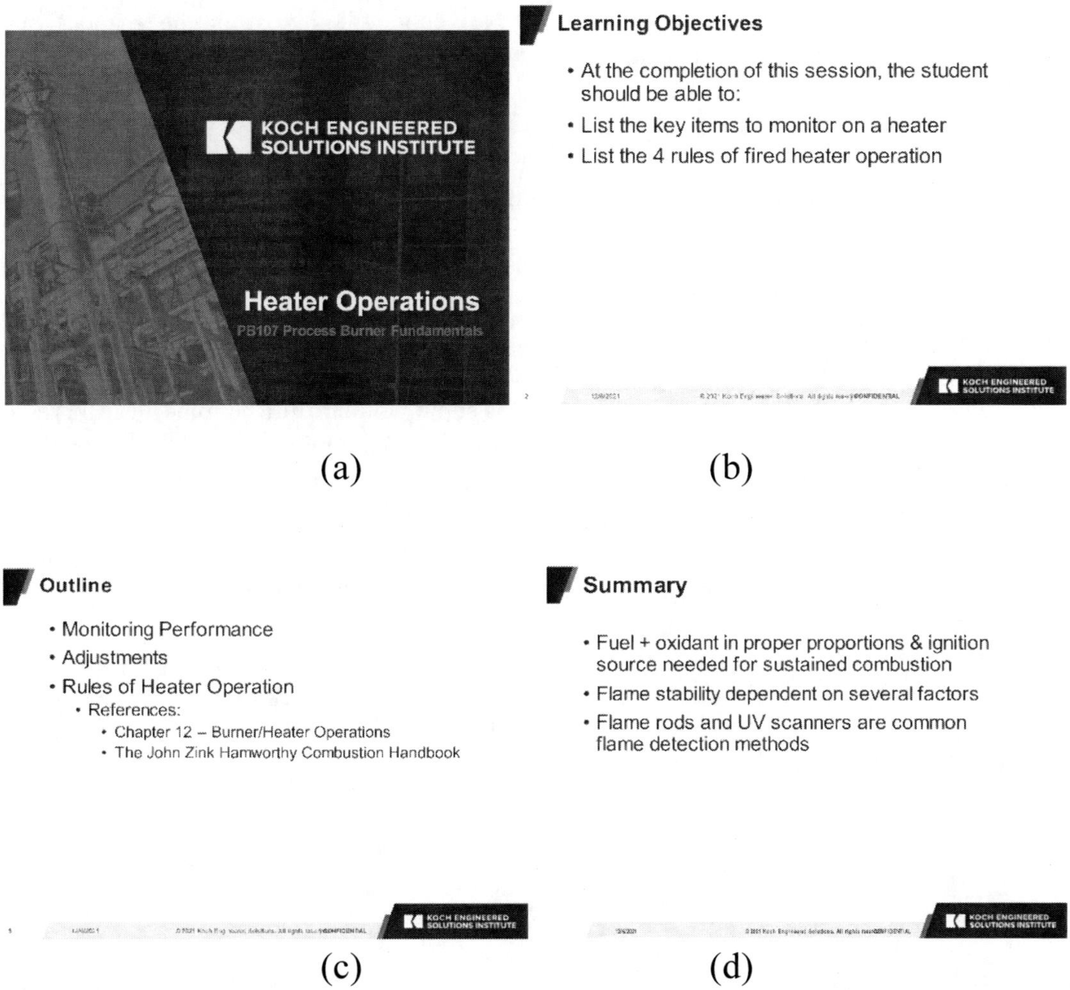

Figure 7. Example of various types of slides (a) title, (b) learning objectives, (c) outline, and (d) summary.

Remember that handouts reflect the speaker and their company. They are takeaways so it is important that they look professional, are easy to read and are of good printing quality. An acronym to remember is *CRISP:* Colorful, Readable, Interactive, Simple, Pictures (graphics, diagrams, etc.) if possible (Michigan State University, 2021).

Apparel

Many work environments allow employees to dress business casual and some even allow casual attire, perhaps jeans on Fridays. It is recommended that the trainer maintain a professional look. Dress for Success! With consideration of providing combustion training in a refinery, the presenter may provide hands-on training outside with the operators on specific equipment, the "preferred" attire will be different. The presenter can still make sure their hard hat is clean, boots are presentable and Nomex or coveralls are clean. There may be an opportunity to "dress for success" during classroom training and change into the proper PPE attire before going outside. In the classroom where several instructors are presenting, the decision may be made to wear identical shirts during training sessions. Not only will this better the chances of a professional look worn by all presenting, but it also allows participants to easily identify the company representatives when they have a question or concern. It's important to select shirts that are a solid color and not too graphic, as to cause a distraction while presenting.

Sometimes trainers may become nervous or find themselves in an uncomfortable situation and reach for items to jiggle or toss. Because of this, it is recommended that trainers remove all coins or small items from their pant pockets before stepping up to the podium. This avoids the opportunity for participants to hear the jingle tune as they play with the coins in their pockets. In an effort to appear fashionable, women may desire to wear bulky and noisy jewelry. This fashion statement is not recommended. This may sound minimal, but any movement or the chance of jewelry getting caught on equipment could cause maximum distraction or become a safety hazard. Today, high heels are a fashion for some women. Just be mindful if there are plans to interact with participants and how quickly the presenter will need to do so; in this case lower heels may work better. A good technique is to take the attention off the speaker early and involve the participants in a class activity or discussion. Evaluate your attire in advance. Ask family or friends for their opinion. It may not be a bad idea to practice your presentation wearing your planned attire to make sure you can move about comfortably and naturally.

Checklist

Ready, set, speak! Langevin Learning has provided a sample checklist that can be used to evaluate trainers during a presentation skills workshop (Langevin Learning Services, 2021). Perhaps this checklist better summarizes attributes instructors may review as they strive to become good presenters.

Presentation Skills:

- Clarity
- AV usage
- Avoidance of speech filters
- Expertise in subject
- Confidence

Organization:

- Appropriate amount of material
- Objective clearly stated
- Pace
- Time management
- Handling of materials

Lesson Structure:

- Introduction
- Feedback to learners
- Transition from phase to phase
- Reviews/summaries

Climate/Rapport:

- Enthusiasm
- Has fun
- Friendly/warm
- Casual/relaxed
- Use of learners' names
- Paid attention to all learners

Response to Learners:

- Responds to signs of difficulty
- Flexibility
- Noticed learner reactions
- Listened

Questioning:

- Questions are clear
- Handling of incorrect answers
- Answered questions well

Involvement:

- Encourages participation
- Invites questions
- Held attention
- Participation was spread

Non-Verbals:

- Voice
- Enunciation
- Eye contact
- Posture
- Facial expression
- Gestures

- Pauses and silences
- Nervousness controlled

Results:

- Objectives met
- Learners receptive

This checklist may appear long and fearful, but it is possible to achieve 98% of the tasks after an instructor's second or third round at the podium. It is important to relax and enjoy presenting to the class. Giving a presentation is not about being perfect as much as personable (Fripp, 2021). If several of your colleagues identify a particular area as room for improvement, that should be an area to focus on before presenting again.

Introduction and Closing

The old phrase: *first impressions are important* still rings true. Not only should an instructor dress professional but one must try to grab the attention of the attendees during the first 3 sentences of their presentation. Start your presentation with a story, a statistic or an icebreaker and you will be off to a great presentation. Provide your students with something they will remember. There are many suggestions available to use and tweak to the preferred training style and/or topic; tie it in and get them interested. Often icebreakers or stories come to a person in the middle of the night. Write them down! Darren LaCroix (2021), 2011 World Champion of Speakers encouraged participants at a North Eastern Oklahoma ASTD retreat to keep a story library (LaCroix, 2021). Darren explained how he maintains an electronic folder full of events that he has turned into stories or icebreakers. This way, a story is just a click away and he is always prepared when called upon to deliver. It is recommended that you stay clear of sharing a joke unless you've tried it before and know that it works, is not offensive and you won't be contacted by HR the next day.

Darren also shared *The Triangle of WOW*. Each side of the triangle represented a goal that every presenter should try to achieve when communicating. They are:

- *Material:* words that are chosen
- *Delivery:* manner in which the topic is communicated
- *Setting:* atmosphere in which the subject is communicated

Icebreakers can be used to arrange the participants in groups to enhance networking with others outside their peers. For example: The Bob Pike Group provides a simple and quick icebreaker entitled, "Line up by Experience." Basically, the instructor will ask participants to stand and move toward a designated wall. The instructor will ask participants to line up according to years of experience, perhaps as an operator or engineer in the combustion field. This activity can be customized to the audience. Perhaps, line up participants by years of teaching. The instructor will need to designate which portion of the line is the lower end vs. the higher end of experience. It is recommended that participants are encouraged to line up quickly and try to do so without talking.

Once participants have formed a straight line, the instructor will then ask the participants to count off by numbering themselves according to the number of tables available in the room. (e.g., If there are 5 tables, then number themselves 1 thru 5.) After this is complete, identify which table is #1, #2, etc. Then ask that all those who are #1, please gather their belongings and move to table #1, and proceed until #5 participants are reached.

The outcome of this icebreaker forces the participants to sit with others outside their comfort zone and provides different skill levels at each table. The more advanced operators or engineers are able to mentor the less experienced ones throughout the training session, during break conversations, lunch or class activities. This creates a great start to their learning experience. It also gives participants another clue that they are not attending a "normal" training session. Perhaps they may enjoy themselves along with learning a new skill or technique (Meiss and McCallum, 2007).

It is important for presenters to share their credentials and experience in the field. Be brief if presenting it verbally and a written copy is not provided to the students. Be careful to not bore them with reciting the entire resume! A key point to remember: the quickest way to connect with the audience is to share personal failures and flaws. Sharing success after success tends to lead the participants to think, "Sure, he can do it, but I doubt that I can. He's perfect!" Begin by identifying with the audience in both personal success and failures. Give the impression that the entire room is open to share knowledge and, possibly gain a few tools and tips that will enhance their daily activities. Although one may be the SME, remain open to hearing experiences and ideas from others in the room. Find an opportunity to connect with your participants. Reminding them that you are there to help them succeed as well.

If introductions have not taken place, this is a good time to have participants introduce themselves. To save time and lower the nervousness of the participants, prepare a slide of questions that will remind them of what information to provide (see Figure 8). Once this is done, inform them of why they are present; WIIFM (What's in it for me). Share the topic objectives and inform them that, in addition to learning, they are encouraged to enjoy the time and have fun.

It is just as important to close the presentation or training session close to how you started. The introduction and closure should tie together. Timing is important, especially if there are several presenters to follow. If a presenter exceeds their allotted time, it may take time away from another presenter.

Often the closing moments are used to wrap up all discussions, gather take away items, handouts and materials and move the focus to returning to their workplace, hotel or catching their flight home. Participants are generally eager to leave and are not 100% tuned into any discussions. This should not persuade the opportunity for the instructor to ask if all the learning objectives and expectations have been met.

One closing activity may be for participants to verbalize to a table partner one or two action items that they will take away from their training session and implement immediately or in the near future at their workplace. Allow enough time for participants to share their experience and exchange contact information with those they have met during their stay. Another closing activity is to provide postcards for the students to write three take away thoughts or action items to preform once they return to their workplace. Have them address them to themselves and leave them as they exit. What a nice way to remind them of their action items the following week and perhaps a reminder to share their training experience with their colleagues.

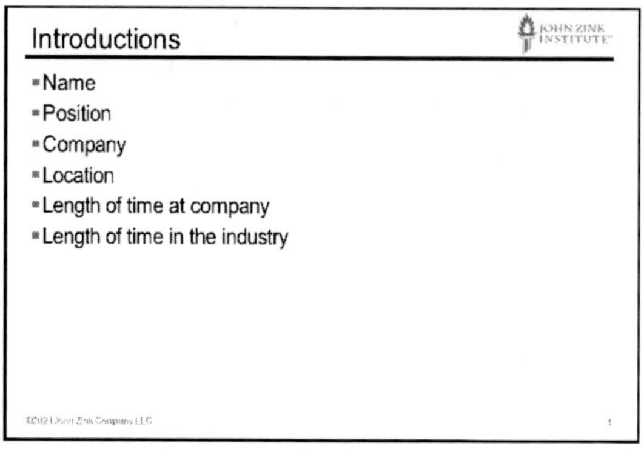

Figure 8. Example slide for participants (students and instructors) introductions.

Closing with another story is recommended. This calms the participants as they quiet down to hear the end of the story. Especially if the training session was started with a great story. For closing activities and stories visit www.ebbweb.com.

It is also important to allow participants to offer feedback, positive as well as negative in some type of writing format. Comments provided by participants are an advantage to the instructors and facilitators providing the training. They allow feedback, suggestions and room for improvement areas that will enhance the next training session. KESI offers feedback time after each presentation or topic is presented while it is fresh on the learner's mind. Allowing them to not enter their name or personal information encourages honesty and completed evaluations. The comments are viewed immediately after the course is complete to see if any immediate actions need to be taken and also reviewed at the next kickoff meeting for an upcoming course.

If possible, encourage a company representative to stand by the exit doors and offer thanks to each participant, encouragement to continue to strive for excellence and wish them well. Thank them for coming and ask that they tell others about their experience and refer them to the next training session. This provides one last positive experience for them upon leaving.

Accreditation

Accreditation refers to an independent third-party giving approval to some aspect of a training program. There are numerous accrediting organizations as well as different types of accreditations. These will be briefly discussed in this section. Accreditation should be considered by an organization hiring an outside party to do training. Accreditation is a measure of quality that shows the training organization meets the rigorous standards set by the accrediting body. For example, most accrediting bodies require that instructors disclose any conflicts of interest. This may be particularly important when hiring a supplier to do training as the course could be a glorified sales presentation if the supplier is not required to meet any accreditation requirements.

Some accrediting organizations, such as the American Petroleum Institute under their Training Provider Certification Program (TPCP),[1] accredit individual courses. Some

[1] https://www.api.org/products-and-services/individual-certification-programs.

accrediting organizations, such as the International Association for Continuing Education and Training (IACET),[2] accredit entire programs. There are also some accreditors that accredit a group of courses leading to a certificate, such as the American National Standards Institute (ANSI).[3] Note that some accrediting bodies, such as the Accrediting Council for Continuing Education and Training (ACCET)[4] are officially recognized by government agencies such as the U.S. Department of Education (2021).

Case Studies

British Petroleum (BP)

This case study is taken from Valencia et al. (2008). The BP Texas City refinery (see Figure 9) is one of the largest and most complex in the world. Located on a 1200 acre site, it has a crude oil capacity of 460,000 barrels/day and nearly 2,000 direct employees. This section describes a partnership between BP and the John Zink Company to deliver classroom training on fired heater operations to process operators.

Figure 9. BP Texas City, TX refinery (Valencia et al., 2008).

Operator training can and should be done in a variety of ways. Some types of refresher training may best be done on-line. Training on a specific unit is usually best conducted on-the-job where an experienced operator trains a less-experienced operator. Training on upset conditions that would be difficult, if not impossible, to simulate with the actual operating equipment is usually best conducted using a process simulator. Training on general theory and best practices is often preferred in the classroom where students and instructors can interact and discuss principles that can be applied to their processes. The last type is of interest here.

---

[2] https://www.iacet.org/.
[3] https://anab.ansi.org/credentialing/certificate-issuers/.
[4] https://accet.org/.

Importance of Training

There are many reasons why operator training is important. One survey found that the three top objectives for operating plants are to improve safety, increase process knowledge, and improve plant profit (Hill, 2005). The most important objective is safety, to ensure that operators can properly respond to abnormal situations (Wilkins, 2006). This includes handling situations they may never have actually encountered before, but which could happen, similar to airline pilots who train for many possible situations that rarely, if ever, occur. This means the operators must have a deep enough understanding of the process to enable them to safely handle abnormal situations and identify potential hazards.

Sustaining and improving the performance of process plant operators is a key element of operational excellence (Pankoff, 2005). This is accomplished in a variety of ways including, for example, proper procedures, documentation, audits, mentoring/coaching, automation and controls, maintenance, and training (both off- and on-the-job). There are many elements of operational excellence including safety, efficiency, productivity, and environmental compliance. It is therefore critical that operators are adequately trained to meet the demands of operational excellence.

Plant operators must be able to quickly and accurately troubleshoot and diagnose process upsets and equipment problems (Smith, 2002). Failure to do so can lead to lost production, off-spec product, equipment damage, and possibly even catastrophic events. Related to troubleshooting is process optimization, which means maximizing production while minimizing costs and meeting environmental regulations. This means that operators should be trained both to handle problems and to look for ways to improve processes even when nothing appears to be wrong. Proper training can sensitize operators about warning signals and what adjustments are needed in the circumstances encountered.

Training Partnership

BP's unit specific operations training department partnered with the John Zink Company to develop a general one day training course for fired heater operations. This course is one component of a more comprehensive operator training program that BP put together which also includes a classroom-lead process fired heater simulator and unit-specific fired heater training conducted by the BP unit training coordinators. Zink conducted its first fired heater seminar in 1950 at its facility in Tulsa. The Koch Engineered Solutions Institute (KESI), the training organization for Zink, now delivers these courses all over the world, including at plant sites. While there is a standardized curriculum for a typical one-day course, most on-site classes are customized to the equipment in the plant. BP's unit specific operations training department developed their desired course outline, learning objectives, and outcomes including a student learning assessment. A pilot course was developed by Zink and delivered to a cross functional group of BP personnel, including training general personnel, engineers, and operations supervisors. The materials were fine-tuned to meet BP's requirements before rolling it out to the Texas City Site operations organization.

One of the benefits for a plant partnering with an experienced training organization is the accumulation of industry-wide best practices. Zink personnel have worked in virtually every refinery in the world. They have seen a wide range of both problems and best practices. This experience is invaluable for training because it exposes a particular plant to potential issues that even their experienced operators may rarely, if ever, have encountered.

Most JZI courses are accredited by the International Association for Continuing Education and Training (IACET) which means that Continuing Education Units (CEUs) are given to students who can demonstrate that they have met the course requirements, which includes successfully passing a post-course examination. One of the requirements of accreditation requires student privacy where test results can only be given to others with written permission from the student. BP decided to forego CEUs for this training because they wanted to see all test results. BP also chose to make the course requirements more stringent than those required by the IACET accreditation.

Customized Operator Training

The objectives for the course are: improve overall understanding, define common fired heater terminology, follow proper and safe operating procedures, increase heater thermal efficiency, reduce pollutant emissions, and provide tools for troubleshooting problems. The agenda shown in Table 1 was developed for the course to meet those objectives. The primary thrusts of the training are safety, reducing pollution emissions, maximizing thermal efficiency, and providing troubleshooting tools. These are addressed throughout the course. For example, while section three specifically addresses combustion safety, it is designed to provide the foundation for other discussions on safety in later sections. Detailed theory and equations are kept to a minimum because the course is designed for operators, so the information must be practical and easily applied to daily operations.

Table 1. Course outline

Section	Title	Length (hr)
1	Introduction	1
2	Combustion Fundamentals	1
3	Combustion Safety	1
4	Process Heater Fundamentals	1
5	Burner Fundamentals	2
6	Heater Operations	2
		8

Sixteen classes were conducted in 2007, with an average class size of 13, to ensure good interaction between students and instructors. Each student received a color copy of the PowerPoint slides used during the training. Figure 10 shows an example of a typical slide which has a picture of actual equipment and a drawing showing more details of free-standing diffusion (raw gas) burners. One copy of the *John Zink Combustion Handbook* (Baukal, 2001) was provided to each class for use as a reference. That book is used as a text in other JZI classes, and although it includes practical information, it is designed primarily for engineers rather than operators. Each student also received a laminated flowchart, designed for use in the plant, for setting the oxygen and draft levels for natural draft and balanced draft heaters.

The students were introduced to each course by a BP Manufacturing Department Leader, followed by Dennis Link, the unit specific operations training Project Manager. They helped provide context expectations and motivation for the class. The students then took a closed-book pre-test consisting of 15 questions requiring fill-in-the-blank answers. No multiple choice or true-false questions were used where students might be able to correctly guess the answers even though they may not have actually known the answers a priori.

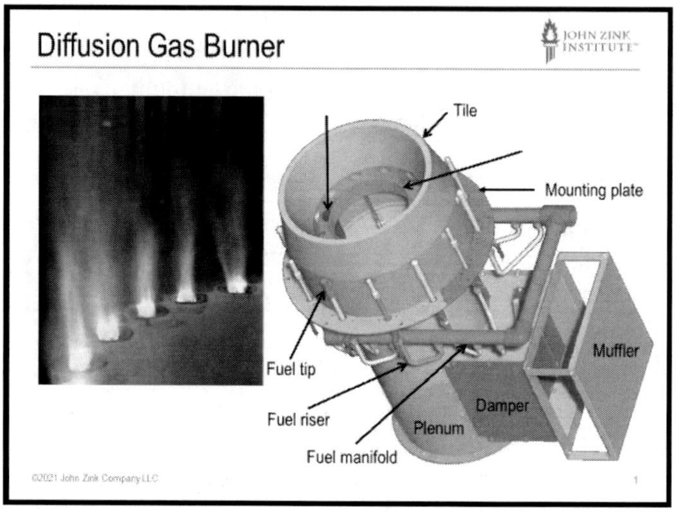

Figure 10. Example content slide.

Figure 11. Irregular flame patterns in an (a) operating process heater and (b) closed air registers on the two improperly adjusted burners.

The classes were designed to be highly interactive, emphasizing the Socratic teaching style where the instructor asks the students several questions and instructs based upon the responses. For example, Figure 11a shows three burners firing horizontally in a process heater (not at BP). The students were asked to determine what might have been wrong with the burners. Then they were shown a picture of the back of the burners Figure 11b which shows that the air registers were closed on the two burners with bright yellow flames. Throughout the course, many photos, animations, and short video clips were shown of process heaters that had problems. The students were asked to determine the problems and recommend solutions. This format exposed them to many potential problems they may never have encountered, but which they might encounter or which they could avoid in the future by taking the proper actions. For example, video clips of burners flashing back were shown during the class. These clips were taken at the John Zink Test Facility under highly controlled conditions. Those clips exposed students to the sights and sounds of flashbacks in case they ever encounter them in their own heaters.

The course was designed to show operators both good and poor heater operations. Many examples were shown of a variety of problems found in actual operating heaters in many plants. Figure 12a shows an example of improperly adjusted burners in a process heater (not at BP). Both the furnace draft and excess O_2 were well out of the desired specifications. The flames were irregular and not producing the desired heat flux pattern on the floor. Figure 12b shows the same burners in the same heater, after they were properly adjusted. These adjustments can be easily made by operators once they have been properly trained. Figure 13 shows a schematic of flame rollover in a cabin heater. Burners are firing up along the wall, but the flames are rolling over into the process tubes, which is very undesirable and can cause a premature shutdown. Possible causes and solutions were discussed for this and other related problems. Students received practice troubleshooting in the classroom using the tools they learned. They also were able to discuss the problems with their colleagues and the instructors.

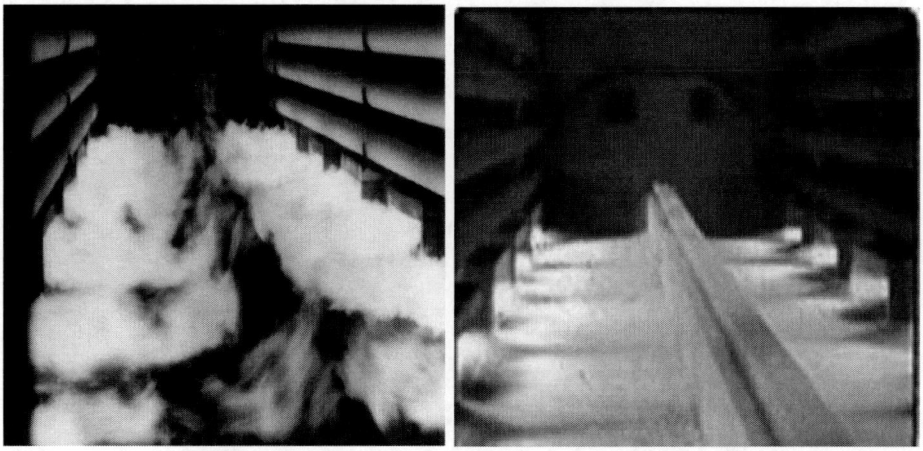

(a) (b)

Figure 12. Burners firing across the floor in a process heater (a) before adjustment and (b) after adjustment.

An important element of the class is that it was designed to be fun. While there is no doubt that a lot of new material was covered over the course of the day, humor and games were used to help engage the students. For example, as a review for the post-test, students were divided into two teams that competed in a game of Zeopardy (see Figure 14). The winning team members received a nice prize, so the competition was often fierce.

Results

The students taking this course ranged in experience from as little as a few months to more than thirty years. Very experienced operators still benefited from the training because they gained a deeper understanding of the reasons for various procedures. For example, a burner that is pulsing or huffing is unsafe and needs to be corrected immediately. The operator may know to reduce the fuel flow to the burner, but may not know why or what could happen if the wrong action is taken. Knowing the why behind the procedures helps the operations personnel to better handle abnormal operations they may never have encountered.

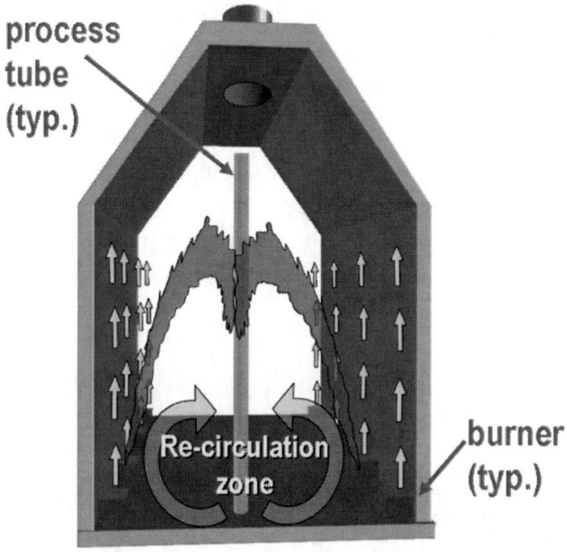

Figure 13. Drawing of a flame rollover.

Figure 14. Main screen for Zeopardy game.

The post-test questions were identical to the pre-test questions, so a direct comparison could be made between what students knew before and after taking the course. After 16 classes and over 200 students, there was a sizeable improvement between pre-test and post-test scores as a result of the course. The average pre-test score was 33% and the average post-test score was 94%. Student were given anonymous course evaluations and asked to rate their interest in and the benefit of each topic using a Likert scale ranging from 1 to 5 where 1=none and 5=great. The overall average rating for all topics was 4.3. One student wrote, "Learning was made fun and interesting." Another wrote, "I know more about heaters than when I came to class!" Figure 17.12 shows a summary of the overall student evaluations for the course. The blue bars show interest in the content and the green bars show how much the student benefited from the class. The vast majority of students found the course interesting and beneficial. The course materials were continuously revised based on student feedback. For example, students in the earlier classes felt there was too much material, so duplicate and superfluous information was deleted in later classes.

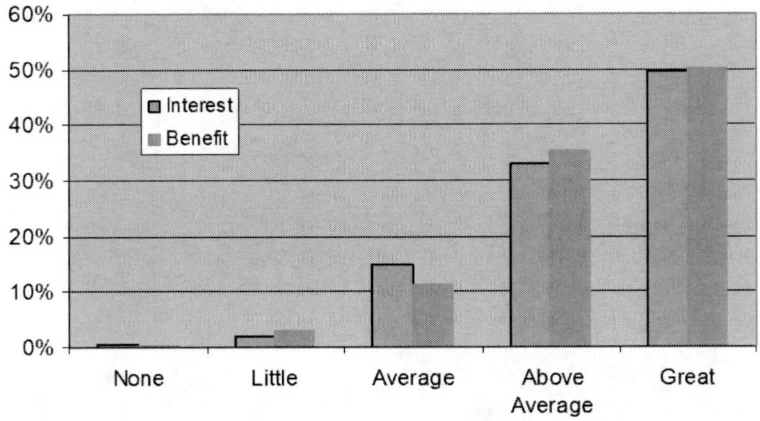

Figure 15. Overall ratings by students on their interest in and benefit from the course.

Conclusion

Operator training is critical for the safe, environmentally sound, and efficient operation of a plant. Partnerships between plants and external training organizations can effectively help meet some operator training requirements. Specific objectives should be developed so the materials and learning experiences can be designed to meet those objectives. Certain types of training are better done in the classroom with a limited number of students to promote interaction and engagement. Some type of assessment tool, such as a post-test, should be used to quantify learning and ensure that key concepts have been grasped. Ideally the training should be fun to fully engage the students, particularly operators who may not be as comfortable in the classroom as, for example, degreed engineers. Some type of course evaluation is important so adjustments can be made as appropriate to continuously improve the content and instruction methods.

Shintech

This section is based on the paper by Gilder et al. (2010). Shintech began construction of its new Plaquemine Louisiana manufacturing facility (see Figure 17.13) in October 2005. The facility is located on a 1,725-acre site and manufactures chlorine, caustic soda and vinyl chloride monomer (VCM). Historically, Shintech has manufactured only polyvinyl chloride (PVC). This new plant is Shintech's first integrated complex. The new manufacturing facility employs state-of-the-art environmental technologies and is subject to the most stringent environmental controls in the country.

Figure 16. Shintech plant in Plaquemine, Louisiana.

John Zink Company, LLC (JZC) manufactures thermal oxidation systems used to destroy unwanted wastes (Karan et al., 2013). Thermal oxidizers (TOs) are commonly used to treat VOC and CO emissions because they have very high destruction and removal efficiencies (Schnelle and Brown, 2002). Thermal oxidation can be defined as, "the process of oxidizing combustible materials by raising their temperature above the autoignition point in the presence of oxygen and maintaining it at high temperature for sufficient time to complete combustion to carbon dioxide and water" (Moretti, 2002). The Koch Engineered Solutions Institute (KESI) is the training group for JZC and delivers training both at its U.S. headquarters in Tulsa, OK and at customer plant sites. The latter type of class is typically from one to three days long and is customized to the equipment in the specific plant. JZI works with the plant to determine a suitable course agenda.

Shintech partnered with JZI to offer a customized TO training to 37 of its plant operators for the start-up of its new facility. A photo of part of the thermal oxidization system is shown in Figure 17. The course was offered with optional Continuing Education Unit credits (CEUs) that were available to any students meeting the following criteria: take (not pass) a pre-test, attend at least 80% of the course contact time, pass (at least 80%) post-test, and complete an anonymous course evaluation. JZI is authorized to offer CEU's through its accreditation by the International Association for Continuing Education and Training (IACET).

Figure 17. Photo of part of the thermal oxidation system during installation.

Course Design

The course was designed to give operators at the new facility a good idea of both the *what* and the *why* of operating the equipment. All too often, operators are trained, sometimes hurriedly and haphazardly, by existing experienced operators. The new operators may learn *what* to do, but not the *why* behind the *what*. It is also fairly common for long-time operators not to understand some of the basics because these basics were never taught to them. The *why* is important because it helps operators better understand cause and effects that can impact safety, thermal efficiency (and therefore operating costs), productivity, and pollutant emissions. For example, Figure 18 shows a slide describing the potential problem of "blow-off" if a burner is over-fired. Moving from left to right shows what happens as the air/fuel mixture velocity is increased. When the last furnace on the far right appears in the slide animation, the sound effect of an explosion is heard to help reinforce the point of the danger of going beyond the design firing limit for the burner. The *why* also better prepares operators to react to new situations that may not have been covered in formal training sessions. The course content included basics that apply to any equipment of this type, along with very detailed and specific information on the equipment in this particular plant.

The course had the following sections:

1. Combustion & thermal oxidizer basics
2. Safety overview & warnings
3. Overall equipment familiarization
4. Detailed walkthrough of equipment
5. Detailed blower, boiler, absorber, scrubber, demister detail
6. Detailed walkthrough of P&IDs
7. Drawings review
8. Pre-start up & refractory cure out
9. Normal start-up & shutdown
10. Logic demonstration & DCS screens

11. Normal maintenance
12. Troubleshooting
13. Drawings

Each student received a three-ring binder containing the color PowerPoint slides. Adequate room was provided for them to make notes in the manual as desired. Some of the students received their manuals prior to the start of the class and came prepared with questions to ask. Statistics for all types of training show that retention of the material diminishes fairly quickly after the training has been completed. The student manual can be quickly and easily referenced as often as needed to refresh previously studied information. Although operators do not generally receive their own copy of the Operation and Maintenance Manual, the student manual contains much of the same information including many of the written operating procedures.

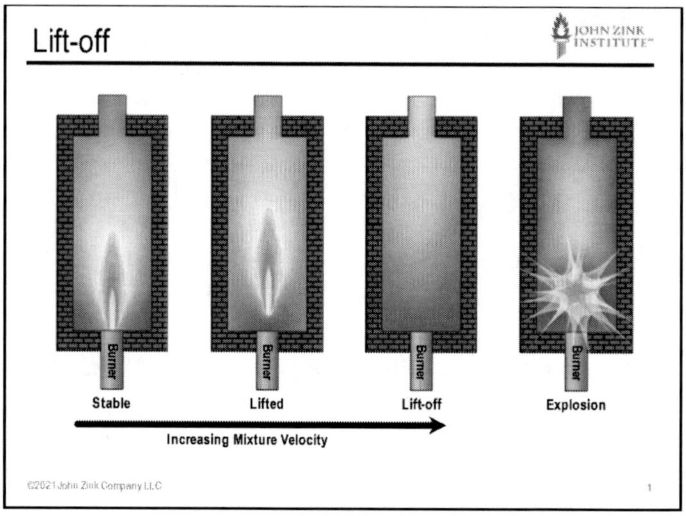

Figure 18. Series of furnace drawings showing the possible progression to an explosion.

Training

The training was conducted over three consecutive days, followed by a fourth day about six weeks later, on a couple of specific pieces of equipment. Although most of the time was spent in the classroom, there were many short sessions spent outside at the equipment going over the specifics after reviewing the basics in the classroom. The plant had not been commissioned yet so the equipment was installed, but not yet operational. While this did not allow the operators to do live training, it did permit operators from all shifts to attend classes together during normal working hours. This produced a great deal of interaction and feedback between participants and with the instructors. Another important aspect of the training was that supervisors were present during most of the sessions, which sent a strong message about the importance of the class.

The format of the training was designed to be very interactive. While colorful PowerPoint slides (see Figure 19 for example) were used to guide the discussion, operators were encouraged to ask questions and make comments at any time. This was encouraged in part through fun games of poker and bingo using poker cards. Every time a participant asked or answered a question, they were given a random card from a poker deck. For the poker game, the student with the best poker hand at the end of the day received a nice prize. For the bingo game, cards were drawn from a deck until someone had enough matching cards to win. The more cards a student had, the more chances of winning, so this encouraged continuous and frequent participation. Other token gifts were also given out during the training as deemed appropriate by the instructors (e.g., to a student asking a particularly good question).

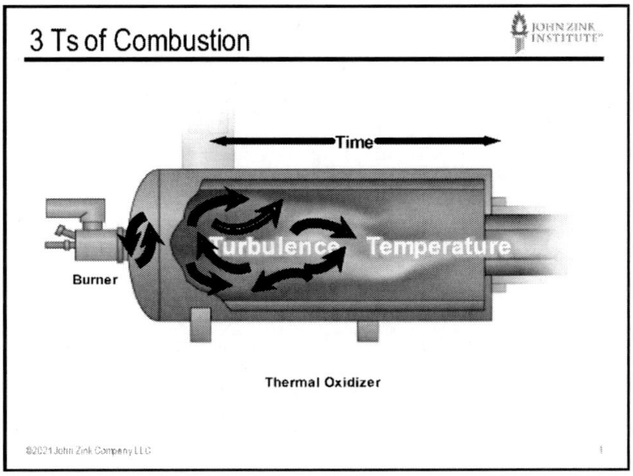

Figure 19. Example of a colorful slide.

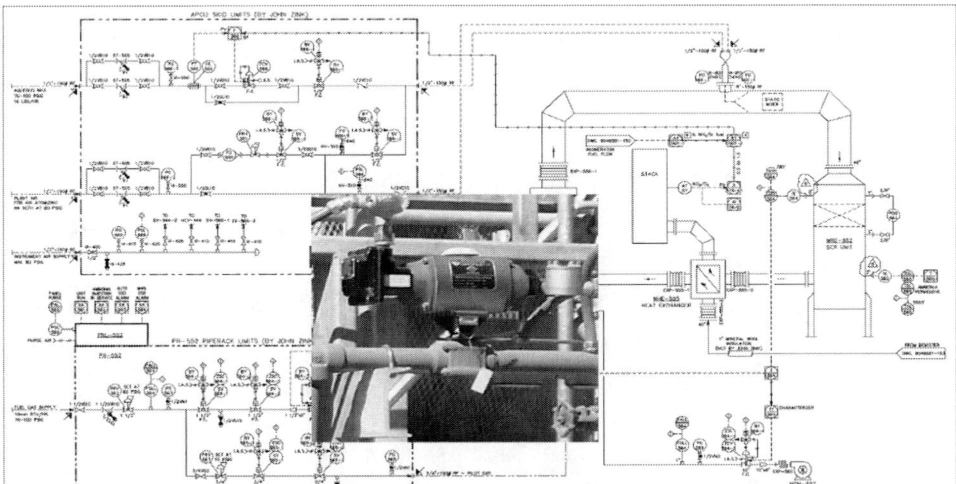

Figure 20. Example of a colorful slide.

Short video clips and brief plant visits were used to break up the lecture periods and to help keep students engaged in the materials. Videos are particularly powerful when demonstrating potential problems, such as flashback from a burner, that may not have been previously experienced at a particular plant, but which could happen under certain circumstances. This is analogous to airline pilots who train in simulators to react to situations they hope they never encounter, but for which they are prepared to handle just in case.

The actual equipment drawings for this plant were used during the training to help familiarize the operators with the equipment and with the operating procedures. To make it even more realistic, photos were taken of individual components that would display on the drawings when clicked. For example, clicking on the symbol of a valve would show a picture of the actual valve in the plant (see Figure 20). This helped personalize the training and made it easier for the operators to connect the drawings to the actual equipment.

Results

Identical fifteen question pre-tests and post-tests were given to the students to measure learning. The pre-test assessed students' knowledge prior to taking the class. The average pre-test and post-test scores were 52% and 99%, respectively. The difference between the scores is an indicator of what was learned in the training.

Students were also given a questionnaire at the end of the course to assess their level of satisfaction with the course. Students did not put their name on the forms, although their name was checked off a list to show they completed the evaluation which is one of the requirements for receiving credits for the course. A five point Likert scale was used, where 1 = none, 2 = little, 3 = average, 4 = above average, and 5 = great. Students rated each section of the course according to their interest in the topic and its benefit to them. There was also a space to write in any comments they may have had on the topic. Figure 21 shows the averaged results by interest and benefit for each topic. The results show that on average students found all topics to be of "above average" interest and benefit.

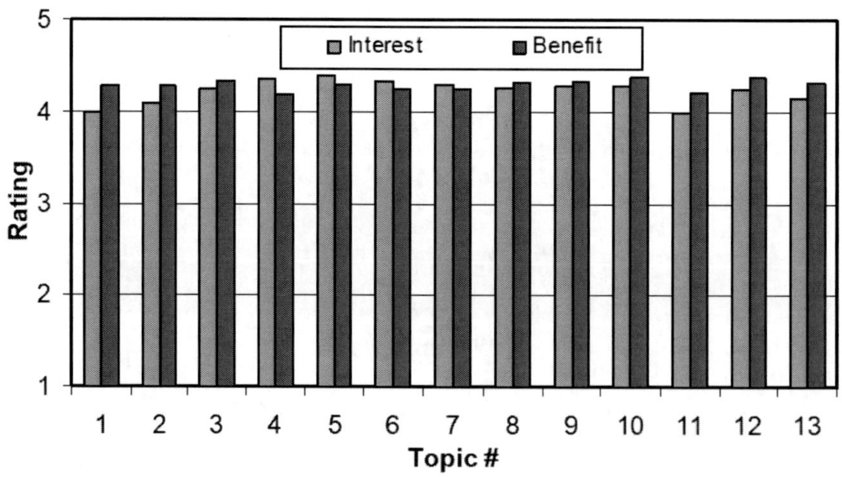

Figure 21. Student ratings of interest and benefit of each course topic.

Another part of the questionnaire asked students for written comments on the instructors and material. Some of the instructor comments included "Very Knowledgeable," "Excellent," and "Very Thorough." Some students felt more time should have been spent on start-up, shutdown, and troubleshooting and less time on the drawings. Content and coverage are always the challenge with a group of students having a wide range of backgrounds and experiences. Note that all students met the necessary requirements and received CEUs for the class.

Conclusions

Properly training plant operators is critical to ensure equipment is operating safely, while maximizing efficiency and productivity and minimizing pollution emissions. Operators need to understand some basic information about the equipment, as well as the details on their specific installation. Although not always possible, it is particularly beneficial to have all operators together in the same class to enhance discussion and mutual learning. Training should be customized to the needs of the plant and incorporate techniques such as fun games to promote interaction among the participants and instructors. Ideally, there should be a "hands-on" portion of the training where instructors use the actual equipment during demonstrations. Pre-testing and post-testing are effective tools to show that operators have learned the key points in the training.

Conclusions and Recommendations

There is growing concern about the effectiveness of CPE, most of which focuses on abstract knowledge rather than on practical application (Cervero, 1992). To improve program effectiveness, the best solution for an organization depends on a number of factors but is often a combination of providers. Organizations need to select the best way of sourcing their training needs (Derose, 1999; Chapter 11), which means selecting the best provider for the specific type of continuing education. Internal providers are often best suited for an organization's core competencies where there are available SMEs who have specialized and proprietary knowledge that may only be available inside the organization and because the organization may not want to share this information which gives them a competitive advantage. This is particularly important for processes and products that are not protected by patents, but are instead trade secrets where the organization must closely protect the information to maintain that competitive advantage. Training that is strategic to the organization should preferably be developed and delivered by internal providers (Blanchard and Thacker 2004). The more complicated the topic the more likely it will need to be taught by internal technical experts (Combs and Davis 2010).

For some types of training, external providers may be best suited to deliver certain courses. This might be for technologies or skills where the expertise lies outside the organization and where this knowledge does not by itself give the organization a competitive advantage. It might also be necessary when the training resources lie outside the organization, such as specialized hardware or software. For example, the operation and maintenance of a pump in a chemical plant is usually not considered to be knowledge that offers a competitive advantage. For this type of non-competitive training, the plant may decide it is best delivered by an external provider such as a supplier. In the past, it was much more common for plants to have specialists who knew a great deal about particular aspects of the operation, such as particular processes or types of equipment. However, the trend toward leaner operations has eliminated many of these specialists, so plants rely much more on suppliers and consultants for that specialized knowledge. Another example where an external training provider would be the best choice is when an organization purchases a new piece of equipment that it has not previously used and for which there are no experienced users in the organization. Specialized training may even be included as part of the purchase (Combs and Davis 2010). JZI often trains clients at the plant site when the clients purchase technology that is new to the plant. These are often short, half-day sessions covering safety, operation, maintenance, and troubleshooting of the equipment.

For certain types of training, a combination of providers may be the best solution. One example might be when competitors collaborate to train on non-competitive topics such as safety, health, and the environment. It is in the best interests of an industry to keep people safe and healthy and not to harm the environment. Incidents that injure plant personnel and damage the environment often bring bad publicity to all organizations in an industry, even those that were not involved in the incidents. Another reason why a hybrid provider might be the best choice is when the specific knowledge of a given organization is combined with industry best practices. An SME in the organization provides the former, while an SME in the industry provides the latter. The internal SME can teach what has happened in their organization, while the external SME can teach what has happened in other organizations in the industry.

One example of using multiple types of continuing education providers is BP (Texas City, Texas, USA.) which used a combination of solutions for a four-day heater operations training course (Valencia et al. 2008). BP hired a supplier to teach a one-day class using a customized computer simulator to train on specific operating units in the plant. They hired JZI to teach a one-day class on process burners and heaters. BP used experienced operators to conduct two-day hands-on training sessions on the actual equipment in the plant. This arrangement of multiple providers received high marks from plant personnel because the best providers were used to teach each portion of the course.

It is unrealistic to expect that all continuing education required by an organization can be provided using only internal resources. "Due to increasing focus on cost control and complex learning technologies in educating adults in the workforce, companies realize that in-house development and management alone are insufficient" (Malloch et al., 2011). External continuing education providers should be used as appropriate, particularly when the expertise on a certain subject is outside the organization. In some cases, the best option may be a hybrid combination of instructors from both internal and external sources. The primary goal for the organization should be to select qualified instructors, whether they are inside or outside the organization, for the given continuing education subject.

There are many potential areas for further research regarding formal continuing education. One important area would be to compare student learning and costs for the various providers. This would likely need to be done in a larger organization with a sufficient number of students to compare learning for the same course taught by the three types of providers. Both quantitative and qualitative studies would help provide depth to such a comparison. It would be particularly valuable to study the effectiveness of hybrid providers such as the effects on learning using multiple instructors from multiple organizations. Research comparing continuing education under a variety of contexts such as organization size and geographic location would help determine the importance of various demographic variables such as age, gender, and engineering specialty. Another potential research area is to compare formal continuing education of engineers to other professions such as accountants, doctors, and lawyers. Best practices from each could be combined to enhance continuing education for all professions.

References

Bartleby Research (2021). Training is the organized procedure by which people learn knowledge and/or skill for a definite purpose. https://www.bartleby.com/essay/Training-Is-The-Organized-Procedure-By-Which-P36JCJ2XY9CXQ. Accessed December 26, 2021.

Baukal, C. (2001). *The John Zink Combustion Handbook.* Boca Raton, FL: CRC Press.

Baukal, C. and Bussman, W., (2009a). Ambient Condition Impact CO and NOx Emissions: Part I, Petroleum Technology Quarterly 14(3), 93-99.

Baukal, C. and Bussman, W., (2009b), Ambient Condition Impact CO and NOx Emissions: Part II, Petroleum Technology Quarterly 14(4), 37-41.

Baukal, C. and Bussman, W., (2009c). Simulator for teaching process heater operating principles, Paper presented at the 2009 American Society of Engineering Education conference, June 14-17, 2009, Austin, Texas.

Baukal, C., Colannino, J., Bussman, W., and Price, G. (2010), Industry instructors for a specialized elective course, Paper AC2010-67, proceedings of the American Society for Engineering Education Annual Conference & Exhibition, Louisville, KY, June 20-23, 2010.

Baukal, C., Price, G., Matsson, J., Bussman, W., and Olson, S. (2011), Industry Adjuncts: Lessons Learned, Paper AC2011-2248, proceedings of the American Society for Engineering Education Annual Conference & Exhibition, Vancouver, Canada, June 26-29, 2011.

Biech, E. (2005). *Training for Dummies*, Wiley http://adulted.about.com/od/teachers/a/teachingadults).

Blanchard, P. and Thacker, J. (2004). *Effective Training: Systems, Strategies, and Practices*, 2nd edn. Upper Saddle River, NJ: Pearson.

Cervero, R. M., (1992). Professional practice, learning, and continuing education: an integrated perspective, *International Journal of Lifelong Education* 11(2), 91-101.

Combs, W. and Davis, B., (2010). *Demystifying technical training: partnership, strategy, and execution*. San Francisco: Pfeiffer.

Darkenwald, G. and Merriam, S. B. (1982), *Adult Education: Foundations of Practice* (New York: Harper & Row).

Derose, G. J., (1999). *Outsourcing Training & Education* (Alexandria, VA: American Society for Training & Development).

Dirkx, J. M., (2011), Work-related learning in the United States: past practices, paradigms shifts, and policies of partnership. In M. Malloch, L. Cairns, K. Evans, and B. N. O'Connor (eds.), *The Sage Handbook of Workplace Learning* (London: Sage).

Fripp, P. (2021). Patricia Fripp. fripp.com. Accessed December 25, 2021.

Gilder, T., Campbell, D., Robertson, T. and Baukal, C. (2010). Customize Operator Training for Your Thermal Oxidizers, Hydrocarbon Processing 89(11), 55-59.

Gronstedt, A. (November 29, 2021). Get Ready for the Metaverse of VR Learning. https://www.td.org/atd-blog/get-ready-for-the-metaverse-of-vr-learning. Accessed December 26, 2021.

Groover, D. (2021). Datta Groover. dattagroover.com. Accessed December 25, 2021.

Hill, D. (2005). Operator training practices revealed, Hydrocarbon Processing 84(6), 17.

Jakes, T.D. (2021). *Don't Drop the Mic: The Power of Your Words Can Change the World.* Brentwood, TN: FaithWords.

John Zink Institute (2011-2012), *Process Technology Training Guide.*

Karan, J., Reese, B., Zschorsch, K-D, and Klaus, W. (2013). Thermal Oxidizer Basics. In C. Baukal (ed.), *The John Zink Hamworthy Combustion Handbook,* Vol. 3: Applications. Boca Raton, FL: CRC Press, 159-209.

Knowles, M., Holton, E., and Swanson, R. (2005), *The Adult Learner* (Boston: Elsevier).
Kornikau, R. and McElroy, F. (1975). *Communication for the Safety Professional.* National Safety Council: Chicago (1975).
LaCroix, D. (2021). Darren LaCroix. Darrenlacroix.com. Accessed December 25, 2021.
Langevin Learning Services (2021). Langevin.com. Accessed December 25, 2021.
Malloch, M., Cairns, L, Evans, K, and O'Connor, B. N., (eds.), (2011). *The Sage Handbook of Workplace Learning.* London: Sage.
Maxwell, J. (2021). https://www.brainyquote.com/quotes/john_c_maxwell_451128. Accessed December 26, 2021.
Meier, D. (2000). *The Accelerated Learning Handbook.* New York: McGraw Hill.
Meiss, R. and McCallum, D. (2007). *S.C.O.R.E Super Closers Openers Review Energizers for Enhanced Training Results,* The Bob Pike Group.
Merriam, S., Caffarella, R., and Baumgartner, L. (2007), *Learning in Adulthood: A Comprehensive Guide* (San Francisco: Jossey-Bass).
Michigan State University (2021). Michigan Approved Trainer of Registered Technician. https://fdocuments.net/document/introduction-to-the-manual-michigan-approved-trainer-of-registered-technician.html. Accessed December 25, 2021.
Moretti, E. (2002). Reduce VOC and HAP Emissions, Chemical Engineering Progress 98(6), 30-40.
North American Process Technology Alliance (2021). Process Technology: An Education that Pays (video). https://www.youtube.com/watch?v=CCMc2y W2Ng8&t=2s, accessed December 25, 2021.
Padfield, C. and Schaufelberger, W. (1998), Lifelong learning in engineering education: a call to action (Brussels: Societé Européenne pour la Formation des Ingénieurs).
Pankoff, J.A. (2005). Improve and sustain process plant operator performance, Hydrocarbon Processing 84(7), 97-99.
Pike, R. (2003). *Creative Training Techniques Handbook, Third Edition Tips, Tactics, and How-To's for Delivering Effective Training,* HRD Press.
Praeger, R., Ponticell, J. and Gordon, E. (1998), Enhancing Learning in Training and Adult Education (Westport, CT: Praeger).
Schnelle, K. and Brown, C. (2002). Thermal Oxidation for VOC Control, Chapter 13 in *Air Pollution and Control Technology Handbook.* Boca Raton, FL: CRC Press.
Smith, K. (2002). Refine your approach to process troubleshooting and optimization, Hydrocarbon Processing 81(6), 63-66.
Strachan, C. (2021). Craig. craigstrachan.com. Accessed December 25, 2021.
Thomas, P. (2021). Premier Presentations Services. http://www.premier-presentation.co.uk/ TIPS.htm.
U.S. Department of Education (2021). Accreditation in the United States. https://www2.ed.gov/admins/finaid/accred/accreditation_pg6.html. Accessed December 25, 2021.
Valencia, R., Link, D., Baukal, C., and McGuire, J. (2008), Consider classroom training for plant operators, Hydrocarbon Processing 87(11), 55-59.
Wilkins, M. (2006). Benchmarking consortium includes operator training, Hydrocarbon Processing 85(11), 15-16.

Chapter 11

Sourcing Continuing Engineering Education

Charles E. Baukal, Jr.*

Koch Engineered Solutions Institute^{SM}, John Zink Hamworthy Combustion, Tulsa, Oklahoma, USA

Abstract

Many are calling for increased continuing education for engineers. However, few details are provided for how to deliver that education. This chapter recommends a strategy for delivering continuing engineering education (CEE). CEE providers are categorized here as internal (the organization itself), external (universities, professional/trade organizations, commercial education, government, vendors, and customers), and hybrid (a combination of internal and external). The focus here is on who delivers the content, not where the training is held, who pays for it, or who organizes it. Typical forms, advantages, and disadvantages are discussed for each type of provider. The objective of this chapter is to compare the various types of CEE providers and to provide a framework for organizations to develop their CEE programs. The best strategy for an organization is usually to use a combination of providers, depending on the training needs.

Keywords: sourcing, continuing engineering education, providers, hybrid training

Introduction

Professions such as law, medicine, and engineering are often characterized by "rigorous professional knowledge" (Schön, 1987, p. 3) that changes, sometimes rapidly, with time. Lifelong learning is required to maintain proficiency and in many cases licensure. Learning may be informal or formal where informal is usually done on one's own without an instructor and formal usually includes an instructor. Examples of informal learning, sometimes referred to as *workplace learning* (Malloch et al., 2011), include reading trade magazines and journals and searching the Internet. Formal learning includes taking classes that may be either for-credit or not-for-credit. The "classes" also include one-on-one on-the-job mentoring by more experienced colleagues.

The Institute of Continuing Professional Development (ICPD, 2011) defines continuing professional development (CPD) as "the systematic maintenance and improvement of knowledge, skills and competence, and the enhancement of learning, undertaken by an individual throughout his or her working life." The on-going education of professionals sets them apart from other jobs, leading to elevated status and prestige. This dynamic process is

* Corresponding Author's Email: charles.baukal@kes.global.

called *professionalization* where professionals continuously upgrade their knowledge and skills (Houle, 1980). Continuing education includes "all those processes that contribute to the advancement of an individual's knowledge, skill, understanding, competence, and general professional and personal development" (Padfield & Schaufelberger, 1998, p. 8). Continuing education, CPD, and professional development are all related terms that generally concern learning outside the normal workplace (Dirkx, 2011). This chapter is specifically limited to professional development and does not include personal development unrelated to the workplace.

Training can be defined as "a planned effort by a company to facilitate employees' learning of job-related competencies" where the competencies include "knowledge, skills, or behaviors that are critical for successful job performance" (Noe, 2005, p. 3). The continuing education and training of engineers is of specific interest here. Engineering can be defined as "the field or discipline, practice, profession and art that relates to the development, acquisition and application of technical, scientific and mathematical knowledge about the understanding, design, development, invention, innovation and use of materials, machines, structures, systems and processes for specific purposes" (Marjoram & Zhong, 2010). CEE can be defined as (National Science Foundation, 1977):

- the substantive content deals with engineering knowledge,
- taken after initial employment,
- addresses updating and diversification rather than advanced education,
- is structured, goal oriented, and intentional, and
- has a duration adequate for a significant increment of training.

CEE tends to be much more practical and applied than university education (Boër et al, 2013). Vanajakumari et al. (2015) argue the most crucial factors for effective CEE are:

- bidirectional interactive learning between students and instructors,
- opportunities to examine real life scenarios in the classroom,
- improved student confidence, and
- communication of the overall objectives of the course without slowing the learning process with theory.

Many institutions have identified the importance of CEE. A number of U.K. government agencies have recommended CEE (Galloway, 1998). The European Society for Engineering Education sponsored a report which argued action is needed to improve the continuing education of engineers (Padfield & Schaugelberger, 1998). A Report by The U.S. National Academy of Engineering (2005) recommended that engineering institutions teach students how to be lifelong learners. The American Society of Mechanical Engineers' 2028 Vision includes lifelong learning as one of the four areas for focused improvement (ASME, 2008). The U.S. National Academy of Engineering (2008) identified fourteen grand challenges for engineering, many of which will require CEE as some of the subjects (e.g., prevent nuclear terror) are not typically covered in university training. "To successfully integrate process and knowledge, engineers must not only stay informed about new and emerging technologies but also be aware of knowledge and skills from other domains" (Sheppard et al., 2009, p. 6). A comprehensive and international report sponsored by UNESCO on engineering also emphasized the need for CEE (Jones, 2010, p. 8):

> With the huge explosion of knowledge, engineers have been forced to become more specialized in their professional skills, which in turn forces judicious choices to be made in the topics taught during an initial education. The only way by which the tussle between breadth and depth can be reconciled is the recognition that a first degree provides just an initial education and that Continuing Engineering Education (CEE) or Continuing Professional Development (CPD) is essential.

The Porto Declaration formalized the ongoing need for lifelong learning by engineers (Soeiro et al., 2016). ABET (2021) is a major worldwide accrediting body for university engineering programs. While lifelong learning used to be an explicitly required Student Outcome, it is now implicit in Student Outcome 7 which states that upon graduation students should demonstrate "an ability to acquire and apply new knowledge as needed, using appropriate learning strategies."

While many have called for CEE, few details have been provided on the specifics of delivering that CEE. The objective of this chapter is to provide some of those details. The goal is to help organizations deliver effective and timely training to their engineers.

The demand for CEE continues to increase. There are many reasons for this including rapid technological advances, globalization, increasing environmental emphasis, and socio-political changes. For example, the fields of nanotechnology and biotechnology are changing so rapidly that continuous training is required to maintain proficiency in those fields. CEE improves job performance and usually leads to increased compensation (Morris, 1978). It is recommended for quality assurance in a profession and in some cases it may be mandated to maintain licensure (Jeris, 2010). Licensure is particularly important in civil and environmental engineering because of the many public works projects done for or approved by governments. Failure to continually learn after receiving an engineering degree leads to obsolescence because of the rapid changes in technology (Barton & Bommer, 1992).

Engineers participate in many types of continuing education, including both informal and formal training (Cervero et al., 1986). Informal training is generally unplanned and initiated by the individual, although it could include sanctioned learning such as mentoring, coaching, and special assignments. For example, reading journals and magazines and attending conferences helps keep engineers current with changes in technology. Formal training is "planned learning activities that are intended to help individuals acquire specific areas of knowledge, awareness, and skills" and "mostly involves institutionally sponsored and endorsed programs, which would include almost all training and development programs that organizations offer" (Jacobs & Park, 2009, p. 140). Engineers may also take formal classes with instructors that may be for credit and possibly even leading to an advanced degree. Formal professional development not leading to a degree is the specific focus here. Because of the time and cost involved, CEE is usually sponsored by the employer rather than paid for by the individual engineer.

Small and medium-sized organizations often need to partner with other organizations to provide training for their employees (Forrester et al., 1995). Even the largest organizations are not able to provide all the continuing education needed by their engineers (Cole et al., 1984). There are several reasons for this. Many organizations no longer have as many subject matter experts (SMEs) in-house as in the past, due to *right-sizing*. This often means greater reliance on vendors to provide needed technical knowledge. Employees at the organization may not be SMEs or trained instructors. There may also not be enough employees needing certain types of specialized training to justify internally developing and delivering the necessary courses. The volume of training may be too much for the available staff within an organization (Anderson, 2011). For these reasons, organizations rely on outside providers to deliver the training they cannot provide themselves. This is often referred to as *outsourcing* (Noe, 2005). Outsourcing in the context of training and development can be defined as (Hale, 2006, p. 2):

> Entering into a long-term relationship with an external resource so as to leverage the capabilities and capacity of that resource. Outsourcing is done to better respond to demands for learning solutions that are technology-based, manage training's costs, decrease cycle time, accelerate development and delivery, and avoid investment in developing the training function.

According to a January 2011 survey by the Human Capital Media Advisory Group, over half of all organizations outsource at least some of their training (Anderson, 2011).

The challenge is how to deliver CEE when and where it is needed, at a reasonable cost. Continuing education must be relevant and should be designed to solve industrial problems (Weimar, 1992). Continuing professional education (CPE) must be related to professional practice to help professionals make decisions about situations they encounter (Cervero, 1992). This means that CEE providers should deliver training that engineers can use in their jobs, preferably as needed.

The individual engineer, employers, academia, professional societies, government, manufacturers, and other education providers all have a role in CEE (Ackermans, 1989; Heinke, 1991; Mott, 2000; National Research Council, 1985; Padfield & Schaufelberger, 1998). Employers are responsible for providing the time and financial support to continually educate their engineers, although this does not always happen as it should (Jeong, 2000). Employers also offer many specialized courses relevant to their own operations. Academia both provides CEE and helps to develop a mind-set in students of the importance of lifelong learning. Professional societies and trade organizations help set guidelines and standards for, encourage, and deliver CEE. Governments set regulations for licensing engineers, develop policies that encourage CEE, and sometimes deliver CEE. Other education providers (usually for-profit) offer a wide range of CEE courses for the on-going training of engineers. Vendors and customers can provide specific training on their technologies and processes.

Employer-sponsored CPE dwarfs that offered by any other type of provider and possibly more than all other providers combined (Cervero, 2001). While much has been written, for example, on CPE offered by universities, relatively little has been written about employer-sponsored CPE. Many reports have identified the importance of CEE, but none have provided specific recommendations on how to provide it. This chapter is intended to fill that gap in the literature.

CEE is provided by many different sources including traditional non-profit institutions (e.g., universities, professional/trade organizations, and government agencies), corporate universities, and commercial content providers (Bourne et al., 2005). Here, the distinction between provider types is made based on who delivers the training, not where the training is held. For example, some refer to *employer-sponsored* continuing education as any courses organized by the employer including in-house seminars, workshops, and courses regardless of who actually delivers the training (e.g., Heinke, 1991). Here, the employer delivering training content is referred to as an *internal* provider. Providers outside the organization delivering training for the organization are referred to as *external* providers. The combination of the organization and an outside provider partnering together to deliver training is referred to as a *hybrid* provider. It is argued here that it is more important who delivers the training, not where it is delivered or who organizes it.

CEE courses include all types of technical (e.g., equipment, processes, software), management (e.g., leadership, employment law, enhancing team performance), and soft skills (e.g., communication, presentation skills, legal training) courses. CEE courses combining engineering and management are the most popular types of courses (Beruvides & Ng, 2009). Cole et al. (1984) broadly defined four categories of continuing education courses for engineers:

- remediating and upgrading basic technical knowledge and skills,
- extending and broadening previously learned scientific and technical skills,
- learning new concepts and skills to keep up with technology advancements, and
- learning new knowledge and skills outside engineering.

Chapter 8 discusses the results of a survey of engineers as to their CEE practices and preferences. Not surprisingly, they strongly prefer technical training over other types such as management, legal, and environment, health, and safety (EH&S) training.

The need for an effective strategy to deliver CEE is becoming increasingly important to address significant global challenges such as poverty reduction, sustainable development, and climate change (UNESCO, 2010).

Context

While there are many stakeholders interested in CEE, the two primary stakeholders are the organization sponsoring the CEE and the individual engineer completing the CEE. Kaufmann and Weaver (1995, p. 34) frame the challenge:

> For training to be effective, it must be quality training for the right people at the right time and tied to the firm's business goals. With proliferation of opportunities, the increased need of continuing education, and declining resources, the decisions become more difficult and more important.

The first aspect of the context is based on human capital theory, popularized by Nobel-winning economist Gary Becker (1962). This involves investing resources, such as training and medical care, into people. This investment is made by an organization in its members. The specific human investment of interest here is education and training (Becker, 1993). When the organization is a company, it generally expects some return on its investment (ROI) in its employees who also usually benefit from a return such as maintaining employment, higher wages, and more responsibilities. A meta-analysis found that human capital relates strongly to organizational performance which suggests organizations should make investments in their members (Crook et al., 2011). A committee investigating career-long education for engineers for the U.S. National Academy of Engineering (1988, p. 29) commented on what is important to employers, "What concerns industry the most is not necessarily the money it spends on education and training, but whether it is getting the best return on the investment."

The second aspect of the context pertains to the needs of the individual engineer. CPD concerns educating adults. Lindeman (1961) believed the curriculum for adult education should be built around the needs and interests of the learners. Many have identified the importance of pragmatism in adult learning. Surveys show that adults prefer practical courses over those that merely satisfy intellectual curiosity (Cross, 1981). Wlodkowski (2008, p. 97) wrote, "adults are highly pragmatic learners." Landis (1989) wrote, "Engineers are primarily interested in continuing education courses that help them do their current job better." Sometimes the need is for immediate training on a given topic which Spak et al. (2011) referred to as "just-in-time knowledge." CEE must meet the needs of engineers (Horvat & Kailer, 1989). Meeting learners' needs particularly pertains to the instructors delivering course content, who must be technically proficient in the content area for instructors of adults (Galbraith, 2004). Wlodkowski (2008) listed five characteristics and skills of motivating instructors: expertise, empathy, enthusiasm, clarity, and cultural responsiveness. The focus of this chapter is on the first skill which Wlodkowski defined as knowing something of benefit to adults, knowing it well, and being able to effectively convey it to adults through an educational process. Therefore, CEE instructors must be technically proficient and be able to deliver practical information engineers can use.

The primary question addressed here is who should deliver CEE that meets the needs of both the organization and its engineers (Horvat and Kailer, 1989)? Secondary questions include: What are the advantages and disadvantages of each type of provider and which should be used under what circumstances?

Internal Providers

Internal providers are organizations delivering training to their employees. Some organizations have set up sophisticated training programs, sometimes referred to as *corporate universities* (Meister, 2001; Lui-Abel, 2011), to provide on-going training for their employees.

Forms of Internal Providers

Training delivered by the engineer's own organization comes in a variety of forms. The least formal and probably most common is on-the-job, sometimes referred to as job shadowing or apprenticeship, where more experienced employees train less experienced employees. For example, new engineers may be paired up with experienced engineers who train the new engineers on the processes and procedures in the organization.

Another type of internal training is where SMEs train fellow employees in their area of expertise. Craig and Evers (1981) referred to these SMEs as *employee educators* and noted that using employee SMEs to teach their colleagues "reflects a common belief of many corporate managers that those who actually performed the job are those best qualified to teach it."

Another type of on-the-job training is referred to as *shared practice* where employees learn from each other while working together (Collin, 2008). This type of continuing education differs from the two described above because it does not involve formal mentoring between more experienced and less experienced employees, nor does it necessarily involve an SME. Working together on projects requires collaboration, problem-solving, and the ability to handle conflict, often leading to a sense of community, which are important skills that can be learned through shared practice.

Internal trainers for an organization sometimes purchase course materials they can either deliver essentially as is or that they can modify according to the needs of their organization. Internal trainers may also develop their own materials, particularly on topics that are specific to the organization. For example, trainers may develop materials for a particular unit in a plant which might include process and instrumentation diagrams, equipment specifications, operating manuals, and other related documentation.

Advantages of Internal Providers

There are many advantages of training provided by the employer, which Gräfen (1991) argued is the most effective. The training is very focused and targeted to the specific needs of the organization, which is in a unique position to know its own knowledge and skill deficiencies as well as those of its employees. For example, training can be provided on specific equipment in a plant, rather than on generic equipment. Training materials may include drawings and manuals of actual equipment in the plant. This makes the training more targeted and time efficient, as well as easier to engage the students since they are learning about the exact equipment they are or will be operating.

Internal instructors usually know the organization and its products, services, and customers. They often also know the students who are fellow employees. This helps instructors tailor the materials and teaching methods to the needs of both the organization and the students.

The training format and duration can be easily adjusted as appropriate. For example, in plants that run 24 hours a day seven days a week there are multiple shifts that may need training. Internal instructors can easily adjust their schedules to provide training to cover the shifts in a timely manner. This may mean teaching certain days and times one week and different days and times another week. While this is possible with external providers, it will likely be much more expensive and more challenging to schedule. Another related factor is that internal courses may be more cost effectively delivered to very small groups of students. The cost of bringing in an external provider, especially if significant travel and living expenses are involved, can be very high per student when training a small number of employees. The more cost effective the training the better the ROI, assuming equal levels of performance.

Students and instructors working for the same organization can discuss topics with less concern about legal and competitive issues. One problem with training classes consisting of students from competing organizations is that detailed and proprietary information cannot be shared. There is also the concern of possible anti-trust violations if certain types of information, such as product pricing and customers, are shared. Even if all students are from the same organization but the instructor is not, sharing proprietary information can be a problem, unless some type of confidentiality agreement has been executed. When students and instructors are from the same organization, they can share important and specific details that can enhance the learning experience and maximize the ROI to the organization.

Disadvantages of Internal Providers

There are some potential disadvantages of internal providers. The first is that training for most employers is not their primary mission, which is usually to earn a profit. This means that training may not receive the support and resources necessary to provide consistent quality continuing education for its professional employees (Cervero, 1988).

Another disadvantage is that training is often limited to the knowledge within the organization. This means that students may not always get the advantage of learning industry best practices. External providers are more likely to know industry best practices if they work with many different organizations across the industry.

While an advantage of this type of provider is the capability to train on the specific equipment the student will be using, a potential disadvantage is that this equipment may not be available for the full range of training exercises. If the equipment has a large impact on production, students may only be able to make limited adjustments to avoid adversely affecting production. This means students may receive little if any training on the actual equipment on those conditions that rarely occur, such as emergency situations, start-up, and shutdown. Some type of simulator may be needed to train for those special operating conditions that occur infrequently (see Chapters 4 and 7). A realistic training environment is important to ensure transfer of training (Grossman & Salas, 2011).

Another potential disadvantage is that internal instructors may be too familiar with the students who are also colleagues, which could lead them to be too easy or too hard on certain individuals. For example, instructors may not challenge students who are superiors for fear of possible job repercussions, while instructors might be overly harsh on colleagues with reputations for not taking their work seriously enough or on colleagues they don't like. External instructors may be less biased toward the students than internal instructors.

Another potential problem with classes taught by internal instructors and consisting only of internal students is that some students may be hesitant to ask questions for fear of how they might appear to their colleagues. While instructors may state the adage there is no such thing as a stupid question, it may be difficult to convince students while they are in the presence of their colleagues, especially supervisors. This can be handled by giving students other alternatives for asking questions such as writing them down anonymously or discussing issues with instructors privately during breaks or outside of class.

Depending on the size and sophistication of the internal training organization, instructors may be asked to teach subjects for which they may not be truly qualified, to save time and money. Instructors may not know or be willing to admit they are not qualified to teach certain topics. A different type of problem related to instructor quality concerns teaching ability. If internal instructors are SMEs, they may be experts in their subject area, but may not be trained on how to teach effectively.

For infrequently taught courses, the quality of internally-delivered courses may not be as good as for external organizations teaching the same courses on a more frequent basis. Because of other priorities, that course is not likely to receive as much attention from internal instructors compared to other courses they teach more frequently.

Very few internally developed and delivered CEE courses are accredited. There are many possible reasons for this. The first is the added expense is not usually justified unless accreditation is required by law or by customers. This is particularly likely for internal courses that are offered infrequently or involve relatively few employees. Another important reason is that most engineering organizations do not have the personnel, policies, and procedures to become accredited and then to maintain accreditation. Initial accreditation is often a rigorous, lengthy, and costly process. Maintaining accreditation is similarly rigorous and requires organizational discipline. Since training is not the primary function of most organizations, accreditation is not usually a priority, especially in an era of *right-sizing* when employees are already stretched to the limit and have little time for lower priority activities. For most engineering organizations, accreditation of their own CEE courses is not viewed as a competitive advantage. Lack of accreditation suggests that internally developed and delivered courses may not have adequate quality assurance.

External Providers

There are many potential reasons why an organization might use external providers to deliver training. Some reasons include gaining flexibility to better control the workload during fluctuating business cycles, outsourcing training for subjects outside core competencies, increasing options for managing the training budget, and using specific expertise to support new strategies or projects (Hale, 2006). Smaller organizations often need to use external providers because they do not have the staff to deliver all their training requirements (Blanchard & Thacker, 2004). Further, using only in-house trainers could cause an organization to become in-bred (Eurich, 1985).

There are many types of external providers that deliver CEE courses. These are broadly categorized here as universities, professional and trade organizations, commercial training organizations, government, vendors, consultants, and customers.

Universities are a major provider of CEE, typically with more emphasis on theory (Cervero, 1992). Industry sometimes partners with universities to provide CEE where the universities deliver the training. Non-credit continuing education courses can be developed by universities in conjunction with industry to meet employee development needs (Anwar & Curley, 2000). These courses are not typically taught in engineering degree programs because they are too technology specific, but they are needed by engineers using those technologies in their jobs. Industry sometimes provides input on what should be included in these university-taught courses and may even supply supplemental materials such as photos, videos, models, or actual equipment.

Community colleges are also a potential provider of CEE (National Academies of Science, Engineering, and Medicine, 2018). In many cases, community colleges have a specific mandate to prepare students to be workers in the immediate region. They may have more flexibility to tailor courses to the specific needs of companies in their local community compared to universities. They may also be able to more cost effectively deliver customized training to smaller numbers of workers.

Universities delivering online courses are sometimes characterized as *virtual universities*, which can be defined as “an electronic representation of the features of a real university” (Bodendorf & Swain, 2001, p. 103). While both employees and employers agree that CEE is important, the real challenge is time for engineers who are busy with work and family responsibilities and often find it difficult to regularly attend classes held on a specific day and time each week (Fink, 2001). This is an important reason why distance CEE has expanded rapidly (Baukal, 2010). These virtual universities help meet the demand for CEE anytime and anyplace (see Chapter 9).

Professional societies are an important source of CPE courses. Engineering societies are generally organized around a specific branch of engineering, such as mechanical, electrical, chemical, and civil engineering. Professional societies are often better suited to provide CEE where the focus is longer term, compared to company-developed training programs that tend to be short-term focused (Van Horn, 1977). Professional societies have some advantages over university and industry training programs such as the flexibility to use the best instructors anywhere a course is needed (Abramson, 1977).

Another category of external CEE providers is non-profit industry trade organizations, which have been largely ignored in most discussions of CEE providers. In contrast to professional societies, trade organizations focus on specific industries or technologies, rather than on particular branches of engineering. These trade organizations are particularly suited to deliver highly specialized courses on topics important to specific industries.

There are many for-profit educational organizations specializing in delivering CEE. These organizations primarily provide topical short courses (Anderson, 1977) and greatly expand the range of topics engineers can learn. A limitation is that they are often limited to subjects of interest to a larger audience to ensure adequate course attendance to generate a profit. They are less likely to cover specialized topics that would generate relatively small course attendance.

Some government agencies offer CEE, particularly related to general health, safety, and environmental topics. They may also deliver courses related to regulations, particularly when there have been recent changes in those regulations, to educate engineers who must help their organizations meet those new regulations.

Many vendors offer CEE related to their products and services, sometimes for a fee and sometimes at little or no cost. Vendors are uniquely qualified to train on their products and services and also have the benefit of industry-wide knowledge of best (and worst) practices related to those products and services. They often have training materials such as photographs, videos, and actual hardware that can significantly improve the quality of the training. They may also have special facilities to demonstrate how to use their products which further enhances the training experience. Delivering technical training not only to its own employees, but to customers and partners, may be a strategic objective for an organization (Combs and Davis, 2010). On the importance of industry delivering training, Bruce et al. (1982, pp. 6-7) wrote, "Engineering faculties cannot by themselves keep up with the knowledge explosion. Close collaboration between engineering faculties and their industrial colleagues is essential if new knowledge is to be distilled from the literature and widely disseminated at the rate at which it is being generated."

Consultants are often available to teach specialized engineering topics. In some cases, they may be retired personnel from a given plant who come back periodically as SMEs to teach on a particular technology. This can be very beneficial as they typically know the plant and personnel very well. The disadvantage is they may not know best practices across the industry. In some cases, individual consultants may be available at a lower cost because of much less overhead compared to an organization. They may also have more flexibility to deliver training when and where it is needed.

Another type of external provider is customers. No substantial discussion of this type of provider was found in the literature. This can be a potentially very powerful type of provider as customers could provide information that may not be readily available within the vendor's organization. The more knowledgeable the vendor is about its customers' needs, the better for both the vendor and its customers. There are some potential challenges with this type of provider as the information may be biased towards a particular customer's operations, there could be some intellectual property concerns, and there would have to be some significant benefits to the customer to dedicate resources to training their vendors.

Forms of External Providers

External providers deliver a wide range of courses in a variety of formats in many locations. Formats include face-to-face, distance, and a combination of both. Locations include at the provider's site, at the client's site, or at other locations such as hotels and university campuses. Providers range from a single person (e.g., a consultant) to complete organizations dedicated to providing training.

Advantages of External Providers

There are many potential advantages of using external providers. The most important is getting the experts in a specific subject area. Pérez-Foguet et al. (2018) describe a European consortium of five universities and four non-governmental organizations that partnered to develop specific online CEE courses related to the subject of sustainability. A particular advantage of professional societies is access to a wider variety of experienced experts compared to other providers (Houle, 1980). Another advantage could be access to custom training resources.

Accreditation of CEE courses has been recognized as a problem (Jones, 2010). While some external providers have accredited courses, this is much less common for internal providers as previously discussed. Accreditation is likely to become more important in the future, particularly regarding licensing of engineers, to ensure that CEE meets some minimum quality standards.

A potential important advantage of an external provider is cost savings which enhances ROI. It is usually not cost effective for employers to match the educational capabilities of organizations dedicated specifically to training on topics outside the employer's core competencies. When selecting external providers, employers often have multiple options to choose from which usually means lower costs due to competition. However, as previously discussed, classes with a small number of students may not be very cost effective when using external providers.

Disadvantages of External Providers

One key disadvantage of external providers is added out-of-organization costs. While there are certainly costs for internal courses, those expenses may be viewed differently as the costs for internally developed and delivered courses do not leave the organization. Company trainers and training facilities are generally part of the operating budget, while costs for externally-provided training usually receive more scrutiny and may be more difficult to justify, especially during times of tight budgets.

Another potential disadvantage of external providers is the course content may be somewhat generic, particularly if the students are from multiple organizations using somewhat different technologies. This means there may be considerable wasted time for students who do not use the technologies being covered at various points in the course. It also usually means that even when the technology being covered is used by the student, it may not be discussed in much depth because of time constraints when multiple technologies must be covered. This problem can usually be overcome by having external providers tailor course content to the specific needs of the organization, assuming this is cost-justified.

A further potential disadvantage may be possible conflicts of interest (Dutta, 2010). This could occur where a vendor offering training may unduly influence students to purchase their equipment which could have negative cost, safety, and environmental ramifications for the organization. This is one reason why organizations should consider using accredited courses where marketing products and services to students during training is usually discouraged if not forbidden by accreditors.

Hybrid Providers

An alternative CEE provider is a combination of both internal and external providers. Two or more organizations create an alliance to develop and deliver training. This form of arrangement has been described as co-sourcing, share-sourcing, strategic sourcing, and in-between sourcing (Salopek, 1998). Note that the collaboration, which Ackermans (1989) calls cooperation, between providers considered here is more narrowly defined than by others. For example, there is increasing collaboration between providers, particularly between employers and universities, to deliver CPE (Cervero, 2001). However, that collaboration primarily concerns employers hiring universities to deliver courses tailored to their needs, where those courses are taught entirely by the university (Darkenwald & Merriam, 1982). Rampazzo and Beghi (2018) describe an example of such an arrangement where the University of Padova delivered specialized training in heating, ventilation, and air conditioning at a company site tailored to the needs of its engineers. As defined here, that would be an external, rather than a hybrid, provider. Beder (1987) describes four forms of collaboration in CPE: co-sponsorship, referral, coordination/control, and donor relationships. These refer more to the administration of and payment for courses, rather than to the delivery. Kost (1980) suggested joint instruction as a potential form of collaboration between industry and academic institutions, although he did not elaborate or give any examples. Evetts (1998) noted that CPE is a collaboration between the individual engineer, the employer, and the education provider. Here, collaboration between providers specifically means *co-teaching*, where both internal and external providers teach a portion of a course. Relatively little is known about cooperative relationships in continuing education (Hohmann, 1980).

While a contractual arrangement is one type of relationship, the more powerful form is where there is a mutual alliance between organizations where each benefits in some way. The objective is to produce more effective training through synergy than could be provided by the individual organizations. An example would be a customer and supplier co-teaching a course. The students learn more about the vendor's product, while the vendor learns more about how their product is used by the customer, including possible areas for improvement and future features. Collaborative relationships among organizations may take different forms, which are discussed next.

Forms of Hybrid Training

One type of hybrid training is where multiple organizations partner together to develop and deliver courses. For example, a university might teach the theory section of a course, while the employer might teach the application of the theory. Another example might be where a vendor teaches about their specific product or service including industry-wide best practices, while the employer teaches how that product or service is used within the organization including specific policies and procedures.

Another form of hybrid training is where competitors work together to offer non-competitive training of benefit to the industry. This is sometimes called *coopetition*, where competitors cooperate on non-competitive areas such as training (Combs & Davis, 2010). For example, teaching safety and environmental best practices is generally good for an entire industry. Incidents can be detrimental to an industry, not just to those organizations involved in the incidents. Preventing incidents is good for the industry.

Advantages of Hybrid Training

Internal SMEs teamed with external SMEs have the potential to deliver customized training that combines the strengths of both. Internal SMEs are experts in their organization's processes and procedures. They can effectively train fellow employees on how things are done in their own organization, but do not necessarily have knowledge of industry best practices or of potential problems that could occur but which have not occurred in their own organization. External SMEs may know industry best practices, but may not know the details of a particular organization's operations, policies, procedures, and experiences.

Another potential benefit of hybrid courses is a better blend of theory and practice in the case when industry partners with academia. Industry is very familiar with the application of technologies used in their facilities but may not be as familiar with the theory or with the most recent developments in a technology area. Academia may be less familiar with actual practice but is well versed in the theory and recent advancements in a field, particularly if the university instructors are actively involved in research in that field.

Disadvantages of Hybrid Training

Hybrid training courses may take more preparation time than those taught by a single organization, whether internal or external. More coordination is required to determine which organizations will cover what topics. Training materials need to be meshed together to eliminate duplication, ensure coherence, and handle any areas where there are conflicts.

Another potential disadvantage of hybrid training is that the training methods and styles of multiple collaborating organizations may be significantly different which could adversely affect learning effectiveness as students may need to continuously adapt to different teaching styles. Course materials may be inhomogeneous. For example, one organization may use one type of template for its slides while another likely uses a different template. If these are not harmonized, this can make the course appear to be disjointed and lack cohesiveness. A related problem is the potential discontinuity of having multiple instructors from different organizations teaching at different times during a course. For some students who enjoy the variety, this may be an advantage. For others who prefer more continuity, this could be distracting. Courses should be designed to minimize the distractions of multiple instructors and teaching styles.

Another potential problem is the legal issues that could arise with multiple organizations, particularly when competitors collaborate. For example, one organization may have a different legal disclaimer than another organization. These may need to be harmonized which may not be an easy task. Another possible legal concern could be ownership of the combined training materials. After a joint course has been completed, can one organization use another organization's materials in future courses that may not include that other organization? These details can usually be worked out but added time may be needed to handle them.

Conclusions and Recommendations

Organizations should invest in CEE with acceptable ROIs (human capital theory). CEE instructors should be SMEs who are trained in how to teach adults and who can deliver practical knowledge and skills engineers can use in their jobs. There is growing concern about the effectiveness of CPE, most of which focuses on abstract knowledge rather than on practical application (Cervero, 1992). To improve program effectiveness, the best solution for an organization depends on many factors but is often a combination of providers. Organizations need to select the best way of sourcing their training needs (Derose, 1999), which means selecting the best provider for the specific type of CEE. Internal providers are often best suited for an organization's core competencies where there are available SMEs who have specialized and proprietary knowledge that may only be available inside the organization and because the organization may not want to share that information which gives them a competitive advantage. This is particularly important for processes and products that are not protected by patents but are instead trade secrets where the organization must closely protect the information to maintain that competitive advantage. Training that is strategic to the organization should preferably be developed and delivered by internal providers (Blanchard & Thacker, 2004). The more complicated the topic the more likely it will need to be taught by internal technical experts (Combs & Davis, 2010).

For some types of training, external providers may be best suited to deliver CEE courses. This might be for technologies or skills where the expertise lies outside the organization and where this knowledge does not by itself give the organization a competitive advantage. It might also be necessary when the training resources lie outside the organization, such as specialized hardware or software, particularly for those that are new to the organization. It may be more cost effective for the organization to purchase some training rather than developing and delivering it themselves (Cervero, 1988).

For certain types of training, a combination of providers may be the best solution. One example might be when competitors collaborate to train on non-competitive topics such as safety, health, and the environment. The industry is best served by keeping people safe and healthy and not harming the environment. Incidents that injure plant personnel and damage the environment often bring bad publicity to all organizations in an industry, even those that were not involved in the incidents. Another reason why a hybrid provider might be the best choice is when the specific knowledge of a given organization is combined with industry best practices. SMEs in the organization provide the former, while SMEs in the industry provide the latter. Internal SMEs can teach what has happened in their organization, while external SMEs can teach what has happened in other organizations in the industry.

It is unrealistic to expect that all CEE required by an organization can be provided using only internal resources (Malloch et al., 2011). There are many factors involved in the selection of an appropriate CEE provider including cost, technical expertise, teaching ability, accreditation requirements, availability, relationship (e.g., partner, customer, competitor), course content, facilities, and proximity. External CEE providers should be used as appropriate, particularly when the expertise on a certain subject is outside the organization. In some cases, the best option may be a hybrid combination of instructors from both internal and external sources.

Jones (2010, p. 329) wrote, "the nature of the challenges for engineers are (sic) changing. Engineering has to be seen in a very much broader context in terms of its role and impact on society, and engineers need to have a very broad set of skills in addition to their engineering expertise." This likely means future growth for external and hybrid providers as it will become more difficult for organizations to provide training for their employees in this complex and dynamic environment. This also likely means growth in delivery of distance CEE to meet individual and organizational needs anywhere and anytime.

Much research is needed regarding CEE providers as no systematic studies were found on this subject. This is not surprising as there are some factors that make this challenging. The first is that non-academic organizations fund most of the CEE. Since educational research is not usually a product or service of the organization, there is little incentive or capability to conduct this type of research. Even if an organization did conduct such research, it would likely be challenging. To fairly compare different types of providers, the same CEE would need to be delivered by different providers to large enough groups of students where learning can be measured, preferably quantitatively. The topic could not be too specialized as there may not be enough participants otherwise. Since this type of research would inevitably require substantial funding, there is little incentive for an organization to share the results of such research if they provided all of the funding and if they believed the results could give them a competitive advantage. A comprehensive research study of CEE providers would be most likely need to be organized and possibly at least partially funded by the government, a professional society, or an industry trade association, all of which have a vested interest in the outcome beyond individual organizations.

There is a relative lack of qualitative studies in engineering education (Borrego et al., 2009; Case & Light, 2011). Qualitative methods such as case studies, ethnography, and narrative analysis could significantly enrich the knowledge base related to CEE. For example, case studies of courses taught by hybrid providers could be used to develop best practices for collaboration between different types of providers.

Another important area of research is quantifying the ROI for organizations investing in CEE. Cost-benefit analyses for CPD in general are scarce (van Loo and Rocco, 2004), let alone for CEE. Morris (1978) investigated the ROI of CEE. A survey of 369 engineers working in government and industry showed that CEE participation was the dominant predictor of job performance, unexpectedly even more important than inherent ability and drive. Performance was measured by compensation and growth in supervisory responsibility, which are indirect measures of ROI that was not measured monetarily. Kaufmann and Weaver (1995) listed seven key questions decisions makers should consider regarding CEE but did not provide any type of cost analysis. Kuprenas et al. (2000) proposed three criteria for evaluating the effectiveness of an engineering training program: availability of training courses, participation in those courses, and comprehension of the courses. While these are useful measures, they do not measure ROI which is difficult to quantify because there are often many factors that impact it (Cranch, 1987). Comprehensive studies quantifying ROI are needed to help organizations better decide who should provide CEE under what circumstances to meet the complex and dynamic challenges faced by both engineers and their employers.

References

ABET, Criteria for Accrediting Engineering Programs, 2021-2022. https://www.abet.org/accreditation/accreditation-criteria/criteria-for-accrediting-engineering-programs-2021-2022/. Retrieved December 21, 2021.

Abramson, H. I. (1977). How one professional society provides continuing education. *Engineering Education*, 64 (4), 287-289.

Ackermans, S.T.M. (1989). Who needs whom in continuing education? A search for patterns of co-operation. *European Journal of Engineering Education*, 14 (1), 5-14.

Anderson, C. (2011). Outsource training to transfer knowledge. *Chief Learning Officer*, 10 (5), 52-55.

Anderson, R. M. (1977). An educator views the limitations of professional societies. Engineering Education, 67 (4), 289-291.

Anwar, S. and Curley, W. G. (2000). Partnering with industry to provide continuing and distance education programs in engineering & technology: a case study. Paper presented at the 2000 American Society of Engineering Education Annual Conference & Exhibition: Engineering Education Beyond the Millennium, June 18-21, 2000. St. Louis, Missouri.

Barton, B. and Bommer, M. (1992). Myths about engineers maintaining professional technical competence. *International Journal of Continuing Engineering Education and Lifelong Learning*, 2 (1), 22-30.

Baukal, C. (2010), Continuing engineering education through distance learning. *European Journal of Engineering Education*, 35 (2), 225-233.

Becker, G. S. (1962). Investment in human capital: a theoretical analysis. *Journal of Political Economy*, 70 (5), 9-49.

Becker, G. S. (1993). *Human Capital: A Theoretical and Empirical Analysis, with Special Reference to Education*. Chicago: University of Chicago Press.

Beder, H. (1987). Collaboration as a competitive strategy. In C. Baden (Ed.), *Comptetitve Strategies for Continuing Education* (pp. 85-95). San Francisco: Jossey-Bass.

Beruvides, M. G. and Ng, E-H. (2009). The continuing education of engineers and engineering managers. *International Journal of Continuing Engineering Education and Lifelong Learning*, 19 (1), 4-18.

Blanchard, P. N. and Thacker, J. W. (2004). *Effective Training: Systems, Strategies, and Practices*, (2nd ed.). Upper Saddle River, NJ: Pearson.

Bodendorf, F. and Swain, P. H. (2001). Virtual universities in engineering education. *International Journal of Engineering Education*, 17 (2), 102-107.

Boër, C. R., L. Canetta, P. Pedrazzoli and C. Redaelli (2013). Academic-industrial international cooperation for engineering education. *J. Intell. Manuf.* 24, 433-439.

Borrego, M., Douglas, E. P., and Amelink, C. T. (2009). Quantitative, qualitative, and mixed research methods in engineering education. *Journal of Engineering Education*, 98 (1), 53-66.

Bourne, J., Harris, D. and Mayadas, F. (2005). Online engineering education: learning anywhere, anytime. *Journal of Asynchronous Learning Networks*, 9 (1), 15-41.

Bruce, J. D., Siebert, W. M., Smullin, L. D., and Fano, R. M. (1982). *Lifelong Cooperative Education.* Cambridge, MA: Massachusetts Institute of Technology.

Case, J. M. and Light, G. (2011). Emerging methodologies in engineering education research. *Journal of Engineering Education*, 100 (1), 186-210.

Cervero, R. M., Miller, J. D., and Dimmock, K. H. (1986). The formal & informal learning activities of practicing engineers. *Engineering Education*, 77 (2), 112-114.

Cervero, R. M. (1988). *Effective Continuing Education for Professionals*. San Francisco: Jossey-Bass.

Cervero, R. M. (1992). Professional practice, learning, and continuing education: an integrated perspective. *International Journal of Lifelong Education*, 11 (2), 91-101.

Cervero, R. M. (2001). Continuing professional education in transition. 1981-2000. *International Journal of Lifelong Education*, 20 (1/2), 16-30.

Cole, H. P., Moss, J., Gohs, F. X., Lacefield, W. E., Barfield, B. J., and Blythe, D. K. (1984). *Measuring Learning in Continuing Education for Engineers and Scientists*. Phoenix: Oryx Press.

Collin, K. (2008). Development engineers' work and learning as shared practice. *International Journal of Lifelong Education*, 27 (4), 379-397.

Combs, Wendy L. and Davis, Bettina M. (2010). *Demystifying technical training: partnership, strategy, and execution.* San Francisco: Pfeiffer.

Craig, R. L. and Evers, C. J. (1981). Employers as educators: the "shadow education system. In G. Gold (Ed.), *Business and Higher Education: Toward New Alliances* (pp. 29-46). San Francisco: Jossey-Bass.

Cranch, (1987). Continuing engineering education in the United States: An overview and assessment. *Engineering Education*, 77 (7-8), 657-668.

Crook, T. R., Todd, S. Y., Combs, J. G., Woehr, D. J. and Ketchen, D. J. (2011). Does human capital matter? A meta-analysis of the relationship between human capital and firm performance. *Journal of Applied Psychology*, 96 (3), 443-456.

Cross, K. P. (1981). *Adults as Learners: Increasing Participation and Facilitating Learning*. San Francisco: Jossey-Bass.

Darkenwald, G. G. and Merriam, S. B. (1982), *Adult Education Foundations of Practice*. New York: HarperCollins.

Derose, G. J. (1999). *Outsourcing Training & Education.* Alexandria, VA: American Society for Training & Development.

Dirkx, J. M. (2011). Work-related learning in the United States: past practices, paradigms shifts, and policies of partnership. In M. Malloch, L. Cairns, K. Evans, and B. N. O'Connor (Eds.), *The Sage Handbook of Workplace Learning* (London: Sage).

Dutta, D. (2010). *Lifelong learning imperative in engineering: summary of a workshop.* Washington, DC: The National Academies Press. http://www.nap.edu/catalog/12866.html.

Eurich, N. P. (1985). *Corporate Classrooms: The Learning Business*. Princeton, NJ: The Carnegie Foundation for the Advancement of Teaching.

European Society for Engineering Education (2010). http://www.sefi.be/. Accessed July 9, 2021.

Evetts, J. (1998). Continuing professional development for engineers: UK and European dynamics. *European Journal of Engineering Education*, 23 (4), 443-452.

Fink, F. K. (2001). Integration of engineering education and productive engineering. *8th World Conference on Continuing Engineering Education*, May 12-16, 2001, Toronto.

Forrester, K., Payne, J. and Ward, K. (1995). *Workplace learning: perspectives on education, training and work*. Aldershot, U.K.: Avebury.

Gräfen, H. (1991). Continuing education: an investment for the future. *International Journal of Continuing Engineering Education and Lifelong Learning*, 1 (2), 121-128.

Grossman, R. and Salas, E. (2011). The transfer of training: what really matters. *International Journal of Training and Development*, 15 (2), 103-120.

Hale, J. (2006). *Outsourcing Training and Development: Factors for Success*. San Francisco: Pfeiffer.

Heinke, G. W. (1991). Continuing professional education for engineers in Canada: whose responsibility is it, and how do we get it going?", *International Journal of Continuing Engineering Education and Lifelong Learning*, 1 (4), 319-333.

Hohmann, L. (1980). Professional continuing education. In H. J. Alford (Ed.), *Power and Conflict in Continuing Education: Survival and Prosperity for All?* (pp. 83-94). Belmont, CA: Wadsworth.

Horvat, M. and Kailer, N. (1989). Needs analysis in continuing engineering education - matching individual and organizational needs. *European Journal of Engineering Education*, 14 (1), 97-104.

Houle, C. O. (1980). *Continuing Learning in the Professions*. San Francisco: Jossey-Bass.

Institute for Continuing Professional Development (2010). http://www.cpdinstitute.org/, Accessed 21 October 2011.

Jacobs, R. L., and Park, Y. (2009). A proposed framework of workplace learning: implications for theory development and research in human resource development. *Human Resource Development*, 8 (2), 133-150.

Jeong, J. (2000). Skill formation of engineers in large Korean firms: an analysis from a comparative perspective. *International Journal of Training and Development*, 4 (1), 66-78.

Jeris, L. H. (2010). Continuing professional education. In C. E. Kasworm, A. D. Rose and J. M. Ross-Gordon (Eds), *Handbook of Adult and Continuing Education* (pp. 275-283). Los Angeles: Sage.

Jones, M. (2010). 7.2.6 Continuing engineering education and professional development. In *UNESCO, Engineering: Issues, Challenges and Opportunities for Development* (pp. 329-332). Paris: United Nations Educational, Scientific and Cultural Organization) (http://unesdoc.unesco.org/images/ 0018/001897/ 189753e.pdf).

Kaufmann, K. and Weaver, C. (1995). Appraising relative worth: continuing-education opportunities for engineers. *Journal of Management Engineering*, 11 (4), 34-38.

Kost, R. J. (1980). Competition and innovation in continuing education. In H. J. Alford (Ed.), *Power and Conflict in Continuing Education: Survival and Prosperity for All?* (pp. 38-51). Belmont, CA: Wadsworth.

Kuprenas, J. A., Haraga, R. K., DeChambeau, D. L., and Smith, J. C. (2000). Performance measurement of training in engineering organizations. *Journal of Management Engineering*, 16 (5), 27-33.

Landis, F. (1989). What is the future of continuing engineering education?" *Engineering Education*, 79 (8), 1005-1009.

Lindeman, E. C. (1961). *The Meaning of Adult Education*. Montreal: Harvest House.

Lui-Abel, A. (2011). Identifying and classifying corporate universities in the United States. In M. Malloch, L. Cairns, K. Evans, and B. N. O"Connor (Eds.), *The Sage Handbook of Workplace Learning*. London: Sage.

Malloch, M., Cairns, L, Evans, K, and O'Connor, B. N. (Eds.) (2011). *The Sage Handbook of Workplace Learning*. London: Sage.

Marjoram, T. and Zhong, Y. (2010). 1.1 What engineering is, what engineers do. In *UNESCO, Engineering: Issues, Challenges and Opportunities for Development*. Paris: United Nations Educational, Scientific and Cultural Organization (pp. 24-26). (http://unesdoc.unesco.org/images/0018/001897/189753e.pdf).

Meister, J. C. (2001). The brave new world of corporate education. *Chronicle of Higher Education*, 47 (22), B10-B11.

Morris, A. J. (1978). *The Return on Investment in Continuing Education of Engineers*. Palo Alto, CA: Genesys Systems, Inc.

Mott, V. W. (2000). The development of professional expertise in the workplace. *New Directions for Adult and Continuing Education*, 86, 23-31.

National Academies of Sciences, Engineering, and Medicine (2018). *Adaptability of the US Engineering and Technical Workforce: Proceedings of a Workshop*. Washington, DC: The National Academies Press. https://doi.org/10.17226/25016.

National Academy of Engineering (1988). *Focus on the Future: A National Plan for Career-Long Education for Engineers*. Washington DC: National Academy of Engineering.

National Academy of Engineering (2005). *Educating the Engineer of 2020*. Washington, DC: The National Academies Press.

National Research Council (1985). *Engineering Education and Practice in the United States: Continuing Education of Engineers*. Washington, DC: National Academy Press.

National Science Foundation (1977). *Continuing Education in Science and Engineering*. Washington, DC: National Science Foundation.

Noe, R. A. (2005). *Employee Training and Development* (3rd ed.). New York: McGraw-Hill.

Padfield, C. and Schaufelberger W. (1998). *Lifelong learning in engineering education: a call to action*. Brussels: Societé Européenne pour la Formation des Ingénieurs.

Pérez-Foguet, A. B. Lazzarini, R. Giné, E. Velo, A. Boni, M. Sierra, G. Zolezzi, and R. Trimingham (2018). Promoting sustainable human development in engineering: Assessment of online courses within continuing professional development strategies. *Journal of Cleaner Production*, 172, 4286-4302.

Rampazzo, M. and A. Beghi (2018). Designing and teaching of an effective engineering continuing education course: Modeling and simulation of HVAC systems. *Computer Applications in Engineering Education* 26 (4), 739-748.

Salopek, J. J. (1998). Outsourcing, insourcing, and in-between sourcing. *Training and Development*, 52 (7), 51-56.

Schön, D. A. (1987). *Educating the reflective practitioner: toward a new design for teaching and learning in the professions*. San Francisco: Jossey-Bass.

Sheppard, S. D., Macatangay, K., Colby, A., and Sullivan, W. M. (2009). *Educating Engineers: Designing for the Future of the Field*. San Francisco: Jossey-Bass.

Soeiro, A., E. la Grange, and A. Smith (2016). Continuing Engineering Education and Sustainability: IACEE Contribution with SERINA and Porto Declaration. In F. J. García-Peñalvo (Ed.). Proceedings of the 6th International Conference on Technological Ecosystems for Enhancing Multiculturality (TEEM 2018) (Salamanca, Spain, October 24-26, 2018), ACM: New York, NY, USA. https://doi.org/10.1145/ 3284179.3284266.

Spak, G.T., P. Schmitt, and C. Bandera (2011). M-Outreach for engineering continuing education: A model for university-company collaboration. Proceedings of the 2011 Annual Conference and Exhibition of the American Society for Engineering Education, Vancouver, June 2011.

UNESCO (2010), Engineering: Issues, Challenges and Opportunities for Development (Paris: United Nations Educational, Scientific and Cultural Organization).

Van Horn, J. H. (1977). Industry/professional society interaction. *Engineering Education*, 64 (4), 285-287.

Van Loo, J. B. and Rocco, T. S. (2004). Continuing professional education and human capital theory. Paper presented at the Academy of Human Resource Development International Conference (AHRD) in Austin, TX, March 3-7, 2004, pp. 98-105.

Vanajakumari, M., K. Johnston, F.B. Lawrence, and R. Menon (2015). An effective teaching methodology for continuing education programs. *Journal of Engineering Technology*, 32 (1), 18-29.

Weimar, B. (1992). Assumptions about university-industry relationship in continuing professional education: A re-assessment. *European Journal of Education*, 27 (4), 379-396.

Wlodkowski, R. J. (2008). *Enhancing Adult Motivation to Learn* (3rd ed.). San Francisco: Jossey-Bass.

Chapter 12

Building a Sustainable Administrative Organization for Delivering Professional Programs to Working Adult Learners

Mitchell L. Springer*

Purdue University, West Lafayette, Indiana, USA

Abstract

Building a sustainable administrative online organization is not singular in nature. It is a multi-faceted endeavor which requires a breadth of perspective and discussion. This chapter is essentially a story. It is an evolution of thought with significant breadth and depth to convey the magnitude of such an endeavor.

Keywords: online, administrative organization, restructuring, professional learners

Introduction

Envisioning, building, and growing a sustainable administrative organization for the purposes of delivering varying modalities of academic programs or courses to professional working adult learners is multi-faceted and highly complex. This is especially true, with complexities compounded, when running the administrative organization as a profit-loss entity within a non-profit public university.

From the onset, there are cultural and highly emotional challenges to the centralization of such an organization. Structurally, there are organizational design model implications, which type of organizational design model is best suited as the entity evolves from inception through growth and sustainability as an on-going concern. Attendant to an organizational design model is the definition of a series of processes from which execution is measured against. Each of these many processes has a series of seemingly sequential, yet highly recursive activities, where each activity has a number of attendant outputs. Throughout these many processes there are defined measurement points collectively creating both leading and lagging indicators. On a periodic basis then, measurement of these indicators against a pre-defined baseline provides opportunities for mid-stream adjustment and correction. All of which requires a deep understanding of organizational structuring, process modeling and key performance indicators to provide maximum efficiency and overall lower expense ratios.

* Corresponding Author's Email: mlspring01@Outlook.com.

In higher education, courses and curriculum are purportedly the intellectual property of the academic departments who provide faculty to teach or approve of non-academic department professors. This directly implies there needs to exist some level of financial incentive for the home department of those faculty, as well as direct incentive to the faculty who are involved in teaching. After all expenses are accounted for, then, what model for faculty and faculty home department incentive is most applicable and will provide maximum incentive to participate in professional studies? Incentivizing both the involved faculty member and their academic department is a critical first step to initiating and maintaining a successful program administration capability.

As to the students themselves, correlations exist between the number of webpage "hits" and the subsequent fall enrollments. While this information is important, it reflects a lagging indicator; that is, it does not tell us why there may have been more or less webpage hits. In contrast, a leading indicator will provide information as to why something has come to be.

Drawing from historical contexts of student data, focus has to center on who these cohort learners are and how they compare to previous cohort groups of comparable demographic age. The information assimilated, focuses on generational cohort differences that impact, or have the potential to impact the offering of professional degrees and other fee-based continuing educational opportunities.

Specifically, consideration must account for:

- Who are the new students.
- The implications of ever-increasing college costs.
- The impact of increasing debt loads.
- Moving back home after graduation and its implications.
    - Postponing marriage.
    - Postponing the buying of homes, cars, and other material possessions.
- Educational and income implications of having children later in life.

In the final analysis, how do we make program and cost changes to our many offerings to most readily satisfy the personal and professional needs of the students and their frequently paying companies and organizations, and yet, remain competitive and sustainable?

These topics and many more are the topics of this chapter. From identifying the need for, to building, to growing and to creating in real-time a sustainable administrative organization for the purpose of delivering academic credentialed programs and courses to professional working adult learners.

Cultural Implications of Running the Academy as a Business

"…We don't need to be in the distance education game. The programs are watered down, the faculty are business people/adjuncts, the students aren't like normal students, the whole distance education concept is for limited audiences and unnecessary, and, we perhaps most importantly we are not a business…"

This, and other similar tunes are the words of the fight songs of those who argue the academy is not a business and should not be engaged in non-traditional distance-oriented initiatives or opportunities.

Within our colleges and universities there are significant inefficiencies, replicated resources and no attention to return on investment, all creating woefully wasteful, exaggerated and inflamed costs. Some of the more obvious problems stem from simple problems to solve such as having multiple mail server applications or over three dozen web design tools. These are simply low hanging fruit when it comes to efficiency gains.

Colleges and universities are under increasing pressure to accomplish four things: increase revenues, decrease expenses, improve quality, and strengthen reputation (Dickeson, 2020, pg. 1). This section deals directly with increasing revenue which feeds and promotes changes in program quality and strengthening a college's or university's reputation.

This section will dig deeper into the cultural implications of running the academy as a business, and the implications for historical thresholds for adult learner entry. This section will examine the implications of offering distance learning programs to professional working adult learners within existing cultures. The section will examine the underlying premise of business versus non-business entities through definition and differentiation of businesses, nonprofits, and public institutions of higher education. Additionally, this section will address the question of the "burning platform" or "call to action" demanded in a competitive environment and quite possibly the greatest missing asset of public institutions of higher education.

The focus of this section is strictly limited to public institutions of higher education. Some will ask why not private institutions of higher education? The answer resides in taxpayer contributions. Public institutions are partially funded by the taxpayers. These are the same individuals who suffer through our many recessions, borrow against their retirements and homes to put their children through college, and have now been forced to work past normal retirement age. Inflated college costs, having the greatest impact on our country's wellbeing, are those of our public institutions of higher education and therefore the target of this section. As a public institution, we have a responsibility to be good stewards of taxpayer dollars and student tuition.

This section will close with an integrative perspective, through a heightened awareness, of the key elements that represent the fundamental underlying premise for suggesting public institutions of higher education are a business and should be empowered and encouraged to engage in those initiatives and opportunities that support the advancement of higher education in today's highly competitive educational marketplace.

How Business Works

At their very basic level, how do businesses work? To answer this requires a discussion on competition for funding, why cash is king, why there are mergers and acquisitions and the role of stakeholders.

Why are for-profit businesses in business? Is it to serve the greater good? In some cases, and perhaps to some degree, yes. But, more generally, for-profit businesses are in business to make money.

What keeps businesses in business? The answer is simple: demand for their products and/or services. This demand in turn produces …again, money.

Why do businesses right-size, down-size, create efficiencies of scale, cut back, delayer, smart-size, redeploy, force shape, reduce in force…and lay-off? The answer: because the general populous demands it! One might ask "who demands it?" Any individual representing a stakeholder in the economy of the world is the answer, you, me, and every other individual who purchases goods and services from business and industry. In every instance where the above people actions have taken place, the price of the organizations stock has gone up. Why, might one ask? The answer is because any individual as a shareholder, wants the greatest return on their investments. Money not paid to salaries goes directly to the bottom-line.

In 2011, Google announced plans to hire more than 6200 new employees. On the surface most concerned with our U.S. economy applauded this. Even the President of the United States stated how great this was for the economy and the many people of our great nation. It is unimaginable anyone not being happy about this, especially given the then state of the economy. But there were those not happy about the decision. Those not happy were the shareholders. Why wouldn't the shareholders be happy about the announced hiring as was everyone else who was made aware of it?

The answer was quite simple; the proposed increase in salaries reduced the profits to the shareholders of the company. This seems preposterous on the surface. Who are these greedy shareholders so they might be tarred and feathered? The answer surprisingly again, is any individual shareholder of the organization; you, me and most every other taxpayer.

To fully appreciate the situation, further explanation is depicted in the below steps.

- Individuals work and as a result make money.
- Given there is extra money available for a rainy day, it is placed in the investments with the highest returns; perhaps a savings account, checking account, money market fund at a local lending institution or wherever.
- The place the investor invests their rainy-day funds has to compete with other institutions for their money; this to provide them with the greatest return on their money and therefore gain on their investment.
- The institution the investor invests in, must find investments to invest the investor's rainy-day funds and make a profit, however small, for themselves as the investing institution. To this end, the investor's investment organization looks high and low for companies that produce the greatest returns for their individual investors.
- Companies that compete for investment institutional money have to create efficiencies to differentiate themselves from other companies wanting the same investment money from the same investment companies.
- To create the required efficiencies these many companies yield higher profits by outsourcing, reducing headcount, streamlining operations and the like.
- In doing the above, these many companies yield higher profits than their counterparts and gain investment company funds to allow these organizations to continue their operations.

In the end, the cycle begins with each individual investor wanting the greatest return for hard earned dollars. We the populous, then, are the ones who demand efficiencies and cost frugality. We are the ones who demand right-sizing, down-sizing, creating efficiencies of scale, cutting back, delayering, smart-sizing, redeploying, force shaping, reduction in force…and laying-off. To quote a saying, "we have met the enemy, and he is us (Walt Kelly, Pogo author, 1970)."

An example of things happening which are not intuitively obvious as to why include the following:

On April 1, 2012, CBS News ran a segment. They were talking about Brevard County, Florida, the home of the space shuttle from the Kennedy Space Center. In this segment they discussed how the last shuttle launch was July 2011, after 50 years of liftoffs. On closing, Kennedy Space Center reduced heads by nearly 7,000. This in turn triggered another 7,000-person impact on the surrounding community because of mass business closings. The town was devastated by all of this to fly into space is to buy a seat on a Russian rocket," or "…there is enormous pride in doing for America what no other workers in the world could even dare…"

The question was why? Why did the Kennedy Space Center have to close? Why did so many Americans have to lose their jobs? The answer is not pretty but is the essence of what drives businesses and allows them to stay in business, money! The Kennedy Space Center closed to save $3 billion dollars a year in taxpayer money! They were quoted as saying "…it's not a personal thing; it's a financial thing…"

Why do mergers and acquisitions occur? Mergers and acquisitions are the coming together of two or more companies. But why? The answers are things we have already broached, to:

- Create shareholder value.
- Create efficiencies.
- Target future markets with additional profitability.
- Serve combined markets more effectively.
- Reward stakeholders with higher returns.

What does create efficiencies and being more effective mean? Generally, in mergers and acquisitions it means reductions in plants, equipment, human capital, processes and all the elements of efficiency gains.

One final question is who are the stakeholders we've been referring to? Stakeholders are:

- Shareholders who hold stock, options, and other financial instruments
- Served populations; individuals, companies, states and Government
- Employees; responsibility to local population, city and state
- Competitors; who are occasionally our strategic partners in financial pursuits

Accommodating stakeholders frequently requires making decisions that benefit some at the expense of others. So, in the end, we have to ask in every business decision, "what side do we take?" and "whose interest will we serve?"

Given all of this, for-profit businesses are motivated to a large extent to stay in business, which requires money. The key to gaining money is quite simple, make more in terms of gross revenue, spend less in terms of expenses, or alternatively, return less to rainy day accounts.

Differences in Mission

Some of the many characteristics that identify nonprofit organizations are strongly related to tax implications. Other of those characteristics have to do with how profit is handled, the mission of the organization, the oversight of the board of directors, and the source of funding.

Nonprofit organizations charge for products, services, or memberships. Some of these nonprofits generate significant amounts of profit; this is allowable and ideal as is the scenario for any business. The difference is the profits are not passed on to shareholders; instead, the profits are reinvested back into the nonprofit organization to advance the mission. This is how nonprofits are allowed to grow. All their costs are covered as expenses and the remaining profits (commonly called residuals in nonprofit organizations) are simply reapplied as deemed acceptable for continued growth of the organization.

The mission of nonprofits is typically different from for-profit organizations in business/industry. In nonprofits, their mission is typically to serve the greater public or social good. This is exhibited in organizations that work in health care (hospitals), education (K-12, colleges, and universities), churches, community programs or foundations with one or more purposes (Nahavandi, 2009, 249).

Business and industrial organizations have a board of directors that is made up of paid individuals. These many individuals are typically highly experienced and bring some form of discipline-specific knowledge to the organization through their involvement. Nonprofit organizations, on the other hand, have board of directors who are not paid. And, although they may have significant discipline-specific knowledge they bring to the table, they may also simply have a burning passion to be part of the organization, which, in the end, creates strong advocates for the nonprofit's mission.

Like businesses, nonprofits may charge a fee for their products or services. But generally, nonprofit primary funding sources are through contributions, grants, donations, agency and industry funding and foundations.

There are hundreds of organizations world-wide that meet the definition of being nonprofits. A few of the top 100 include National Public Radio (NPR), United Nation's Children Fund (UNICEF), Smithsonian Institute, World Food Program – USA, American Cancer Society, Save the Children, Feed America and many more.

On leadership challenges of nonprofit organizations, Nahavandi (2009, 252) notes:

> The leadership of nonprofit organizations involves the same principles as other organizations. Their leaders must help individuals and groups set goals and guide them in the achievement of those goals. The public-good mission of nonprofits, along with the voluntary participation of many of their employees, contributors and other stakeholders create a particular burden on leaders of such organizations to lead through a collaborative and trust-based style. In most cases individual donors, except for their tax benefits when applicable, do not get tangible benefits from their donation and the resources they contribute do not always stay in their community…
>
> As much as integrity, trustworthiness, and self-sacrifice are elements for all leadership situations, they are even more so in the nonprofit organizations. Without the profit motive, which legitimately guides business organizations and the rewarding of its leaders (e.g., top leaders being compensated with company shares), the nonprofit organizations are likely to attract leaders with a stronger focus on civic contribution…
>
> One of the major challenges that leaders of nonprofit organizations face is how to recruit, retain and motivate employees, many of whom are volunteers, without having access to

> substantial monetary rewards. Even in the case of paid employees, salaries are often lower than comparable positions in business organizations. The leaders of nonprofits, therefore, require considerable skills in motivating and inspiring their followers. In many cases, followers have joined the nonprofit because they are passionate about its mission; however, passion alone does not always lead to effectiveness. An additional factor is that the structure of many nonprofits is relatively flat with few employees and few layers of management. Effective leadership requires empowerment, effective use of all resources available, often by harnessing the power of teams and participation to creatively solve problems without many resources.

There are numerous subtleties from the above observations. First, relative to individuals who participate in nonprofit organizations, they are not typically paid or in the least not paid very highly.

Second, those individuals who wish to participate in nonprofits, which assumes colleges and universities as well, wish to do so because they are serving the greater good.

Third and perhaps most notably, whether in for-profits or nonprofits there is the opportunity for leadership to react to a burning platform. This means, moving resources where needed, increasing support in areas of weaknesses, eliminating those activities or initiatives that are a monetary drag on the interests of the whole organization and other efficiency gaining opportunities in an effort to serve their stated missions. This is not the case in public institutions of higher education which have shared governance and a slower reaction time. Leadership in public institutions of higher education cannot make the required changes for efficiency when shackled by the vote of those not inclined to participate; those who feel compelled to maintain the status quo. Therefore, leadership in those organizations appears to be hand-tied to simply management versus true leadership and facilitators of change.

Recognizing the Burning Platform

> "… [if we all knew we were going to get into heaven regardless, then what incentive would we necessarily have to live by God's commandments, be kind to others or avoid any other ills]…"
> Fr. Patrick – St. Thomas Aquinas.

Throughout business/industry the impetus for immediate change is frequently called a "burning platform." In fact, the phrase is so common; an image can readily be acquired from a search of Google images. The burning platform is a visual image which instills a sense of urgency, a sense of immediacy.

The burning platform creates a call to action. It shocks us into wanting to do something, recognizing no action would certainly be to our detriment and potential demise. The burning platform rallies our collective efforts toward that which will make us once again safe and secure. We recognize in its image, a catastrophic and cataclysmic fate.

Higher tuition rates, poor social or economic conditions, increasingly greater debt loads of graduating seniors and the like are beginning to have an impact on colleges and universities. Some, however, don't see, or perhaps more appropriately "feel" the immediacy of the situation.

There is danger in the comfort zone. By this, becoming comfortable creates many ills. When too much is provided to individuals or organizations and there is no expectation of accountability it is easy to feel a sense of entitlement. In business and industry, entitlement can develop because of poor management. This happens when managers want to avoid the confrontations that result from holding people responsible for outcomes.

In any work environment, people feel a sense of entitlement when they do not have to earn or work for their rewards or benefits. The root cause of entitlement stems from many things, most notably (Nahavandi, 2009, 18):

- When employees collect paychecks on a regular and expected basis whether they produce a lot of output or a little
- When raises are given for nothing more than time in position
- When there is a sense of guaranteed lifetime employment
- No formal process for measuring performance or one that has no real impact
- An emphasis on traditions
- A compensation system that does not really reflect an individual's contributions
- A formal hierarchy where differences in status dictate overall behavior

When people do not have to earn what they get, when it is already there, they very soon take for granted that which they receive. The tragic human irony is that these individuals are not grateful for what they have. Instead, they want something more than what they already get; this is the premise of entitlement.

In her book "Danger in the Comfort Zone," Bardwick (1991, 19) states:

> [Too much security] ..."I've worked here a long time and have done what you expected. I've earned my security." That is the working presumption of a majority of people who work in our various government bodies, in all levels of our schools, in our large and powerful unions, and in our mature and prosperous corporations. It amounts to total job security. Too much security is what entitlement is all about.
>
> Some groups of people have formal tenure. They set a precedent for others to argue their right to have the same certainty. In recent years those precedents became transformed into court decisions so it can become legally difficult to fire people"

Increasingly, when job security is absolute, there is no reason to have performance reviews. To this end, organizations that provide this level of lifetime employment stray further and further from these performance related measurements, quite simply, because they are instruments to measure performance. "This is the truly insidious danger of entitlement: it settles in everywhere and becomes part of the mortar that holds everything together."

In discussing what entitlement costs, us, Bardwick (1991, 29) explains:

> The dollar cost of entitlement packages to entitled employees is only one reason we cannot afford to tolerate the conditions of entitlement. The real cost to our nation is the loss of creativity; innovation and creative thinking... people are not at their keenest when life is too safe. When people receive without having to achieve, they are protected from failure. There is no punishment for not achieving.... By protecting people from risk we destroy their self-esteem. We rob them of the opportunity to become strong competent people. Facing risk is the only way we gain confidence, because confidence is the result of mastering challenge. Confidence is an internal state. It cannot be given, it can only be earned... thus, too little anxiety is destructive. It deprives people of the experiences that create confidence.

> Instead of strength, courage and confidence, people trapped by entitlement are cautious and avoid risk. When security is very high and people shouldn't be afraid, they are. Entitled people cling to entitlement, fearing they couldn't earn or compete or survive on their own. The long-term result of too much protection is an endless search for protection because we are afraid of losing it."

When people are not held accountable for performance, they do not perform. Instead, they can become complacent; this given no performance criteria is required of them. Productivity is always lowest among those who feel a sense of entitlement. "We must shake loose the psychology of entitlement. We must empower people by giving them responsibilities and we must hold them accountable for their actions. Only in this way can we engender the achievement, growth, and confidence that are necessary for maximum productivity."

Unfortunately, whenever entitlement exists in an organization, it is usually because it has been established as part of the culture. This is why it usually takes a significant shock or call to action to gain the appropriate motivation to act. Given, moving away from entitlement creates a risk, there will always be an element of fear to overcome.

Bardwick goes on to describe the relationship between entitlement, earning and fear. Bardwick depicts productivity as being very low when either anxiety is very low or very high. Relative to earnings, when security is dependent on producing, or output, productivity is at its highest. On entitlement, Bardwick suggests avoiding risk and creating safety are institutionalized in rules and procedures. In this scenario, appearance is more important than achievement. When fear is present, there is no sense of having any control, resulting in people panicking and creating a sense of every person for themselves.

Generally, people resist leaving the comfort of entitlement. When they are pushed out of it, they will try to return. The only real way to energize apathetic organizations is to push them into the psychology of earning. After years of avoiding risk, most people will find this distasteful. Only a call to action or burning platform effect will move people and organizations out of entitlement.

There has been a lot of work done that discusses the relationship between competition, self-control, and innovation. Bronson and Merryman (2013), in their book Top Dog address this very natural phenomenon.

In their book the authors analyzed the saliva samples of those who skydive and those who ballroom dance. In looking first at the skydivers, they discovered skydivers had a huge rush of stress on the first jump but succeeding lower levels of rush stress on the second and third jumps. The authors concluded "…the true "high" of skydiving, and other edgeworks, stems from the way skilled performance brings control to a situation most people would regard as uncontrollable (Bronson and Merryman, 2013, 5). This sense of control stems from the skydiver's capacity to focus attention on the actions necessary for survival. "The feeling of self-determination they get from conquering the risks is the real payoff. It is not pure thrill they seek, but the ability to control the environment within a thrilling context (ibid.)."

When the authors looked at ballroom dancers, they found something unexpected. The ballroom dancers did not experience the fall-off of rush stress the skydivers had. This no matter how many times they competed; the rush stress was always high.

> "…according to what science tells us, dancing at that point in their lives should have required very little cognitive control. All of the muscle memory should have been driven down into the

> cerebellum region of their brains, where it was automated. There should have been no worry over forgetting to vary the inside and outside of their feet to create style and line… but that wasn't the case. The intense stress reaction was no different between the [groups tested]. The cutthroat world of ballroom dancing remained terrifying no matter how long they'd been at it. The contestants did not habituate. But how is it that someone can get used to skydiving but not ballroom dancing? Because the real difference between skydiving and ballroom dancing isn't defined by the physical environment of the activities. It is not even about the actual jeopardy to life and limb. The real difference between skydiving and ballroom dancing was the psychological environment. The expert dancers were in competition and the skydivers were not. To be more precise, it isn't the dancing that is stress inducing, it is being judged. It was winning and losing *(ibid., 8)*."

This argument suggests competition creates a sense of competitiveness. Competitiveness is what causes people to come out swinging, to do the hard things, to put their noses to the grindstone and produce when required.

Competitiveness is that which turns external energy into internal energy; it is the visualization of the purpose; the thing which makes most individuals want to be part of the solution and put forth maximum effort. Even if there is no burning platform, competitiveness and the internalization of the stated purpose provide unparalleled impetus to do more and to get ahead of our competitors. Competition spurs motivation (Ibid., 18). It is even true competition increases creative motivation. Competition "…doesn't kill creativity; it facilitates creative output by supplying motivational drive. Competition teaches people to be comfortable with conflict and opposition, which is a necessary building block for developing the creative psyche" (Ibid., 19). Success in competition requires taking risks, which are normally held back by fear.

The authors take this relationship between competition, self-control, and innovation to the next level by postulating how competition leads to self-improvement which in turn leads directly to greater innovation and performance. They state "…the real benefit of competition is not winning – it is improved performance… competitors discover an extra gear… competition facilitates improvement [p. 27]." The Matthew Effect and the Mark Effect have been coined to address an interesting dichotomy of assisting those who are competitively oriented from those who do not do as well under such stress.

> "…The term Matthew Effect was coined by sociologist Robert Merton in 1968; it refers to the dynamic that the early leaders in a competition tend to get showered with resources that make them even better, increasing the gap on weaker competitors over time. For instance, the best students get sent to the best schools, where they have the best teachers; in the same way the best players get sent to the best teams, where they get trained by the best coaches. The term was inspired be the Gospel According to Matthew: "For to everyone who has, more will be given and he will grow rich; but from the one who has not, even what he has will be taken away."
>
> Whenever we try to counteract this, whenever we try to distribute resources to prop up the weaker competitors – we're applying the Mark Effect. The term was coined in 2009 by the University of Chicago's Matthew Bothner, drawing from the Gospel According to Mark: "But the first will be last, and the last will be first." In our society, there's almost an unlimited number of ways we try to assist, or intervene, when competitors are unequal. We take it as a given that competition is predicated on a level playing field – that the rules apply to all, and if some redistribution isn't done now and then, the rich will just get richer, to the point there is no competition left [too big a gap between those with and those without, thus eliminating the psychological effect of competition]" *(Ibid., 40)*.

Daniel Paquette (, 119) changed the perceptions of social psychology by proposing against the micro-monitoring of helicopter parents. He argued attachment theory has overvalued a parent's role in providing comfort when children feel insecure and undervalued the parent's role when fostering exploratory behavior. He noted that studies of animals deprived of rough and tumble play show they grow up unable to be successfully aggressive: they perceive threats when there are none, and they do not perceive any when they should. Paquette explains this kind of play, where the parent can escalate or reduce this kind of aggressiveness, teaches children how to express their aggressiveness but in a modulated and controlled way. This is not the scenario with those who possess security in the comfort zone. Modulation and control are seldom exercised in favor of ignoring the behavior. Ignoring the behavior, therefore, simply acts not to instruct or counsel, but to reinforce bad behavior. The important thing about roughhousing, explains Paquette, is that the parent maintains control, animating children but de-escalating when kids are on the brink of anger or frustration.

On April 11, 2014, ASEE First Bell, an on-line electronic newsletter for members of the American Society for Engineering Education, identified one of their daily top articles as:

> Community Colleges Begin Offering Bachelor's Degrees Across US.
>
> The Hechinger Report (4/10, Marcus) reports that community colleges in 21 states are now able to issue bachelors degrees and have started offering degrees in high demand subjects. In Florida, the number of students entering community colleges has grown to 30,000, many attracted by the tuition, which is significantly cheaper than four-year universities. The trend has been opposed by four-year universities, who have lobbied against community college degree expansion in Colorado, Michigan, and California. Florida Sen. Joe Negron (R-28 District) proposed halting the expansion of community college bachelor programs leading Kenneth Walker, founder and former chairman of the Community College Baccalaureate Association, to say the debate "comes down to the money. And from my perspective, the focus ought to be on meeting the needs of the students, meeting the needs of business and industry." The article notes that California will begin allowing its students to receive bachelor degrees at community colleges to address a predicted shortfall in 2025.

This article led to numerous others of similar content. The topic throughout the many articles described how 21 of the 50 states in the United States had provided their respective community colleges the right to offer bachelor's degrees. Community colleges by their nature usually stop at an associate degree. Participants in these associate degrees then have the opportunity to transfer to a four-year college or university should they choose to do so.

What is most interesting about these series of articles and the changes to state regulations to allow this to happen is the similarity this has to work done by Clayton Christensen. Christensen et al. (2011, 2017) have written about disruptive initiatives that will change the face of higher education. Until this series of articles had heightened awareness, most had always thought, and still do, that on-line education of increasingly greater quality will be the disruptive technology to change the face of higher education. This new movement, however, might relegate technology-oriented degrees to the lesser-priced community colleges, while the higher-priced four-year degree colleges focus their efforts on research-oriented education. Research versus applied is a big differentiator when examining these two models. In the final analysis, this movement may very well be what separates these two foci: research versus applied and their respective college/university offerings.

Providing a More Efficient and Effective Organization

Depending on the discipline and the efficiency focus of the product life cycle, there are hundreds of opportunities to reduce redundancy and cost; therefore, providing a more effective and cost-efficient output for the betterment of the organization, shareholders and its customers. Sometimes these many efficiencies come in the form of centralizing policies, procedures, methodologies and practices, other times in the form of common software and hardware, and still others in the form of inventory control and production efficiency. No matter where you look in the product or service life cycle, there are ways to increase efficiency and reduce overall costs; this has been proven in business and industry many times over hundreds of years.

The bottom line is even if the mission of higher education is perceived as being different from business/industry, that doesn't mean higher education can't act like a business. This, perhaps best summarized by Selingo (2013, 4):

> "...Despite the tear-filled nostalgia that the image of college evokes – tree-shaded quads, ivy-covered neo-Gothic buildings, and fall football weekends – the truth is that in the last two decades higher education in the United States has evolved into a big business...there are some *5,300 colleges* and universities in the US, everything from beauty schools to Harvard. They bring in *$490 billion in revenue* each year. They employ more than *3.5 million people.* They hold more than *$990 billion in assets,* including cash, investments and campuses that are essentially mini-cities. And they spend *$440 billion on goods services and people each year."*

Organizational Design Models – An Initial Perspective

The Center for Professional Studies in Technology and Applied Research (ProSTAR) was approved by Purdue University under the College of Technology as an academic Center in February 2009. At that time, the underlying foundation for ProSTAR's professional education activities was a Master of Science degree with a primary focus in technology leadership and innovation skills including tools for process improvement and quality management.

This section addresses the many complex and potentially controversial aspects of designing and implementing an infrastructure for an academic center to centralize professional studies within a major educational/research institution. Topics to be addressed are the creation of the underlying infrastructure, the alignment and implementation of a time-phased 18-month pro forma rolling financial budget, issues associated with course availability and ownership, trials and tribulations of master's and PhD level program offerings, and faculty availability.

Initial Infrastructure

On June 11, 1998, the College of Technology requested through internal University administration the approval of a non-traditional, fee-based weekend alternative to a traditional campus tuition-based Master of Science with a major in Technology degree. Below is an excerpt from the originally submitted internal request to administration.

> "Technology is transforming all aspects of our life, both at home and at work. Therefore, it is important that we consider the future of those individuals completing undergraduate degree programs in technology and engineering technology. The number of students completing an undergraduate degree in technology-related programs continues to increase. The demand for graduate technology education opportunities is a growing population that will continue to increase. Also, as this demand continues to grow, the need for greater flexibility in delivery will also expand. Competition for this potential customer base will encourage higher education administrators and faculty to evaluate new and innovative delivery systems to serve the educational needs of these customers. A sincere commitment from all parties involved will make this initiative work. This is not a new degree program, just a different delivery approach to an existing one. Technology and engineering technology programs continue to change and evolve in striving to meet society's technological expectations and needs. It is imperative that graduate education be considered as an important element. A survey study conducted at Purdue University reported that 92% of the alumni and faculty indicated that graduate education in technology is important for the professional development of individuals working in industry and that there exist a perceived demand for graduate education in technology and engineering technology"

Subsequently, on October 13, 2000, the Indiana Commission on Higher Education (ICHE) approved the University request for delivering a full distance-based alternative to the traditional program. The entire process from conceptualization to full final approval took two years and four months.

Within the context of this newly approved master's Program, prior to a named center for professional studies, the Graduate Studies Office took a lead role in offering the first Weekend Master's Program (WMP) in Technology on the campus of Purdue University in West Lafayette, Indiana. In early 2004, budget responsibility was assigned to the department of Industrial Technology, as they provided the majority of faculty in the program.

In the fall of 1998, on verbal approval from ICHE, the University offered non-traditional MS Technology program was made available for enrollment, some 129 years after the founding of the University and 50 years after the formation of an independent School/College of Technology within the University organizational model.

The original offering followed a weekend format, meeting from Friday through Sunday. The program met three times a semester, twice in the summer semester, for a total of five semesters (Fall, Spring, Summer, Fall and Spring). The cohort format graduated all participants in the May graduation ceremony after 22 months. The price of the program was $325 per credit hour, $975 for a 3-credit hour course for a total of $10, 725 for a 33-credit hour program. Three of these credit hours were for three 1-credit hour, pass/fail directed project courses.

Since that initial program offering, there have been 17 unique offerings of this basic program, some with minor variation for company specific interests in areas of concentration. Not all cohort-based programs were open for public enrollment. Some programs were created specifically for companies who wished to fund an entire cohort themselves and subsequently make basic underlying modifications for concentrations more applicable to their company specifics.

Differences between Tuition-Based and Fee-Based Programs

The predominate differences between tuition-based, residential programs and fee-based alternative format programs resides in three categories: namely, financial, geographical and ICHE approval.

From a financial perspective, under the University umbrella practices,

- Fee-based programs do not provide assistantships or fee remittance for participants
- Fee-based programs do, however, qualify for financial aid, e.g., Stafford loans
- Organizations providing fee-based programs are self-supporting and therefore must be entrepreneurial; i.e., they receive no University-based funding
- Fee-based pricing is bounded by University guidelines; this means the fee charged must be at least that of traditional tuition-based programs, and, out-of-state rates for fee-based programs must be at least that of traditional tuition-based programs. This is actually not an issue, given the infrastructure to support an organization is entirely self-funded, meaning, the infrastructure receives no State general funds or traditional student tuition assistance
- Traditional tuition-based students provide no funding for fee-based programs should they participate in a fee-based program or course; this tends to reduce incentive on the part of fee-based program host organizations to mix student types

Geographically speaking, fee-based programs are considered non-traditional in that they are not necessarily on-campus programs; they may be on-campus utilizing an alternate format; weekend, etc., or may be off-campus in approved locations. The Graduate School ensures fee-based programs do not compete with traditional on-campus programs. This latter point is seldom an issue as most students wishing to participate in on-campus programs are residential and anticipate a traditional experience. Alternatively, most student s who participate in non-traditional weekend programs are already in the professional ranks and not able to participate in a traditional program format. In fact, the greatest single stated advantage to a weekend format program is that it is a weekend format which allows the working professional to actively participate in normal family and job-related activities, and yet, takes those few weekends required to participate in class; recognizing there exists required non-class activities typically handled through other distance learning employed methodologies.

From a program approval perspective, the approval process is very similar to the traditional program approval process in that both internal university as well as ICHE approval is required. This point may be restated to say the desire to offer a non-traditional program requires yet another ICHE full committee approval comparable to that required for any other traditional new program offering. A department, school or college cannot simply repurpose an existing traditionally provided program in an alternate non-traditional format, or, alternatively, cannot holistically change an existing curriculum without full ICHE Committee approval.

Budget Control

After February 9, 2009, on approval of the College of Technology Academic Center for Professional Studies in Technology and Applied Research (ProSTAR), an effort was afoot to create an underlying infrastructure which would promote the current and future growth of professional studies. The intent was to organize facilities, human capital, processes, practices, and methodologies in support of this anticipated growth. Additionally, a new and more advanced budgeting capability was required.

The solution space involved attacking the stated objectives from two primary fronts: budget control and workload organization. ProSTAR methodically followed a program and project management methodology for achieving the objectives.

Recognizing the need for strong budget control, ProSTAR senior leadership immediately formed a tightly connected relationship with the College of Technology Business Office. Tight coupling with the Business Office was critical, recognizing the many changes to come and the need for their support with college Dean approvals, coupled with the ever-present recognition of being self-funded. Their first effort was to create an 18-month rolling pro forma financial budget. It began simplistically identifying basic revenue, expenses, and resultant residual (profit). Each month, then, as actual financial data evolved, PROSTR senior leadership updated the budget in real-time, maintaining an 18-month rolling window of known revenue and expenses.

The immediate value of a credible budget, being maintained and monitored by the Budget Office is unparalleled in the eyes of senior administrative leadership. The pro forma allowed expenses to be incurred based on revenue predictions of future opportunities. This is tremendously important given the chicken and the egg scenario, where you need resources, both human and equipment, before you can bring in new business, and can't hire new resources without already having the business. A pro forma of future residuals (profits) of expenses against revenue was critical for future growth and allowed for current incurrence of expenses against anticipated and planned revenue.

One of the most important decisions during the formation of the financial model was how residuals (revenue-expenses) would be distributed and to which departments. This is a very sensitive area to be considered. In the end, courses and curriculum are owned by the academic departments who provide faculty to teach. This directly implies there needs to exist some level of financial incentive for the home department of those faculty, as well as direct incentive to the faculty who are involved in teaching. Additionally, there are other costs, as part of overall expenses, which must be paid to organizations such as central university administration, central administrative bodies that support fee-based programs and many more. After all expenses are accounted for, then, what model for faculty and faculty home department incentive is most applicable and will provide maximum incentive to participate in professional studies? This becomes the area of most interest to all parties and will support or alternatively potentially kill forays into centralized professional studies.

There are many models for compensating faculty and incentivizing departmental participation. In the beginning ProSTAR senior leadership selected a very simply model where faculty were compensated a variable amount based on an algorithmic determination of weighted criteria. This later evolved into a fixed incentive, regardless of faculty specific criteria. Department incentive, however, becomes a much bigger issue and one ProSTAR senior leadership learned to not take lightly. Insufficient incentive, as mentioned above, undermines the noblest attempt to build a centralized organization. The distribution must appear fair, but there has to be recognition from all parties that "taxes" to outside interests must be paid; in other words, a portion of all residuals must go to the governing hierarchical organization; to be read "College." This scenario is no different than paying taxes to the U.S. Government. The "taxes" are applied to support a greater infrastructure and are the only source of funds for non-revenue generating organizations. In the end, a centralized functional model for professional studies is more efficient and effective than a more distributed model. This holds true given the required redundancy in decentralization and inherent increased costs.

Roles and Responsibilities

In defining the roles and responsibilities, the basic work definition questions to be addressed following a program management process are (Springer, 2010):

- What is the target organization being asked to do?
- What work is there to be done?
- Who will be performing the work?

While there are numerous other related questions to further defining an executable program plan, the initial definition of the work is most critical to subsequent activities.

What is the professional studies organization being asked to do?

Successfully defining the work to be performed requires an analysis of the request for work to be performed, most generally referred to as a request for proposal, or in the case of a target professional studies organization, the mission statement. The mission statement for the Purdue University Center for Professional Studies in Technology and Applied Research (ProSTAR) reads as follows (Kerzner, 2006):

The ProSTAR mission is to provide education, and engagement in applied research, for business and industry.

- To provide fully accredited degrees, certificates, courses, and workshops to professionals in industry at the local, state, regional, national, and international levels
- To actively pursue the scholarship of Professional/Technology Education
- To actively pursue industry-based engagement in applied research through the College of Technology.
- To deliver programs on campus, by distance, on site, or a combination utilizing applicable industry-specific education mediums.

This mission, then, provides the framework for identifying the stated and derived requirements for the work to be performed.

What Work Is There to Be Done?

Work to be performed may be defined as "stated" and "derived" requirements (Beatty, 1996; Simonson et al., 2008). A stated requirement is one stated directly and explicitly in the provided mission statement. In this scenario, the stated requirements, although many, may be defined generally as:

- Provide accredited degrees
- Provide certificates
- Provide workshops
- Service populations at the local, state, national and international level
- Provide industry engagement through applied research
- Deliver programs by distance, on-site or a combination of education mediums

While the stated requirements provide the explicit desires as stated in the mission statement, the derived requirements are those things that must be done to satisfy the stated requirements. Derived requirements are those requirements not explicitly stated but deemed necessary to fully satisfy the stated requirements. The derived requirements could be nearly infinite. It is only necessary to identify those high-level derived requirements that pose significant cost, schedule, or performance implications. In this scenario, derived requirements may be defined as:

- Accredited degrees are defined as Bachelor or Master's degrees
- Accredited degrees are credit-based versus non-credit certificate courses
- Certificates may be defined as either credit or non-credit certificates
- Workshops may be defined as non-credit courses
- Business office support is required
- Marketing support is required
- Applied research is required

Again, while there may be literally hundreds or even thousands of derived requirements, only those at the highest level need to be defined; those that are cost, schedule, or performance drivers.

The management of the process used to evolve the requirements of the effort is referred to as requirements management. As a process, requirements management has five basic steps:

- Identification
- Analysis
- Allocation
- Verification
- Traceability

Identification is the process of collecting both stated and derived requirements as reflected above. Analysis is the step of the requirements management process that separates similar requirements into chunks or groupings of higher-level requirements. Allocation is the process step that allocates the requirements to an individual or organization to perform the work and subsequently satisfy the identified requirements, either stated or derived. Verification is the process step of determining how well the identified requirements were satisfied. There are generally four basic types of verification: analysis, inspection, demonstration, or test. Each of these types of verification has cost and schedule implications. Traceability is the process step of tracing the stated or derived requirement through the evolution of the target instantiation and therefore ensuring it has been addressed and subsequently satisfied or not.

Four-Phase Natural Evolution of an Administrative Organization Serving an Online Population

February 2019 marked 10 years since the formal creation of the Purdue University, College of Technology's Center for Professional Studies (ProSTAR). Ten years of positive growth in the number of professional working adult learners served. What a great year it was to be celebrating this milestone; aligned to Purdue University's 150th year sesquicentennial celebration!

Given the significance of this milestone, this section provides a rich historical account of the evolution of the College's efforts to serve the many professional working adult learners of the great State of Indiana and Nation. This section addresses the growth of enrollments and gross revenue as each of the four phases evolved and matured. It will further examine the positives and negatives of each evolutionary milestone; specifically addressing the challenges of each phase, operating as a profit/loss center within a public non-profit research university.

Phase I, in the fall of 1998, the College's Department of Industrial Technology took a lead role in implementing the first weekend master's program (WMP) on the campus of Purdue University in West Lafayette, Indiana. The original offering was cohort-based, and it employed a weekend format, meeting from Friday through Sunday. The cohort met three times a semester, twice in the summer semester, for a total of five semesters (Fall, Spring, Summer, Fall and Spring). Because of its non-traditional approach, the state's authorization included the establishment of a different fee structure than normal on-campus classes which resulted in a program cost that was higher than traditional on-campus equivalent programs. Hence, the beginning of tuition-based versus fee-based program offerings.

Phase I, the initial on-campus distance-hybrid offering in the fall of 1998 spawned a comparable off-site instantiation of this sole modality in 2005.

Phase II, the 2005 instantiation was delivered in an off-site format at the location of a target corporate partner. This industry- and corporate-specific outreach instantiation provided for the first time, significantly increased outreach, and enrollments outside of the main campus and the prior and on-going, on-campus, distance-hybrid baseline programs.

Phase III, began in 2010 when the online administrative organization, in collaboration with the academic departments, entered the distance education market with three semi-asynchronous distance programs. Semi-asynchronous distance offerings created a significantly greater outreach, one not bounded by geography.

In 2021, the online administrative organization entered Phase IV, the offering of 100%, globally available asynchronous delivered programs with highly evolved marketing and subsequent outreach. These new scaled program offerings place the unit at the precipice of great change once again. The targeted emphasis on asynchronous global outreach was expected to realize challenges, as well as opportunities not previously experienced.

Phase I

On June 11, 1998, the Purdue University College of Technology (COT) initiated the process for university, and subsequently, Indiana Commission for Higher Education, approval of a non-traditional delivery medium, fee-based weekend alternative to Purdue traditional on-campus tuition-based Master of Science with a major in Technology degree (Springer et al., 2011b).

On October 13, 2000, the Indiana Commission on Higher Education (ICHE) approved the University request for delivering a hybrid distance-based alternative to the traditional on-campus tuition-based classroom-only programs. The entire process from conceptualization to full final ICHE approval took two years and four months. Noteworthy, from the proposal excerpt, the concept of evaluation [quality of the program] was integral to the program proposal from the onset.

In the fall of 1998, the COT's Department of Industrial Technology took a lead role in implementing, pursuant to ICHE authorization, the first weekend master's program (WMP) in Technology on the campus of Purdue University in West Lafayette, Indiana. The original offering was cohort-based and employed a weekend format, meeting from Friday through Sunday. The cohort met three times a semester, twice in the summer semester, for a total of five semesters (Fall, Spring, Summer, Fall and Spring). After 22 months, all members of the initial cohort format graduated in the May 2000 graduation ceremony. Because of its non-traditional approach, the state's authorization included the establishment of a different fee structure than normal on-campus classes which resulted in a program cost that was higher than traditional on-campus equivalent programs.

The Center for Professional Studies in Technology and Applied Research (ProSTAR) was approved by Purdue University under the College of Technology as an academic Center in February 2009. At that time, the underlying foundation for ProSTAR's professional education activities was a Master of Science degree with a primary focus in technology leadership and innovation skills including tools for process improvement and quality management.

In addition, the program incorporated other innovations beyond its delivery system, schedule, and fee structure. To be consistent with its goal of developing practical skills and knowledge immediately, or at least quickly, applicable to business and industry, its plan of study (Springer et al., 2011a) incorporated a base of essential core studies, flexible and easily tailored courses to ensure relevance to emerging technologies, and a guided, industry focused applied research and development project simply called the Directed Project (DP). The latter DP was deliberately designed to require work commensurate to what is typically expected of a master's degree thesis (Springer et al., 2011b).

Phase II

The initial on-campus distance-hybrid offering in the fall of 1998 spawned a comparable off-site instantiation of this sole program in 2005. The 2005 instantiation was delivered in an on-site format at the location of a target corporate partner. This industry- and corporate-specific instantiation provided for the first time significantly increased enrollments outside of the main campus and the prior and on-going on-campus distance-hybrid baseline programs.

The geographical limits of on-campus programs, even taking into consideration the distance-hybrid aspects of the programs, created a self-imposed constraint. Moving the programs to a customer's location, geographically distant from the main on-campus programs, spurred enrollment growth, but it too created enrollment limitations based on corporate sponsorship and geographical specificity.

Phase III

In 2010, ProSTAR, in collaboration with the academic departments, entered the distance education market with three 100% semi-asynchronous online distance programs. Distance offerings created a significantly greater market, one not bounded by geography. It was anticipated that most future growth providing the greatest opportunity for sustainment of an on-going administrative organization would materialize through distance delivery models.

The 2010 offering of online distance programs came in two noticeably different delivery methodologies: namely, synchronous, and semi-asynchronous.

Synchronous delivery of a given distance program implies the recipient of the instruction is receiving the instruction in real-time as the instruction is being provided. While this methodology supported the distance element thereby not requiring the student to attend class on the university campus or, in a previously discussed format, at an employer's location, it still limited student participation by requiring a set time by which the instruction was to be given and therefore a set time by which the student must be available and prepared to receive said instruction. This concept created yet another limitation to full enrollment possibility or potential for maximum enrollment opportunities.

Distance programs offered through the latter delivery modality of semi-asynchronous delivery, fixed, or removed, the barriers other predecessor synchronized program delivery models created. This delivery modality allowed the students to view the recorded modules, yet "sync" on when individual course assignments were due.

While the on-campus distance hybrid, on-site distance-hybrid, and 100% distance semi-asynchronous delivery models still exist in a multitude of program offerings, the 100% distance asynchronous delivery model creates the greatest promise for maximum enrollments; only limited by the desire to obtain the offered degree program.

Figure 1 depicts the enrollment growth from varying delivery modalities.

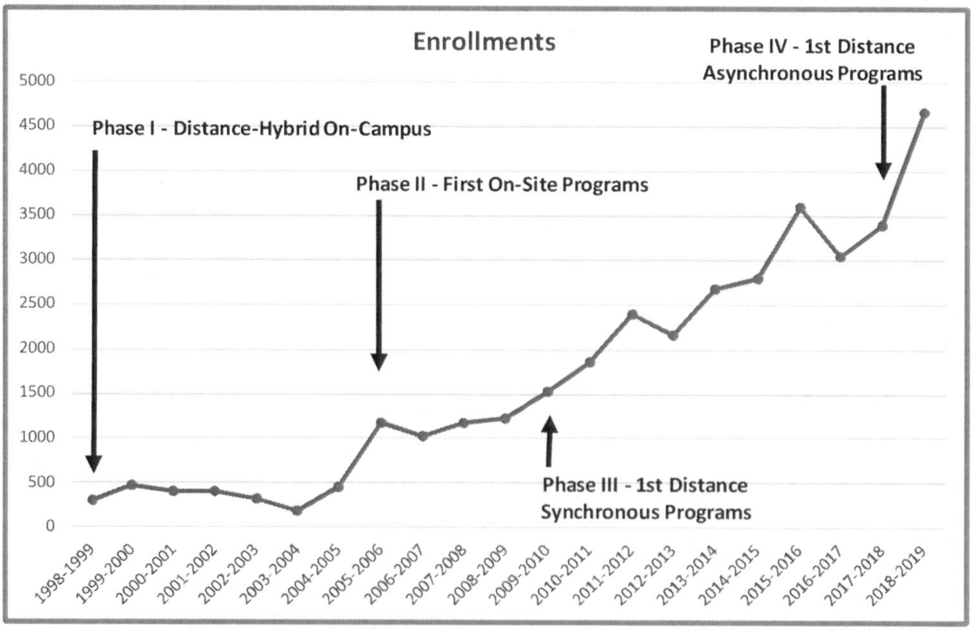

Figure 1. Delivery Modality Induced Enrollment Growth.

Table 1 depicts the delivery modes employed and their respective limitations.

Table 1. Enrollment limitations by delivery modality

Delivery Mode	Limitations		
	Geographical	Employment	Real-Time
1998 Distance-Hybrid; on campus			
2005 Distance-Hybrid; on-site, off-campus			
2010 Distance Synchronous Programs			
2018 Distance Asynchronous Programs			

Phase IV

To collectively wrap university arms around online education, the collection of previously independent and interdependent college-level administrative organizations administering online academic programs were centralized under a single organizational entity. This new entity was organized as a matrix organizational design model. Whichever model is employed, it is critically important an overall strategy exists aligned to the model. Strategies include such things as information system infrastructures, physical plant and equipment, financial systems, and certainly human resources.

As an organization matures through increasingly greater evolutions of growth, it typically evolves from a traditional model to potentially a product structure, and then into a matrix structure. The matrix structure is organized as functions along the vertical axis and programs, projects, colleges, or other organizational entities along the horizontal axis, hence the matrix concept. The idea is money is made along the horizontal axis units, with resources supplied along the vertical functions. In short, functions provide resources to meet the needs of the fiscally responsible horizontal units.

From this perspective, program/project managers, business unit leaders or other horizontally organized profit/loss entities perform all the typical management functions of planning, organizing, staffing, directing, and controlling. However, being assigned the actual staff is the responsibility of the functional organizations. In other words, the horizontal profit/loss unit identifies the functions required to be successful but makes these requests for human resources through the functions. Whether the function assigns a junior resource, experienced resource or senior resource is the function's call; but, if the tasks are not performed adequately, there will be a most expected conversation between the functional lead and the profit/loss unit lead on the human resource assigned to perform the work.

Table 2 below depicts an online organization superimposed over a matrix organizational design model.

In this online organization instantiation of a matrix organizational design model, the college leads (responsible for horizontal profit/loss centers) were made part of the online organization; meaning, they were administratively hardline to the head of the online organization.

Table 2. Matrix structure with online instantiation functional depiction overlay

Online Organization							
Colleges	**Function #1**	**Function #2**	**Funtion #3**	**Function #4**	**Function #5**	**Function #6**	**Function ...#N**
College #1	Functional Resource Assignment	Functional Resource Assignment	Functional Resource Assignment	Functional Resource Assignment	Functional Resource Assignment	Functional Resource Assignment	Functional Resource Assignment
College #2	Functional Resource Assignment	Functional Resource Assignment	Functional Resource Assignment	Functional Resource Assignment	Functional Resource Assignment	Functional Resource Assignment	Functional Resource Assignment
College #3	Functional Resource Assignment	Functional Resource Assignment	Functional Resource Assignment	Functional Resource Assignment	Functional Resource Assignment	Functional Resource Assignment	Functional Resource Assignment
College #4	Functional Resource Assignment	Functional Resource Assignment	Functional Resource Assignment	Functional Resource Assignment	Functional Resource Assignment	Functional Resource Assignment	Functional Resource Assignment
College #5	Functional Resource Assignment	Functional Resource Assignment	Functional Resource Assignment	Functional Resource Assignment	Functional Resource Assignment	Functional Resource Assignment	Functional Resource Assignment
College #N	Functional Resource Assignment	Functional Resource Assignment	Functional Resource Assignment	Functional Resource Assignment	Functional Resource Assignment	Functional Resource Assignment	Functional Resource Assignment

Generally, the profit/loss organizations have a hardline to the profit/loss organization leads, while these same organizations are dashed-lined to the online organization itself. In this manner, the profit/loss organization feels some level of assurance their lead has loyalty and commitment to what might be best for the profit/loss organization, and not be intentionally or unintentionally biased by the interests of the overarching online organization. The success of the profit/loss organization has an impact on everyone in the profit/loss organizational unit, in this instantiation the colleges. To this end, the colleges should want to maintain an administrative hardline to their respective leads. Failure to do so, places the colleges themselves in a subservient role to the direction of the online organization lead, which may not be in the college's best interest for long-term growth or sustainment, or, may run counter to branding integrity of the individual colleges.

Being a part of a matrix also requires each function have identified a well-defined process, depicting the many activities of the process as well as the attendant outputs of each activity. These many activities, then, are organized into a hierarchical breakdown of work referred to as the Work Breakdown Structure (WBS).

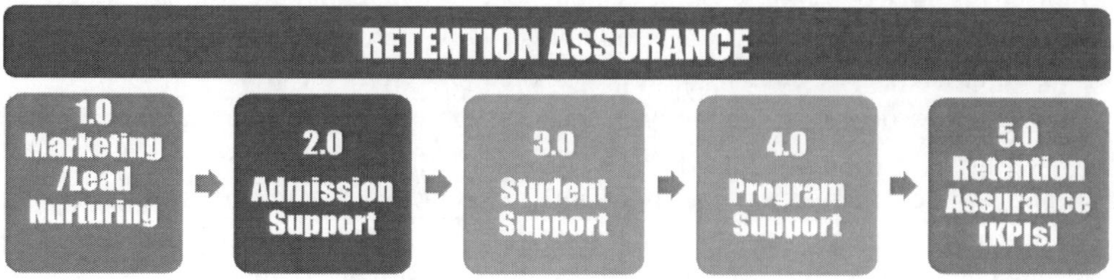

Figure 2. Functional Level 1 Process Overview.

Figure 3. Drill Down of WBS Element 1.0.

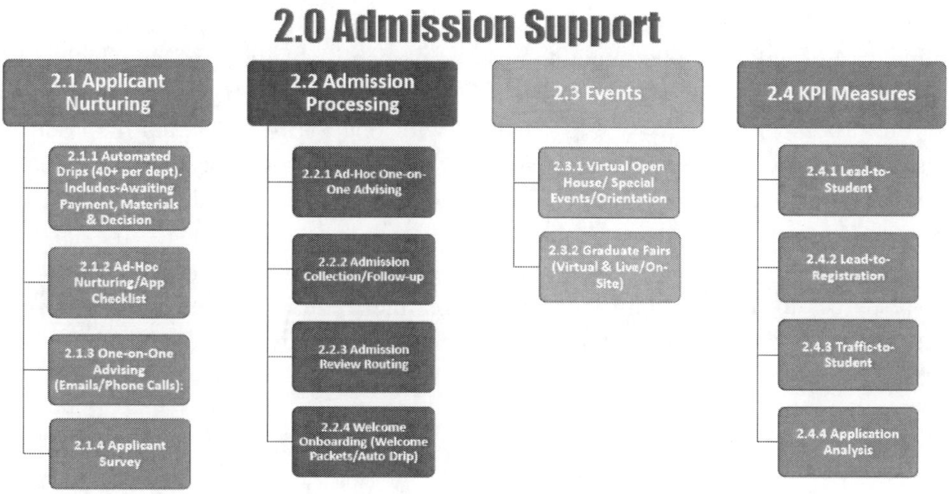

Figure 4. Drill Down of WBS Element 2.0.

Once the WBS has been created, then a matrix can be created depicting the cumulatively defined work on the vertical axis and individuals performing the work within the function on the horizontal axis. Each function member, then, can reflect their fractional full-time equivalent (FTE) in the corresponding tiered elements (activities) of the original WBS, now depicted in the responsibility assignment matrix (RAM).

From these two entities, the WBS and RAM, human resource shortages and over staffing can be seen. Anything short of this process frequently creates what is commonly referred to as "knee-jerking" or "reactionary management"; where work, while well intentioned, is seemingly reassigned without consideration of fractional assignments or the above mentioned job analysis.

A partial example of one of these functional processes may be seen in the following drill down WBS figures.

What Is a Matrix Organizational Design Model?

Creating a matrix organizational design model, at the highest level, is a three-step process. While not difficult to understand in theory, it can be quite difficult in practice. The reasons for this difficulty are the multiple opportunities for failure. Meaning, if the sequence of methodology is not followed, or any of the formal steps minimalized in favor of expediency, then significant rework or possible failure will occur. Failure to create a matrix organization when one is needed comes at the expense of operational efficiency, resource effectiveness, and engaged people. As academic institutions increasingly face complex operational challenges, the benefits of a matrix organizational model will become more and more evident.

The first step in creating an efficient and effective matrix model is to recognize the many disciplines of the organization. In business/industry, these disciplines include engineering, human resources, operations, finance, legal, and as many as 17 others.

Once the disciplines have been defined, the second step is to recognize each discipline as a separate body of knowledge. If the discipline does not have a body of knowledge, then it may be the discipline is not a discipline, but in fact, a subset of subsidiary activities of a parent discipline. If a discipline is determined to possess an identifiable body of knowledge, then it is critical to express the discipline as a process, with logical, sequential, or recurring activities. Within the process, each activity is associated with one or more outputs or products. The process, then, can be defined in terms of a hierarchy of processes, procedures, methodologies, and practices. Each discipline process should have interfaces to other disciplines defined with pro-active and re-active key process indicators (KPIs).

Once disciplines have been identified and defined, the third and final step is to allocate human resources to the vertically identified disciplines/functions. To avoid later significant restructuring rework and cultural upheaval, this step must occur as the final step.

Context

Understanding the basic building blocks of matrix organizational model design and process management is the science and increasingly the art of creating an efficient and effective matrix organizational model. As important, however to the blocking and tackling of understanding the many phases, activities, and processes, are the cultural activities related to people. Even the best at blocking and tackling, will fail if people processes are not fully accounted for during the change management process.

There are many opportunities for introducing inefficiencies and unnecessary cost drivers. Common mistakes are centered not only on an improper matrix model implementation, but the manifestation of power and ego in the process. Below highlight mistakes commonly made within higher education, in what routinely can be a highly emotional experience.

- Failing to recognize the horizontal businesses/colleges are in fact the profit/loss centers, and the vertical functions/disciplines are support organizations.
- Not fully understanding the fundamentals of change management in organizational matrix model development.
- Not fully understanding the science of process management.
- Not fully understanding the art of technical performance measurement.
- Ignoring the people issues of change management.
- Lacking the imagination to identify and plan for the risks involved.
- Overcoming leadership emotional immaturity during the highly emotional organizational design process.

Creating a Matrix Organizational Design Model

Theorists have devised many ways to partition an organization into subunits, with the intent of improving efficiency (Springer, 2019). Additionally, the intent of partitioning an organization is to decentralize authority, responsibility, and accountability. The mechanism through which partitioning is accomplished is called "departmentalization." In all cases, the objective is to arrive at an orderly arrangement of interdependent components.

Many basic management courses refer to the three-variable formula below:

- Accountability = Authority + Responsibility

Authority is the power granted to individuals (possibly) by their position in the company, so they can make decisions for other individuals to follow.

Responsibility is the obligation incurred by individuals in their roles in the formal organization to effectively perform assignments.

Accountability is being totally answerable for the satisfactory completion of a given assignment.

In the above formula, when given any two variables without the third, there is a high probability of some form of failure. For example, the results would be obvious when someone is given responsibility and held accountable but has no formal authority to execute. Likewise, authority and responsibility, without accountability, promote subjectivity in decision making.

The matrix structure is a hybrid organization that attempts to balance the use of human resources as people are shifted from one project to another. It can be viewed as a project organization superimposed over a functional organization. Figure 5 is an example of a typical matrix organizational structure.

The matrix structure is more complex than either the traditional or product-oriented structures. To this end, it requires some basic ground rules to be successful:

- Participants must spend committed time on a project; this ensures a degree of loyalty.
- Horizontal as well as vertical channels must exist for making decisions.
- There must be quick and effective methods for conflict resolutions.
- There must be good communication channels between managers.
- All managers must have input into the planning process.

- Both horizontal and vertical managers must be willing to negotiate for resources.
- Horizontal line must be willing to operate as a separate entity except for administrative purposes.

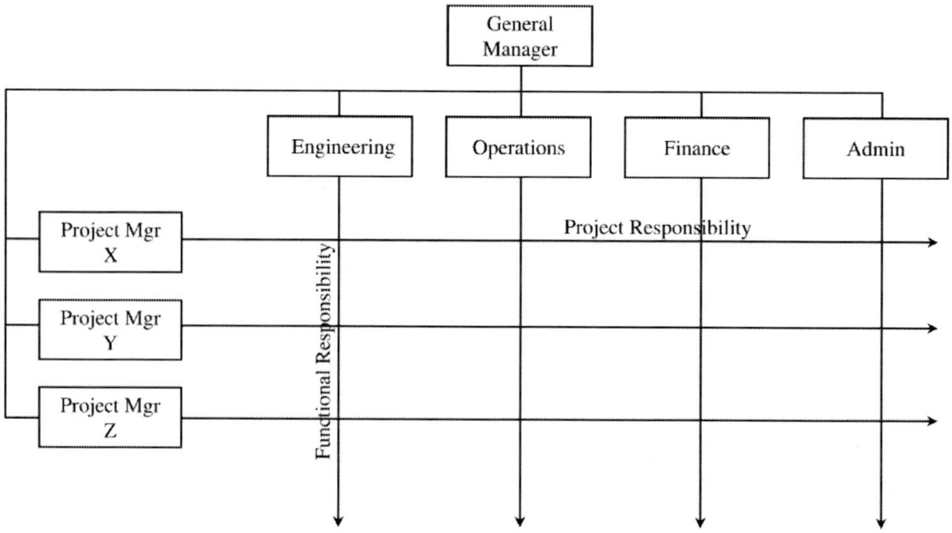

Figure 5. Matrix Organizational Design Model.

Organizational change should be viewed and treated as a formally chartered project. Because of the people involved and the transactional nature of the integrative process, project management is often more behavioral than quantitative. Interpersonal and communicative skills are extremely important attributes of the project manager and should be used as criteria in the selection of the project manager.

In a matrix organizational structure:

- There should, ideally, be no disruption due to dual accountability.
- A difference in functional management judgment should not delay work in progress.

Advantages of this type of organizational structure are:

- Combines the strengths of both project and functional organizations.
- Provides a good interface with the outside customer.
- Promotes effective interdisciplinary task integration.
- Promotes an efficient use of production resources.
- Promotes effective project control, as programmatic concerns are assigned to a single individual.
- Promotes career continuity and professional growth, as each functional individual has a home after project completion.
- Perpetuates technology. By this, functional resources gain the benefit of a functional strength, which can be transferred to the program of the day.
- Functional knowledge is available for all projects on an equal basis.

Disadvantages of this type of organizational structure include:

- Dual accountability of personnel. This is perhaps the biggest threat to this type of structure. Personnel will generally favor whoever it is that completes their performance review and subsequently has control over their income adjustments. Confusion here can derail a unified effort.
- Conflicts between project and functional managers. This issue will be discussed in more detail below.
- Profit and loss accountability is more difficult.
- There are continuously changing priorities, especially on the part of the functional managers, who control the resources.
- The balance of power between functional and project managers must be watched.
- Functional managers might be biased toward their own priorities.
- Because of the duality of authority, employees may not feel a strong commitment to a single source.
- Employees may feel confused about loyalty.

Step #1 – Defining Disciplines

The first step in creating an efficient and effective matrix model is to recognize the many disciplines of the organization. In business/industry, these disciplines are well defined. For example, these disciplines may, include engineering, human resources, operations, finance, legal, and as many as 17 others. Even within a broad-stroke discipline such as engineering, there may exist subdisciplines, for example, software engineering, hardware engineering, electrical engineering, chemical engineering, and many more.

Typically, if a discipline is in fact a discipline, there will be significant support in the form of professional associations, entities dedicated to credentialing, and wide-spread recognition that entity being defined is in fact a discipline. On occasion, in the infancy of an evolving discipline, support organizations as those just mentioned may not yet exist. In these scenarios, the authenticity of the discipline will be authenticated as time passes and other individuals or organizations have had sufficient time to perform the necessary analysis of proposed discipline-specific activities and outcomes.

Historically, disciplines align to existing continuums of study, both research and educational. From a research perspective, one can point to Pasteur's Quadrant. Where research has been framed along two coordinates: its consideration for use and its quest for fundamental understanding (Stokes, 1997).

Pure research is defined as research that is sought for understanding without thought of use: high in quest for fundamental understanding but low in consideration for use. This type of research is typically employed prior to and during the early concept exploration phase and helps to form an underlying premise for further research and exploration.

Use-inspired research, that which is both high in quest for understanding and high in consideration for use, is concerned with advancing basic understanding while seeking opportunity for an application of such research. This type of research is best suited for concept exploration and the early demonstration and validation phases of the product life cycle.

Pure applied research, low in quest for fundamental understanding yet high in consideration for use, may best be described as the application of existing knowledge. This type of knowledge may be thought of as the utilization of third or fourth generation programming languages set on top of an underlying operating system. Pure applied research is first and foremost the application of previous pure research efforts; for example, programming using Oracle as the underlying database versus designing the underlying database.

Step #2 – Identifying Each Discipline-Specific Body of Knowledge

Once the disciplines have been defined, the next step is to recognize each discipline has a body of knowledge. If the discipline does not have a body of knowledge, then it may be the discipline is not a discipline at all, but in fact, a child subset of activities of yet another parent discipline.

If a discipline is determined to possess a body of knowledge, then that body of knowledge should be expressed as a process (Springer and Schuver, 2017), with seemingly sequential or recurring activities, where each activity has attendant to it one or more outputs/products.

The process, then, can be defined in terms of a hierarchy of processes, procedures, methodologies, and practices. Each discipline process should have interfaces to other disciplines defined and pro-active and re-active Key Process Indicators (KPIs).

By definition, the many activities of a process, when executed successfully, produce a consistent end result. Process management is concerned with making sure the defined process is still efficient and effective, in that it minimizes the activities of the individuals performing the process, and that the end result is still what is desired. Process management is simply managing the existing process.

Creating an efficient process involves the elimination of non-value-added activities. In other words, once we identify all the activities to be performed and the order in which we intend to perform them, we must then look to see if some activities:

- Are redundant and can be deleted
- Are best performed in another sequence
- Can be combined with previous or subsequent activities
- Are potentially missing, which could enhance the efficiency of the entire process

In business and industry, process management, as characterized by Choyce (1992) and Gioia (1992), provides management with:

- A way of thinking systematically about the behavior of people at work in an organizational setting.
- A vocabulary of terms, concepts, theories, and methodologies that allow work experiences to be clearly analyzed, shared, and discussed.
- Techniques for dealing with many of the problems that commonly occur in the work setting.

Process management is not a new concept. Process management originated as part of the production-oriented statistical quality control movement in the late 1920s and early 1930s.

What is relatively new, however, is the transition of process management methods from a manufacturing environment to a total company orientation.

Process management is a continuous effort that recognizes that the work done in an organization is accomplished through a series of processes and charges the organization's managers with ensuring that these processes are clearly defined, healthy, and competitive. It is a comprehensive approach, the goal of which is to increase the effectiveness, efficiency, control, and adaptability of a given organization.

Business process management represents a break from some of the traditional concepts of organizational authority (Stinnett, 1992). It requires a new way of looking at, and thinking about, long-established assumptions concerning hierarchies and organizational structure. For instance, in a conventional organization it would be most unusual for the vice president or director of one group or division to become directly involved in the activities taking place in another group or division.

Because process management involves managing processes across divisional and organizational boundaries, as well as within these boundaries, it requires a more flexible management strategy. It also requires close cooperation among managers in diverse functional and operational units to ensure that the process flow is not interrupted by conflicts over lines of authority (King, 1992).

Process management relies on process definition, elimination of non-value-added activities, customer/supplier orientation, and a team approach (Hoban, 1992; Price, 1992).

Process management processes utilize continuous process improvement (CPI), which assumes that a measurement baseline has been established. Through CPI, the process is measured forever. CPI accounts for error elimination, innovation, and business changes. All activities of a process are questioned; nothing is sacred.

Process management offers organizations a means of applying to nonproduction functional organizations the same quality improvement and defect reduction techniques used in manufacturing processes. Many engineering, service, and business processes offer an organization the greatest untapped potential for cost savings through quality and productivity improvement (Welsh, 1992).

Process management, with its emphasis on business process quality, is the most meaningful way to apply the principle of quality throughout an enterprise (Zells, 1992).

The basic steps in creating an efficient process are:

- Determine what end product or result is desired.
- Identify the activities currently used to accomplish this process.
- Determine how the current activities are ordered (this is the interrelatedness of the many activities). Connecting the current activities provides a flow chart depicting the current condition.
- From the current flow chart, identify which activities do not seem to add value, could be merged, or seem inappropriately placed in time.
- Create a new flow chart depicting the ideal scenario (don't worry about who currently does which activities or how).
- Identify measurement points in the new process that will allow one to determine how well the new process is working.

- Test the new process. In a business environment, this may mean making people assignments to the activities. It may further mean reassigning individuals or work in a manner not previously assigned.

As stated above, it is only through proper measurement that we can make required changes to an existing process to increase either efficiency or effectiveness. Proper measurement requires that we identify sufficient measurement points throughout our process, and, that these measurement points are reflective of how the process is running.

Step #3 – The Assignment of People

Once disciplines have been identified and defined, the final step is to allocate human resources to the vertically identified disciplines/functions. To be optimal, this step must occur as the final step, after Step #1 and Step #2 above.

Why is the allocation after step #1 and step #2 above important? The answer resides in the cultural implications of change. Change management is premised on the idea change happens all the time and is no more prevalent than when bringing together a new program, in possibly a new business, and generally often with a new team of colleagues. To this end, it is imperative to discuss change management as it impacts individuals involved in a new endeavor.

The impact of this type of change is relatively the same as in any other change involving emotional reactions—and all change involves emotional reaction. While it is critical to understand the vision, goals, and objectives of the change, and the processes, methodologies, practices, and technologies employed to help ensure the change, there is probably nothing more important than managing the emotional impact of the change.

Change is not the same thing as transitioning. Change and transitions are different in that change is an event, situational, and context bases. Examples include such things as receiving a new boss, new office, and implementing new procedures.

Transition, on the other hand, is a process with activities and attendant products. Transitions can be extremely emotional, and potentially generating resistance and possibly upheaval within the organization. Individuals in the transitional phase of change are likely to experience emotions like those experienced during a grieving process.

The basic process for change includes three concepts:

- Letting go of the current and previous state
- Moving through the transitional aspects of the change
- Forging a new beginning

The starting point of letting go of the current and previous state is dealing with the many challenges of leaving the old behind. This includes such things as emotional ties, new infrastructure changes, and the inevitable organizational changes attendant to people, processes, practices, and methodologies. Things as simple as how an employee might fill out a travel expense report become highly emotionally charged process changes.

During this initial phase, we should recognize the potentially crippling personal effect of the process (likened to Kübler-Ross model of grieving):

- Denial—this cannot be happening
- Anger—why do we have to be going through this?
- Bargaining—what if we simply did . . .
- Depression—loss of things past and loss of future differences to current people, practices . . .
- Acceptance—I understand and recognize my limited ability to change it, so let's make do

It is important during this phase to recognize the emotional context of everyone involved. Leaders must:

- Understand exactly what is changing
- Understand the potential for secondary changes and what they may be
- Understand who is going to be losing something—department heads, line workers
- Recognize the "way things are" is what helps individuals feel comfortable and at home
- Understand that if everyone is going to feel a loss of something big in the "organization's history," for example, mergers, then a major effort to celebrate the past is important
- Understand the loss is extremely personal to everyone — it probably doesn't work to suggest, "I know how you feel."

According to Bridges (Bridges, 2009), it is important that we make known what is over and what is not. There are three potential side effects of not being clear on this:

- People will not stop doing anything; they will try to do it all
- People make their own decisions about what to keep doing and what to stop, causing inconsistency and chaos
- People toss out all the past artifacts; loss of important items, information, processes, practices, and so forth may occur

Bridges says we should celebrate the endings of the past. Treat the past with respect; do not denigrate the past. Let people keep a piece of the past; create a banner of historical accomplishments and provide a vision of continuity from the past to the future.

Major Opportunities for Failure

Failure #1 - Failing to recognize the horizontal businesses/colleges are in fact the profit/loss centers, and the vertical functions/disciplines are support organizations.

Frequently when a new matrix is created, there is a struggle for power between the hierarchical vertical disciplines and the horizontal profit/loss centers. This struggle can produce undesirable consequences.

Horizontal profit/loss centers in business/industry may include business units, divisions, and even company businesses. In higher education, profit/loss centers are typically oriented around colleges, schools, or perhaps academic departments. The horizontal profit/loss centers are where the money, revenue, or operational funds originate.

Vertical disciplines exist to provide the resources and the attendant expertise for each of those resources. This would include discipline-specific processes, procedures, methodologies, and practices.

The vertical discipline component is often without the resources and expertise, because there are few opportunities to generate revenue. The horizontal organizational elements are often without the product or service of the profit/loss center. In the most extreme cases, there would be no need for the discipline resources. Both vertical and horizontal organizational elements are mutually beneficial, needing the other's resources to be successful.

Unfortunately, human nature tends to promote a selfish imbalance in power, authority, and overall sense of self-importance. This imbalance often leads to decisions not in the best interest of the organization and should be avoided.

Failure #2 - Communicating with people without performing Step #1 and Step #2 above, identifying the disciplines and the discipline-specific processes.

This usually happens as a "power grab" to initially convey the power and strength of the vertical discipline leader of the newly formed organization. This mistake often leads to unnecessary resistance between the vertical matrix discipline and the horizontal profit/loss center. This type of problem often leads to resolution by the organizational parent.

In this instance, collecting all people creates two primary and potentially enduring negative reactions: (1) people are already performing in one capacity and may be ultimately asked to perform in another, and (2) once the matrix model begins to evolve through step #1 and step #2, significant negative cultural implications occur.

Failure #3 - Senior leaders not knowing how to create an efficient or effective matrix organizational design model and unwilling to listen to those who do know how to do this. Worse, not wanting to look ignorant or lack knowing how to do their jobs.

The saying here is "...the only thing worse than ignorance, is ignorance in power..." Thousands of books have been written on the topics of management and leadership. Good leadership recognizes the more we learn, the more there is to learn, and the corollary, the more we learn, the more we recognize how little we know. As individuals gain success in life, their perception of rightness increases. This self-perception of rightness, however, is limited to that which the individual has expertise. Unfortunately, all too often, individuals perceive their success in one area to naturally propagate to another, perhaps where their expertise is less. This is the premise for the modern-day saying "...pride comes before the fall."

A large part of emotional intelligence is recognizing what we know and what we do not. To this end, building a team of individuals who can help shape an overall success is critical to achieving organizational objectives. The implications of this phenomenon are riddled in the literature of business and industry success stories. This topic is not the topic of this paper and will not be further addressed.

Failure #4 - Thinking that adding more people to the organization is a solution, when creating an efficient organizational design model more readily supports the growth and continuity of an ongoing concern.

When senior leaders do not fully understand how to create an efficient matrix organizational design model, they tend to look to hiring those who can help to do so. This frequently leads to increasing staff and hierarchy within the newly created model. These additional staff may, on the surface, appear to be aligned to the intent of how work will be accomplished in the newly created model, but without having performed the previously defined steps #1 and #2, this logic is flawed.

Without a solid understanding of the disciplines, their processes, and practices, adding people may not be well grounded. In fact, how does a newly formed matrix model perform, if the newly formed model is being evolved in real-time and does not yet resemble the required outcomes of the two steps defined above?

The act of prematurely adding human resources to an ill-defined matrix structure, unjustifiably increases resources and attendant costs. In the end, this approach starts off wrong and is seldom corrected as time passes and the newly formed matrix model evolves.

The New Professional Working Adult Learner

With greatest respect and reference to Bob Dylan's 1964 song "…the times [students] they are a changin…," there has been a transitioning of professional working adult learners from one generational cohort to another, and now, to yet the youngest of generational cohorts… the Millennials.

Correlations exist between the number of webpage "hits" and the subsequent fall enrollments. While this information is important, it reflects a lagging indicator; that is, it does not tell us why there may have been more or less webpage hits. In contrast, a leading indicator will provide information as to why something has come to be.

From this perspective, this section will examine the leading indicators to the new professional working adult learner; these individuals being the target cohort for fee-based professional education providers offering Master of Science or Master of Arts degrees.

Drawing from sixteen years of student data, this report will focus on who these cohort learners are and how they compare to previous cohort groups of comparable age. The information presented will focus on generational cohort differences that impact or have the potential to impact the offering of Master of Science degrees and other fee-based continuing educational opportunities.

Specifically, this section will focus on:

- Who are the new students?
- The implications of ever increasing college costs.
- The impact of increasing debt loads.
- Moving back home after graduation and its implications.
    - Postponing marriage
    - Postponing the buying of homes, cars, and other material possessions
- Educational and income implications of having children later in life.
- How do we make program/cost changes to our offerings to remain competitive?

Who Are the New Students?

To better understand how best and who is best to teach today's college student, the personality and essence of the student needs to be better understood.

In a book by Levin and Dean titled Generation on a Tightrope (Levine and Dean, 2013), the authors present a snapshot of undergraduate students enrolled between 2009 and 2014. The authors begin by laying the foundation:

> Today's college students are struggling to maintain their balance as they attempt to cross the gulf between their dreams and the diminished realities of the world in which they live. They are seeking security but live in an age of profound and unceasing change. (p. ix)...
>
> They desperately want the economic opportunity their parents enjoyed but are coming of age during a deep recession with reduced career prospects. They want to believe in the America Dream and are optimistic about their personal futures but they are pessimistic about the future of the country. They want to be autonomous grown-ups but seem more dependent on their parents and the adults around them than any modern generation. They want intimacy—a partner and a family—but they are isolated, weak in face-to-face communication skills and live in a hook-up culture. They want to play by the rules but they don't know the rules and the rules are in flux because of the dramatic changes in our economy, the rise of new technologies, the condition of our public and private institutions, and a world growing flatter. They want to live in an n Internet world, a digitally connected globe but the adults and social institutions around them are analog or digital immigrants, including their blackboard universities.

From the onset, although not much of an issue, an understanding of their many names help, such that all may uniformly address these individuals as a cohort; that being born in a similar time period and having experienced similar major milestone events in their lives. Levin and Dean (p. 5) summarize the historical and current names being applied.

> Current undergraduates have been called millennials (Howe & Strauss, 1992) and generation 2K (Zoba, 1999) because they are a part of the first college generation of the twenty-first century. They have also been called generation Y (Tulgan, 2009) and generation iY (Elmore & Cathy, 2010), which is logical because they followed generation X and are partial to the Internet. There is generation Z (Hopkins, 2005) because they are the children of generation X. Building on that Internet theme are the meat and potatoes Internet generation (Milner, 2010), the too-easy-to-confuse-with-basketball net generation (Tapscott, 2008), the insightful digital natives (Palfrey & Gasser, 2010) and the less committal digital generation (Jukes, 2010). There must be an iGeneration somewhere but we haven't found it. Taking an entirely different tack are the names me-first generation (Lipkin, 2009), meaning they are a tad self-involved, and echo-boom generation (Alch, 2000), referring to the fact that these are the children of the baby boomers, not booming like their parents, just echoing, and so it goes.
>
> As of this writing millennials leads the pack in popularity with generation Y following second but seeming to have faded because it may or may not refer to a somewhat older group of young people.

Two of the many characteristics of this study and reported findings are the digital nature of this most recent cohort and the climate of continual change in which they live.

Colleges must educate this generation of undergraduates to thrive in an era of continuing change, to live productive and successful lives. A majority of undergraduates said their courses would be improved if they made greater use of technology, if their professors knew more about how to use technology, and if more of their classes made use of blended instruction, combining online and in-person classes (p. 165). The parallel to this statement resides in the truth of how many know a grandparent, or for that matter perhaps even a parent, that doesn't know how to use Facebook or how to tweet, or even for that matter how to text on a smartphone?

The Millennials, live in an anytime and anyplace world, operating twenty-four hours a day, seven days a week unbounded by physical location. "This causes a divergence between higher education and digital natives on the goals of education. Universities focus on teaching, the process of education, exposing students to instruction for specific lengths of time, whereas digital natives are more concerned with the outcomes of education, learning, and the mastery of content in the manner of games (p. 166)."

Higher education and digital natives' slant toward different methods of instruction.

> This is reflected in a difference between professors and students who approach knowledge in very different ways. Faculty members may be described as hunters who search for and generate knowledge to answer their questions. Digital natives by contrast are gatherers, who wade through a sea of data available to them in the disciplines, focusing on breadth versus depth of knowledge.
>
> Digital natives are oriented more toward group learning and social networking, characterized by collaboration and sharing of content. This causes an ethical challenge for universities, which under certain circumstances view collaboration as cheating and uncited content sharing as plagiarism (p. 167).

Higher education as a provider of services is predominantly driven in belief and practice. "That is, the university through its faculty determines the curriculum, the content, the instructional methods, the study materials, and the class schedule. Digital natives tend to be consumer driven, preferring to choose if not the curriculum and content they wish to study, then the instructional method by which they learn best, the materials they use to learn and the schedule by which they choose to study (p. 167)."

In the December 2012 report titled Knocking at the College Door (WICHE 2012) by the Western Interstate Commission for Higher Education (WICHE, 2012), the authors discuss the declining enrollment in institutions of higher education and the changing face of the college student from a racial and cultural perspective:

> The landscape of American higher education has changed rapidly in recent years and will continue to do so into the future. Simple demographics suggest that some states and regions will continue to see increases in the number of high school graduates, while others will see declines. In addition, the composition of our graduating class will continue to change, with increasing number and shares of the population coming from communities of color… our nation's leadership position is in peril: the U.S. has slipped to 16th in the share of its young adult population with a college education.
>
> The number of high school graduates overall having peaked during the 2010-2011 academic year, all four regions will see short-term declines in their numbers…

Why Are College Costs so High?

As a nation, we are standing at the precipice of seismic shifts in national and international higher education and public institutions of higher education in particular. "The U.S. economy is stuck in neutral since the last recession, tuition prices are skyrocketing, student loan debt has surpassed $1 trillion, parents, who have leveraged their homes through equity loans and first loans - for the second time, are losing faith in the value of education, state funding is dwindling, federal grants are shrinking, and donor dollars are smaller" (Moody's, 2013). These are the times in which we live. Change is inevitable. We can continue to do what we do, until such time when we can't, then, we must do something else. This is the prevalence of the literature today.

There is a growing trend toward college and university mergers. Marcus (2013) states "...it's a kind of private sector-style consolidation that is becoming increasingly common, not only for public institutions, but also for nonprofit, independent ones that can pool their resources and cut their costs in a time of falling budgets and demand for efficiencies in higher education..." Marcus goes on to state "...there have been few mergers of colleges and universities in the past... but the pace of such consolidations is picking up..." What is happening is a very natural next phase in the business life cycle; costs are rising, the number of new freshmen entering into college is flat, and college and universities are experiencing the financial implications of reduced revenue and increased costs (p. 2). In the end, we cannot protect something that does not have an economic right to exist. Market forces will prevail as they always do; free money through taxpayer's indebtedness and rising tuition does not last forever.

Moody's, in their January Industry Outlook report of 2013, was negative about the financial prospects of higher education. They highlighted the consolidation trend as one of the "bolder actions by university leaders" that can "foster operating efficiencies and reduce overhead costs amid declining state support..." by centralizing such services as marketing, fundraising, purchasing, and information technology.

In the report The Next Generation University[7], the nonpartisan New America Foundation stated higher education must adopt business practices to improve efficiency. Writing, the authors of the report state "...in the business world, the prevailing philosophy has long been that efficiencies and savings can be achieved by getting bigger and building economies of scale... which is why companies grow or merge with competitors..."

"It's not an easy thing politically," said Richard Novak, senior vice president for programs and research at the Association of Governing Boards of Universities and Colleges, "...you certainly have academic departments that have on the surface the most to lose, and will be the most vocal critics. And faculty, who have tenure and job security, are the most likely to be outspoken about it (Selingo, 2013, 2)." Marcus goes on to state "everybody is realizing that we're not going back to the way things used to be... change is coming – and you can either get on board or be left by the side of the road (p. 3)."

States have been increasingly rolling back their financial support for higher education, leaving their public universities, which already educate eight in ten Americans, scrambling for cash at a time when students are trying to get in. This leads to the finger pointing of inefficiencies. And, when the finger pointing of inefficiencies begin, it almost always points at the oft quoted "bloated administration" and "overbuilding" across any given campus. Remedies to increasing efficiency and reducing costs suggest (Selingo, 2013, xiii):

- Limiting the number of majors and tying those remaining to the needs of the local economy
- Offering classes year around
- Offering distance hybrid courses

Other suggestions are generally aligned to standard business practices of (Kiley, 2011):

- Simplifying organizational structures by decreasing layers of management
- Increasing the number of direct reports for any given supervisor
- Eliminating redundancies in service organizations such as information technology (IT), human resources (HR), finance or marketing through centralization, and, consolidating purchasing

There is no shortage of ideas on how to fix the growing and currently perceived financial debacle of higher education. Over twenty years of literature reveals hundreds of potential solutions. While these are specific recommendations for colleges and universities to follow, most are really nothing more than basic practices in any business/industry looking to increase efficiencies and reduce costs.

Moving Back Home and Its Implications

Increasingly students, those who graduated and those who did not, are moving back home. While there is no single authoritative source for actual percentages, a proliferation of studies reflects 34% of graduating students moved back home in 2011 and a more recent survey reflected 60 – 85% of graduating students intended to move back home after graduation (Giang, 2012, 1). All of this seems to be due to the slowly recovering economy from the last official recession; December 2007 through June 2009, and a difficult job market.

CNNMoney (Dickler, 2012, 1) reported in 2011 that as many 85% of graduating college seniors intended to move back home for at least a short while; this up from 67% in 2006. PEW Research Center reported in December 2011 that 53% had intended to move back home, at least, again, for a short while.

Whatever data is used for reporting, clearly there is a trend that implies college graduates are in fact expecting to move back home, if not for only a short while after graduation.

Vivian Giang, reporting for Business Insider in 2012 states: "…It's becoming the norm for young people to move back in with their parents. In fact, 53 percent of 18-to-24-year-olds are living with their parents, and 85 percent of college seniors plan on moving back home after graduation (Giang, 2012, 1)."

In March 2012, PEW Research Center reported more generally "…This generation of young adults has sometimes been labeled the "boomerang generation" for its proclivity to move out of the family home for a time and then boomerang right back. The Great Recession seems to have accelerated this tendency. The Pew Research survey found that among all adults ages 18 to 34, 24% moved back in with their parents in recent years after living on their own because of economic conditions." Tracking the number of young people in a given household is more easily measured than relying on voluntary data from those nearing college completion. Although, job prospecting in a nearing college graduate's vision is a very strong indicator of intent; meaning, if a college graduate is not finding gainful employment than their reporting they intend to move back home is likely a positive correlation to reality upon graduation.

February 9, 2012, PEW (2012) reported "…More affluent adults and those with higher levels of education are among the most likely to say that young people are struggling in today's economy. Nearly half (48%) of college graduates say young adults are having the hardest time these days. This compares with 37% of those whose educational attainment is a high school diploma or less. Similarly, adults with annual household incomes of $75,000 or higher are much more likely than those making less than that to say young people have been hit harder than their older counterparts (52% vs. 37%, respectively)."

The PEW Research Center report (2012, 3) leads directly to subsequent discussion on the implications of the slowly recovering economy and difficult job market on the postponement of marriage, having children, and purchasing of material possessions typical of age-specific cohorts.

Postponing Marriage and Children

While the data on how many college graduates is subject to many studies, date on the average and median age at first marriage is a readily available from U.S. Government census data and reporting.

Subsequent to moving back home after graduation is the continuing, and next logical postponement of adulthood, namely, getting married. Again, as with moving back home there is significant literature on the seemingly ever-increasing increase in age at first marriage.

From 1960 to 2010, the median age at first marriage went up nearly six years, from 22.8 years of age to 28.7 for males, and from 20.3 to 26.5 years of age for females (PEW, 2011, 3).

In piecing together the puzzle, established is that college graduates are moving back home after graduation, and, are postponing marriage in record numbers. Aligned to the postponement of marriage are discussions on total fertility rate, education and fertility, income and fertility, and the biological implications of postponing having children.

The total fertility rate is premised on the number of babies the average woman would bear over the course of her life if she were to survive until the end of her reproductive years and age-specific birth rate were to remain constant (Last, 2013).

> "…Demographers measure fertility in different ways. They start with the simplest observation: the number of births each year plotted against the numbers and ages of the rest of the population. From there they calculate the "crude birth rate," which is the number of children born per 1,000 people in that particular year. The crude birth rate isn't particularly useful, but for sake of context, the U.S. population was about 312 million in 2011 and about 4 million babies were born, so the U.S. crude birth rate was around 13. By combining the crude birth rate with other census data, you can determine "completed fertility," which is the number of babies actually born to each woman in American by the time she's 50... And finally, there's the "total fertility rate." The TFR is closely related to completed fertility: It's the number of babies the average woman would bear over the course of her life if she were to survive until the end of her reproductive years and age-specific birth rate were to remain constant *(Last, 2013, 5)*."

A few quick facts on total fertility rate and its implications on the world population [51, p. 11].

- The American fertility rate in 2021 was 1.93
- In order for a country to maintain a steady population, it needs a fertility rate of 2.1

- Which means that the Japanese and Italians (with fertility rates of about 1.4) are on the verge of downsizing their countries. Their cities are dwindling; some small towns are on the cusp of simply closing
- 1979 world's fertility rate was 6.0, today it's 2.52

Relative to education and fertility, the more educated a woman becomes, the less children, on average, she has (Last, 2013, 12).

- U.S. average 1.93
- College graduate = 1.78
- Women with a graduate degree = 1.61

Education, then, is directly correlated with delayed family formation. The drop in fertility among women with college and advanced degrees, then, is in large part a function of delayed family formation.

Specifically, it's not just the length of education that diminishes fertility, or the careers the education makes possible, but the debt load the education incurs. Since 1987, when the Nellie Mae Corporation began keeping statistical track of student loans, the average student- loan burden on college graduates has almost quadrupled, from $7,500 to $29,400.

In looking at income and fertility rates, there is again a direct correlation between increasingly higher levels of income and decreasing fertility rates (Last, 2013). Household income level under $20,000.00 represents the highest fertility rate at 2.038. As household income rises through the range of roughly $75,000 to $100,000 per year, fertility rates decline and stabilize around 1.75%. Household incomes over $100,000 show a slight uptick from the 1.75% to 1.83%, but far less than the highest fertility rate of 2.038.

Given young adult college graduates are assuming increasingly greater levels of school loan debt, are forced into moving back home for a period of time, delay marriage, opting to find gainful employment and establish their careers all prior to having children; when then, do these young individuals have their first-born child and how does that compare to previous generations?

The answer, as might be expected, is the median age of first-time parents has been skewed to the right, meaning, the median age of first-time parents has gone up. Date reflects the median age for first time parents has increased as much as 5% for mothers 30 - 35 years and older; this while 20- to 30-year-old first time mothers decreased in age by 3%.

The Center for Disease Control depicts declining birth rates for three age categories, 15-19, 20-24 and 24-29: while depicting increasing birth rates for age groups 30-34, 35-39 and 40-44.

From another perspective, the skewing of median age of first marriage and children, has its limits. From a biological perspective, between the ages of 24 and 34, a woman's chance of becoming infertile increases from 3 percent to 8 percent. By 35, half of all women trying to get pregnant over the course of 8 months will not succeed. After 35 it gets even more difficult. By age 39, a woman has a 15 percent chance of being unable to conceive at all. And, by a woman's 43 birthday, her chances of getting pregnant are nearly zero. All of which is why today, 1 out of every 100 babies born in the United States is created via in vitro fertilization (Last, 2013, 3).

Postponing the Purchasing of Material Possessions

As we chronologically age, we have very predictable patterns of spending. These patterns of spending are directly linked to what have been coined gerontological life phases. When we refer to gerontological phases we are referring to the scientific study of human development. The scientific study of human development is the science that seeks to understand how and why people change, and how and why they remain the same, as they grow older (Berger and Thomson, 1998, 4).

To help individuals to better understand the developmental changes we experience as we grow older, there have been established three domains of human development: biosocial, cognitive, and psychosocial.

Biosocial development includes all the growth and changes that occur in a person's body, and the genetic, nutritional, and health factors that affect those developments, as well as motor skills, everything from grasping a rattle to driving a car. Cognitive development includes all the mental processes that are used to obtain knowledge or to become aware of the environment. Psychosocial development includes development of emotions, temperament, and social skills. (Berger and Thompson, 1998, 5).

Gail Sheehy was one of the first to authoritatively document the basic consumption patterns of ageing individuals or cohorts of individuals based on age. Harry Dent (2014), more recently aligned these patterns to actual accumulated date from the U.S. Bureau of Labor Statistics Consumer Expenditure Survey. Dent's findings can be summarized in the below. The consumer life cycle presented below, uses U.S. Census Bureau data to reflect the median age by which most Americans participate in a given event; for example, our first starter homes are purchased generally around the age of 31.

- Single (ages 18 – 22)
- Young Married (ages 22 – 30)
  - Average age of first apartment – 26
- Young Family (ages 31 – 42)
  - Average age of 1st starter homes – 31
  - Average age of trade-up homes - 41
- Family, College Kids (ages 46 – 50)
  - Average age, largest consumption of furniture - 46
- Empty Nesters (ages 50+)
  - Average age for college tuition peak – 51
  - Average age for purchases of autos – 53
- Retired (60+)
  - Average age for hospital visits – 60
  - Average age vacation and retirement homes – 65
  - Average age for cruises – 70
  - Average age for predominance of prescription drugs – 77
  - Average age for nursing homes - 84

As we chronologically age, our life demands for shelter, transportation, food and clothing all change. These many changes are a reflection in large part to our changing family makeup at distinct periods in time. From above, it can be construed our U.S. populace generally get married in their 20's. The 30's are typically a time for the collection of material possessions (cars, homes, furniture, appliances, etc.) Our late-40's reflects an empty nester's phase, where our children become, theoretically, independent and move out onto their own; whether this be college or simply moving away from home to begin working. During our 50's, we begin to notice the changes of primary ageing; that is, those changes related to biological changes in hearing, eye sight, and other physical changes that are frequently the underlying premise for sayings such as "...ageing is not for the faint of heart..."

It becomes considerably more apparent when looking through the above lens to see the overall impact delaying adulthood can have on our national economy. Skewing our purchases to the right, places pressure on the entire economic infrastructure designed around historically predictable patterns of purchasing. In summary, delaying entry into adulthood, meaning moving away from home, being married, having children and all of the implications of this shift to the right, has a negative impact on the whole of the U.S.

In 1950, the median age in the U.S. was 30 years of age. In 200, the median age had increased to 35 years of age. By 2050, the median age in the U.S. will be 40. As people age, they consume less in terms of material possessions. Those chronologically aged also reduce their earnings and subsequently their taxes, therefore reducing the overall U.S. tax base.

Complimenting this line of thinking, is the required number of working adults 16-64 who pay into Social Security in support of those who draw down from this Government sponsored social support program. The below table reflects the severity of this ratio from 1940 (the first-year social security checks were issued) to more recent 2010.

> "The Social Security Administration predicts that by 2034, the ratio of workers-to-retirees will fall to just 2.1 workers for every retiree as a result of (1) roughly 809 million Baby Boomers retiring and (2) the declining fertility rates having failed to produce a proportionate number of new workers [p. 108]."

Table 3. Number of workers supporting each retiree

Ration of Workers Paying FICA Taxes to Retirees Collecting Social Security Benefits			
Year	*Workers (in millions)*	*Beneficiaries (in millions)*	*Ratio (number of workers supporting each retiree)*
1940	35.390	0.222	159.4
1950	48.2850	2.930	16.5
1960	72.530	14.262	5.1
1970	93.090	25.186	3.7
1980	113.656	35.118	3.2
1990	133.672	39.470	3.4
2000	155.295	45.166	3.4
2010	156.725	53.398	2.9
Source: Social Security administration (http://www.ssa.gov/history/ratios.html).			

On the Whole

Overall, we have a slowly recovering economy, difficult job market, high student debt, graduates moving back home after graduation, delays in marriage, having children and the purchasing of material possessions. All this demands action for the betterment of our society and the world populous as a whole.

Although one could argue vehemently the root cause of any one of the above triggers of delayed adulthood, the perspective of this section suggests the precipice of change must error on the side of continuing to educate our populous at a fair and reasonable costs.

There are multiple areas for future research. Below reflects a few of these areas.

- A mapping of what we now know about this next generational cohort to the many programs offered to professional working adult learners. Specifically, with the above presented information on economic variables and their impact on life phases, how will organizations offering opportunities for continuing education be better equipped to align program costs to participating students and program applicability to needed employment skills?
- Given the many demands on the new cohort of program participants, what impact might be expected to time to graduation?
- Will the average age of the professional learner go up with the above presented life-phase shift?
- Will there be greater need for career and financial services in support of the many competing economic considerations?

Changing U.S. Age, Racial and Ethnic Demographics and its Impact on Higher Education

The United States of America is undergoing, and will continue to undergo, a demographic transformation the likes of which have never been experienced in this great Nation. The demographic changes which surfaced in the literature and became more pronounced around 2008, are now at the precipice of tectonic change, and its impact on higher education is already being felt.

Three major events will take place over the upcoming decade. Each of which, by itself, may appear harmless and go relatively unnoticed. Together these three transformative changes paint a forever changing face of the demographics of the U.S. The impact of these three primary drivers of demographic change is already being felt in the hallowed halls of higher education. Colleges and universities are scrambling to accommodate these, still to be fully understood, major impacts.

The first of these three major changes is the “graying” of America. The last of the Baby Boomers, born between 1946 and 1964, will turn 65+ years of age in 2030. This is particularly significant because of the financial impacts on social services and safety nets currently enacted into law in support of a generally aging population.

The second of the three major changes is the marked cross-over (2035) where the number of people 65+ years of age outnumber the youths under the age of 18. The manifestation of this cross-over resides in the number of working age individuals for every aged dependency. When youth dependency, those aged under the age of 18, is added to the older-aged dependency, the net effect is a total dependency where there are two dependents for every three working age adults. This cross-over as well represents what has been termed the new minority majority of America, where the non-Hispanic White population becomes the minority overall population for the first time in U.S. history.

The third, and final, of the three major demographic changes is the recognition that the primary driver for population growth in the U.S. will be from international migration. Not because of an increase in international migration, but because of an aging natural population and a declining birth rate of same.

The new demographic of the United States has had a negative impact on enrollments in higher education. New minority populations are not equally prepared, financially, or otherwise, to participate in higher education as the current non-Hispanic White majority population. To this end, 25 years of researched literature materializes into multiple changes currently being implemented by institutions of higher education to accommodate this new minority majority population.

This section extracts from the literature the most recent current demographic changes, the impact of these changes on the enrollments in higher education, the response of colleges and universities to these rapidly changing American demographic realities, and the heightened awareness of these changes and their implications on continuing professional development administrative organizations.

Changing U.S. Demographics

The U.S. population, overall, is expected to grow more slowly, age considerably and become significantly more racially and ethnically diverse.

It is expected the U.S. population will reach roughly 400 million people in the year 2058 (Vespa et al., 2018, 2). At this writing, according to the U.S. Census Bureau's World Population Clock, the U.S. population is 329 million; with one birth every eight seconds, one death every twelve seconds, one international migrant every twenty-eight seconds, for a net gain of one person every twelve seconds. The U.S. population is increasing, at a decreasing rate. Through the year 2030, the population grew at a rate of 2.3 million people per year. However, going forward from 2030, the population is expected to grow at a much slower rate of 1.5 million people per year from 2040 to 2060. The decrease in the rate of growth is predominantly due to an ageing Boomer population and declining fertility rates of the U.S. majority non-Hispanic White female population.

The United States Census Bureau (2018, 9) calculates population between 2017 and 2060. It considers the births by demographic cohorts, native-born mothers versus foreign-born mothers living in the U.S. Accounting for births by individual native-born, foreign-born, racial, and ethnic cohorts is critical given the differences in fertility rates between these populations. Deaths is figured the same way as births; again, accounting for differences between native-born, foreign-born, racial, and ethnic cohorts. Net immigration, discussed later, becomes an increasingly greater percentage of the population; not because of an increase in immigration, but because of an ageing majority non-Hispanic White population coupled with a declining birth rate of same.

The Baby Boomers were born between 1946 and 1964. The last of the youngest Boomers, those born in 1964, will turn 65 years of age in the year 2029; marking 2030 as the first year all boomers will be at least 65 years old. This single fact has enormous implications, sociologically, economically and without any doubt demographically. This change is coming and is inevitable.

The U.S. Census Bureau (2018, 1) reflects the changing age demographic of those 65 years of age and older by calendar year. As they discuss, the U.S. has a generally ageing population with greater percentages of total population being 65 years of age and older as time progresses. By 2020, 17% of the total U.S. population will be 65 years of age and older, by 2030, 21% of the U.S. population will be 65 years of age and older, and by 2060, nearly one quarter of the entire U.S. population will be over the age of 65.

In the U.S., the working age population is between the ages of 18 and 64. Those below the age of 18 are calculated as the youth population, while those age 65 and above are considered the senior, or aged, non-working population. In the ideal scenario, the backfill of an aged population is the youth population. This younger population becomes the working age population of a society and ensures the continuation of social programs such as social security, Medicare, and other like programs for the aged population.

The U.S. Census Bureau (2018, 3) states the year 2035 marks the first time in U.S. history the older (65+) population is expected to outnumber the youth (under 18) population. They suggest, it can be seen in the year 2030, the under 18 population is at 18.4 million, while the 65 and above population is at 73.1 million. By 2040, however, the under age 18 population is at 76.8 million, while the over 65 age population is at 80.8 million. The actual cross-over in population projections occurs in 2035.

The shift from a youth-dependent population to an elderly-dependent population has significant implications as discussed above. The combined youth and old-age dependency, however, is even more revealing. In the year 2020, the total dependency ratio, as a measure of the burden on the working age population, is 64. Meaning, in the year 2020, there will be two dependents for every three working age adults. The combined dependency ratio, with the elderly population taking a higher percentage of the total dependency ratio, increases steadily through 2060, the last of the current estimated years. This dependency is, again, a reflection of a slower growing population, a declining fertility rate and a generally aging population.

Impacting this discussion are adjustments attributed to postponing retirement for those social security age eligible. Current economic, political and social events have caused some of those eligible for "full" retirement, as defined by the U.S. Social Security Administration, to postpone retirement until a later age. While there is significant previously reported data on this topic (Springer, 2019, 236-238), the true impact at this writing is uncertain. And, for purposes of this section, will simply be mentioned to heighten awareness and raise consciousness.

Growing Racial and Ethnic Diversities

The population in general is ageing and growing more slowly. This is especially true for non-Hispanic Whites. The fastest growing populations are two or more races (+197%), Asian (+101%) and Hispanics (+93%). The U.S. Census Bureau (2018, 7) depicts the percent change from 2016 to 2060. The non-Hispanic White population is, again, the only population expected to decline, reflecting a drop of over 19 million people for a percent change rate of -9.6%. This decline in population reflects the general aging of the population, coupled with the declining fertility rates of this cohort.

While the non-Hispanic White population remains the single largest cohort group, by 2045 they will no longer be the majority of the population of the U.S.

Youth Minority Majority Change in 2020

The youth of the U.S. are the bench strength of any country's population. They backfill the aging population and are the primary cohort to sustaining age related social programs, the strength of the working class and a country's capacity to innovate. It is, therefore, imperative this population is sufficiently educated and capable of sustaining an on-going enterprise, or in this case, a country.

In 2020, less than one half of the children under 18 years of age were non-Hispanic White. Meaning, a majority of the youth population will be what has been historically known as a minority population; this cross-over has been coined as either the new minority majority, or alternatively, the new majority minority (when referring to the non-Hispanic White population).

The U.S. Census Bureau (2018, 8) depicts this changing demographic. In the year 2020, the non-Hispanic White population represented 49.8 percent of the total youth cohort. The combined minority populations will exceed the non-Hispanic White population for the first time in U.S. history. By 2060, roughly two in three youth will be non-Hispanic White. This trend is not expected to reverse as significantly greater growth of the combined minority populations outpaces that of the non-Hispanic White cohort.

International Migration

By 2028, the share of the U.S. population that is considered foreign-born is projected to be higher than any time since 1850. The U.S. Census Bureau [55, p. 9] depicts 15.2% of the entire U.S. population as being foreign-born, exceeding the previous high of 14.8%; the actual cross-over occurs in 2028.

The overall share of the U.S. population that is foreign-born is expected to continue to increase; not because of a growing international migration, but because of a shrinking overall natural-born aging population.

By 2030 (U.S. Census Bureau, 2018, 11] net international migration will be the single largest driver of U.S. population growth. In 2030, the net natural-born increase to the population that being the increase due to natural born U.S. citizens, will be roughly 1 million, while at the same time, in 2030, the net international immigration increase in population will be roughly 1.1 million. Through 2060, net international migration population increase will remain stable at 1.1 million, while the natural-born net population increase will continue to decline. This overall population increase reflects an increasing population at a decreasing rate of growth.

College Enrollment Impact

In the Distance Education Learning Report, Allen, and Seaman (Allen and Seaman, 2017) report higher education enrollments, overall, for academic years spanning 2012-2015, are down across public and private for-profit institutions, while enrollments are slightly higher in private non-profit institutions.

The Allen and Seaman data reflects nearly a one million student decrease of -931,317 in students studying on campus.

They (Allen and Seaman, 2017) depict both the percent change from 2012 – 2015 as well as the equivalent student population. Worth noting is that public universities maintain the predominance of total student enrollments, both as a headcount and as a percentage.

Looking at the data presented by the authors, one would be naturally compelled to ask, "where did nearly one million students go?"

The students did not migrate to those institutions with a negative percent change from 2012-2015; namely, they did not go to private non-profit 4-year or above, private for-profit 4-year or above, public 2-year, or private for-profit 2-year. It appears the sole significant increase in enrollment institution type was to private non-profit 2-year institutions.

On reflection, if total loses do not add up to equal total gains, then where did the students go? Further research suggests there are simply not as many students; but why? The simple answer can generally be stated as an aging overall demographic population and declining fertility rates (Springer and Schuver, 2018).

Demographic Deeper Dive

While it can be stated there is a slower growing and generally aging overall population, not every population demographic graduates high school or enters college at the same cohort rate. To be more specific one must look at each location throughout the United States, the demographic make-up of that location and the probability of members of each cohort going to college. This deeper analysis provides a formula for more accurate predictions of college attendance.

The demand for college [59, pgs. 24-25], therefore, may be calculated as follows:

Demand for College (lt) =

- Probability of attendance (lt)
- Number of children (lt)

Where (l) = location, (t) = birth cohorts by year of expected H.S. graduation

College participation may also be viewed as an overlap between birthrate and social-economic factors; in other words, not every demographic cohort goes to college in equal numbers. For example, high-income children participate in some college at a rate of 73%, while low-income children participate in some college at a rate of 30%.

When overlapping birth-rates with racial/ethnic factors, the data suggests the following (Grawe, 2018, 118).

- Asian Americans participate in some college at a rate of 84%
- Non-Hispanic Whites participate in some college at a rate of 75%
- Non-Hispanic Blacks participate in some college at a rate of 60%
- Hispanics participate in some college at a rate of 60%

When overlapping birthrates with the level of education of a child's parents, the data suggests children whose parents have a degree at the bachelor's level, or more are more likely to attend some college than those children whose parents have not completed their H.S. education.

- Child's participation in some college, when parent(s) have a bachelor's degree, or more is 87%.
- Child's participation in some college when parent(s) did not complete High School is 47%.

The Winds of Change

The winds of change are blowing the sands of time through the pages of history as the saying goes. We are standing at the precipice of seismic shifts in national and international higher education and public institutions of higher education in particular. The U.S. economy is stuck in neutral since the last recession (Moody's, 2013, 1), tuition prices are skyrocketing, student loan debt has surpassed $1.5 trillion (Student Loan Hero, 2018, 1), parents – who have leveraged their homes through equity loans and second mortgages – are losing faith in the value of education, state funding is dwindling, federal grants are shrinking, and donor dollars are smaller. These are the times in which we live. Change is inevitable. We can continue to do what we do until such time as we can't. Then we must do something else. This is the prevalence of the literature today; something has to give, something other than addressing the low hanging fruit of administrative cuts. The changes must be institutional and must involve physical infrastructure.

There is a growing trend toward college and university mergers. Marcus states "…it's a kind of private sector-style consolidation that is becoming increasingly common, not only for public institutions, but also for nonprofit, independent ones that can pool their resources and cut their costs in a time of falling budgets and demand for efficiencies in higher education…" Marcus goes on to state "…there have been few mergers of colleges and universities in the past… but the pace of such consolidations is picking up…" What is happening is a very natural next phase in the business life cycle; costs are rising, the number of new freshmen entering into college is flat, and colleges and universities are experiencing the financial implications of reduced revenue and increased costs (Marcus, 2013, 2).

In the end, we cannot afford to protect something that does not have an economic right to exist. Market forces will prevail as they always do; free money through taxpayer's indebtedness and rising tuition does not last forever.

Moody's, in the "January Industry Outlook" report of 2013, was negative about the financial prospects of higher education. They highlighted the consolidation trend as one of the "bolder actions by university leaders" that can "foster operating efficiencies and reduce overhead costs amid declining state support" by centralizing such services as marketing, fundraising, purchasing, and information technology.

In the report "The Next Generation University" (Selingo et al., 2013, 2), the nonpartisan, New America Foundation, stated that higher education must adopt business practices to improve efficiency. The authors of the report state "...in the business world, the prevailing philosophy has long been that efficiencies and savings can be achieved by getting bigger and building economies of scale...which is why companies grow or merge with competitors..."

"It's not an easy thing politically," said Richard Novak, Senior Vice President for Programs and Research at the Association of Governing Boards of Universities and Colleges, "...you certainly have academic departments that have on the surface the most to lose and will be the most vocal critics. And faculty, who have tenure and job security, are the most likely to be outspoken about it" (Marcus, 2013, 2). Marcus goes on to state, "Everybody is realizing that we're not going back to the way things used to be... change is coming – and you can either get on board or be left by the side of the road" (Marcus, 2013, 3).

States have been increasingly rolling back their financial support for higher education, leaving their public universities, which already educate eight-in-ten Americans, scrambling for cash at a time when students are trying to get in. This leads to the finger pointing of inefficiencies. That pointing finger is almost always aimed the oft-quoted "bloated administration" and "overbuilding" across any given campus. Remedies to increasing efficiency and reducing costs suggest (Selingo et al., 2013, xiii):

- Limiting the number of majors and tying those remaining to the needs of the local economy
- Offering classes year round
- Offering distance hybrid courses

Other suggestions are generally aligned to standard business practices of:

- Simplifying organizational structures by decreasing layers of management
- Increasing the number of direct reports for any given supervisor
- Eliminating redundancies in service organizations such as information technology (IT), human resources (HR), finance, or marketing through centralization, and consolidating purchasing (Kiley, 2011, 1):

David Wessel in *The Wall Street Journal* (Kiley, 2011, 1) sited four ideas to fix higher education:

- Use a stick – "...the federal government could link an institution's student aid to its track record, essentially punishing schools whose graduates do poorly in the job market..."
- [The] President could offer extra money to those colleges wishing to "...test new approaches shortening a path to degree..."
- Transparency – "...the administration argues that the higher-ed market would work better if students knew more about the careers and wages of graduates of programs

they're considering and families knew more about the likely return on their big investment."

- Lowering barriers to entry – old rules "including an archaic system of deciding what's a college for purposes of federal aid and what isn't, may be preventing innovation…"

There is no shortage of ideas on how to fix the growing concerns of higher education. Looking back over twenty-five years of literature reveals tens upon tens of potential solutions. While these are specific recommendations for colleges and universities to follow most are really nothing more than basic practices in any business/industry looking to increase efficiencies and reduce costs.

One of the more exhaustive studies of what action should be taken to grow revenue and increase cost efficiency comes from Moody's Investor Service, "Industry Outlook," January 16, 2013 (Moody's 2013). Moody's negative outlook for higher education culminates in the following:

> …in the coming years, all universities will need to fund growing financial aid budgets, increasing healthcare costs, and capital improvement necessary to maintain or build market position. There is also a growing risk that public universities will be forced to assume payment responsibility for post-employment benefits that their respective states have historically paid. In order to do so, administrations will have to focus on traditional cost containment and operating efficiency, as well as take on the bigger issue of tenure and the student service intensive residential experience" *(Marcus, 2013, 16).*"

Their recommendations for change include:

- Centralization and shared services: human resources, fundraising, marketing, financial services, and information technology.
- Consolidation: within public systems; shared leadership and mergers.
- Expense flexibility: increase use of adjuncts; reduced percent of tenured faculty; leasing space for satellites/expansion.
- New market outreach: continuing education and degree completion; non-traditional students.
- Online education: distance learning programs; hybrid classes; fully online degrees; MOOCs for credit.
- Partnerships/collaborations: purchasing cooperatives; dual enrollment; joint BA/MA degree programs.
- Programmatic review: elimination of small, underutilized programs.
- Space utilization: weekend, evening, summer courses; moving administrative functions off core campuses to administrative spaces.
- Tuition pricing strategies: recalibration of the number of credits covered by the standard full-time tuition with per-credit charges for additional coursework (Moody's, 2013, 16-17).

The greatest long-term changes are not the low-hanging fruit, but the structural changes that have lasting impact. Changes that are structural in nature, and the most readily available to create efficiencies and reduce costs are seldom mentioned, and if mentioned seldom acted upon.

These changes include centralization of our many decentralized colleges and elimination of programs which fail the return on investment calculation.

On January 18, 2013, President Daniels's "Open Letter to the People of Purdue" makes explicit reference to higher education as we know it being poised for big change. He goes into great detail highlighting concerns for higher education according to many literary works.

These include:

- College costs too much
- Administrative costs have run up the cost to students without enhancing the value of education
- Rigor has weakened
- The system lacks accountability
- Too many professors are spending too much time writing papers for each other with no real contribution to understanding or human knowledge
- The unique system of tenure promotes "the narrowest form of closed-mindedness and the worst repression of dissident ideas"
- Athletics is out of control as a priority of university attention

President Daniels goes on to say, "…the operating model employed by Purdue and most American universities is antiquated and soon to be displaced…" In response to these many concerns and criticisms, President Daniels offers suggestions for collective thought and action. To name a few:

- Excellence – Purdue is not its buildings, or even its wonderful past or traditions "…this would be a great university if it met in a tent." Purdue is its faculty and students and what happens when brought together effectively.
- Affordability – "…every university community should embrace the shared responsibility to reexamine current practices and expenditures with a determination to keep its tuition and fees within the reach of every qualified student…"
- Shared governance – "...shared governance implies shared accountability. It is neither equitable nor workable to demand shared governing power but declare that cost control and substandard performance in any part of Purdue is someone else's problem. We cannot improve low on-time completion rates and maximize student success if no one is willing to modify his schedule, workload, or method of teaching…"
- Common purpose – "…the widespread duplication of identical functions can work against the common goal we must have of affordability and liberating resources for new investments in faculty and facilities… many choices will necessitate a communitarian outlook that consciously places the interests of the overall university first…"

Time and again, President Daniels's "Open Letter to the People of Purdue" refers to being good stewards, creating efficiencies, becoming more effective, reexamining current practices and expenditures, and addressing the duplication of support function services. These and many other references are indicative of areas for improvement and alignment to our current mission.

President Daniels's strong background in business affords him great understanding of the steps necessary for economic success.

The National Academies of Science produced a report titled "Research Universities and the Future of America: Ten Breakthrough Actions Vital to Nation's Prosperity and Security" (National Academy of Sciences, 2012). In this report, Recommendation #4 focused entirely on increasing productivity and reducing inefficiencies, stating "...the nation's research universities should set and achieve bold goals in cost containment, efficiency and productivity in business operations and academic programs..." Changes relative to increasing efficiency and reducing costs are directly impacted by the level of cooperation and collaboration of the administration and tenured faculty.

And Carlson (Carlson, 2018), penning a report titled Sustaining the College Business Model for The Chronicle of Higher Education states:

> "...The 20th Century was good to higher education as institutions and enrollments grew. But pressures have mounted in recent decades. Rising labor costs, falling public funding, suppressed tuition revenue, and demographic changes are straining the college business model. And the sector isn't known for finding efficiencies. Both buoyed and burdened by years of tradition, acres of infrastructure, and scores of personnel, many colleges and universities are struggling financially" *(Carlson, 2018, 4).*

Carlson goes on to recommend numerous potential changes required to sustain current college and university economic models, naming a few:

- Growing tuition revenue
- Collaborating and consolidating
- Streaming operations
- Reinventing yourself

Implications for Continuing Professional Development

The implications on continuing professional development organizations serving the needs of professional working adult learners are quite similar to those of the wider

- The new student population is changing.
- The economics of the new student population have a greater impact on their ability to participate in continuing education beyond a Bachelor's degree.
- The current supply-demand equation does not necessarily support ever increasing debt accrued as a result of post-Bachelorette educational attainment.

The Current Supply-Demand Equation

Scott and Nightengale (2018), in a published report titled "The Education-Jobs "Mix-Match": How Much Opportunity is there for the College-Educated Workforce in America's

Metropolitan Areas?" discusses at great length the implications of having a highly educated national populace.

The authors report on the increasingly greater percentage of the American population that has continued to obtain higher levels of educational attainment. The data reports:

- 1940 – 5% of the U.S. population had a Bachelor's degree.
- 2016 – 31% of the U.S. population had a Bachelor's degree.
    - 2016 – 29% more had some college but not a formal Bachelor's degree.
- 2018 – 33% of the U.S. population had a Bachelor's degree.

The authors go on to highlight those with higher levels of education experience less unemployment, earn higher wages, but require a wage premium for their knowledge and skills, which is sometimes hard to get.

The author's final point is that the unemployed, or more readily stated "underemployed" has risen over the last few years.

In looking at the supply and demand of higher education, the data suggests more than half of the total jobs in the United States (63%) require a High School degree or less. This, while over 60% of the population, ages 25 and above, have more than a High School degree.

- 63% of jobs require a H.S. degree or less
- 60% of people 25 and over have more than a H.S. degree
- Of adults with "some college"
    - < 1/3 have an Associates degree, certification or not completed any terminal diploma

The authors report the share of jobs requiring college versus the percent of the total U.S. population with this level of education. Worth noting is that:

- 24% of the total jobs in the U.S. require less than a High School degree, while only 13% of the population satisfy this educational requirement.
- 39% of the total jobs in the U.S. require a High School or equivalent education, while only 27% of the population satisfy this educational requirement.
- 11% of the total jobs in the U.S. require "some college," while a full 29% of the population have this minimal level of education.
- 26% of the total jobs in the U.S. require a four-year degree or more, while 31% of the population have this minimal level of education (2016).

From this data, there are more highly educated individuals than jobs requiring this level of education at entry, and for those jobs requiring High School or less than High School, there are more jobs than individuals who satisfy this entry level of education. This directly implies there are jobs that, if filled, would be filled with overqualified individuals.

The authors go on to point out:

> "…If people with postsecondary degrees cannot find jobs where they can use their education, it is reasonable to assume they would need to look for other jobs that require less education

instead, potentially displacing people with less education who could have held those jobs. As the share of people with postsecondary education increases, people may also need to continuously upgrade their skills or education, not necessarily to keep up with new duties required for a particular job, but to compete with other workers in the labor market...[pg. 11]"

Additionally, in the year of the report, only 24% of the employers in the U.S. had a problem hiring individuals due to a lack of degrees or credentials. The continuing thought suggests the oversupply has led to a substantial "up-credentialing," which is supported when viewing the data for the last 15 years. Job up-credentialing means the job has not changed, but now has a higher educational minimum credential than was previously required.

- Only 24% of employers had problem hiring due to a lack of degrees or credentials
- Most important skills
    - Problem solving
    - Interpersonal
    - Communication
    - Teamwork
    - Leadership skills
    - None of which require postsecondary education
- Over supply → substantial "up-credentialing" over last 15 years; jobs require higher education without the job changing

All of this directly implies there is a potential for diminishing returns on postsecondary education. Meaning:

- Countries with low percentages of formal postsecondary education yield increasing income growth and decreasing inequality as the populace becomes more educated
- Countries with widespread college education yields slowing growth and increasing inequality

Conclusion

As stated in the introduction, building a sustainable administrative online organization is not singular in nature. It is a multi-faceted endeavor which requires a breadth of perspective and discussion.

This chapter tells a story. It walks through an evolution of thought with significant breadth and depth to convey the magnitude of such an endeavor.

This chapter demonstrated through this evolution of thought how any one or more of the topics of this chapter can help to maintain and grow an on-going cencern, or, can inhibit the natural evolution of an otherwise high potential organization.

Envisioning, building, and growing a sustainable administrative organization for the purposes of delivering varying modalities of academic programs or courses to professional working adult learners is multi-faceted and highly complex. As this chapter documents, it is especially true, with complexities compounded, when running the administrative organization as a profit-loss entity within a non-profit public university.

References

Allen, I., Seaman, J (2017). Digital Learning Compass: Distance Education Enrollment Report 2017. Babson Survey Research Group.

Bardwick, J. (1991). *Danger in the Comfort Zone.* New York, NY: AMACOM.

Beatty, P.T. (1996). *Connecting With Older Adults: Educational Responses and Approaches.* Malabar, Florida: Krieger Publishing Company. Pgs 69-73.

Berger, K. S., and Thompson, R. A. (1998). *The developing person through the life span* (4th ed.). New York: Worth.

Bridges, W. (2009). *Managing transitions: Making the most of change.* Philadelphia, PA: Perseus Books Group.

Bronson, P. & Merryman, A. (2013). *Top Dog*. New York, NY.; Twelve, Hachette Book Company.

Carlson, S. (2018). *Sustaining the College Business Model.* The Chronicle of Higher Education.

Choyce, R. (1992, January). Why become a project management professional? PM Network, 32–33.

Christensen, C. & Eyring, H. (2011). *The Innovative University: Changing the DNA of Higher Education from the Inside Out.* San Francisco, CA.: Jossey-Bass.

Christensen, M., Horn, M., and Johnson, C. (2017). *Disrupting class: how disruptive innovation will change the way the world learns.* New York: McGraw-Hill.

Dent, H. (2014). *The Demographic Cliff.* New York, NY.: Penguin Group.

Dickeson, R. (2010). *Prioritizing Academic Programs and Services: Reallocating Resources to Achieve Strategic Balance.* San Francisco, CA.: John Wiley & Sons, Inc.

Dickler, J. (2012). Boomerang Kids: 85% of college grads move home. CNNMoney Online May 15, 2012. Downloaded from the internet September 18, 2014. http://money.cnn.com/2010/10/14/pf/boomerang_kids_move_home/.

Giang, V. (2012). Why Recent Grads are Moving Back Home in Droves. Business Insider, September 12, 2012. Downloaded from the internet on September 19, 2014. http://www.businessinsider.com/recent-graduates-heres-some-reasons-to-move-back-home-after-graduation-2012-9.

Gioia, J. (1992, November). Comprehensive program management. PM Network, 5–7.

Grawe, N, (2018). *Demographics and the Demand for Higher Education.* Baltimore, MD., Johns Hopkins University Press.

Hoban, F. T. (1992, August). An overview of training and development strategies for NASA project management. PM Network, 44–49.

Kerzner, H. (2006). *Project Management: A Systems Approach to Planning, Scheduling and Controlling.* Hoboken, NJ: John Wiley and Sons. Pgs 87-127.

Kiley, K. (2011). Where Universities Can Be Cut. Inside Higher Education Online. Downloaded from the internet April 30, 2013. http://www.insidehighered.com/news/2011/09/16/unc_berkeley_cornell_experience_show_where_administrative_cuts_can_be_made.

King, C. G. (1992, August). Multi-discipline teams: A fundamental element of the program management process. PM Network, 13–22.

Last, J. (2013). *What to Expect When No One's Expecting.* New York, NY: Encounter Books.

Levine, A. & Dean, D. (2013). *Generation on a Tightrope.* San Francisco, CA.: John Wiley and Sons.

Marcus, J. (2013). Like Private Businesses, Universities Consolidate to Cut Costs. Downloaded from the internet on July 21, 2013. The Hechinger Report. http://nation.time.com/2013/07/19/cash-strapped-universities-turn-to-corporate-style-consolidation/?xid=newsletter-daily.

Moody's (2013). US Higher Education Outlook Negative in 2013. Industry Outlook, January 16, 2013. Downloaded from the internet on July 19, 2013. http://www.marquette.edu/budget/documents/USHigherEducationOutlookNegativein2013.pdf

Nahavandi, A. (2009). *The Art and Science of Leadership*, 5th Ed. Upper Saddle River, NJ.: Prentice Hall.

National Academy of Sciences. *Research Universities and the Future of America: Ten Breakthrough Actions Vital to our Nation's Prosperity and Security.* Washington, D.C.: National Academies Press (2012).

Paquette, D. (2004). Dichotomizing Paternal and Maternal Functions as a Means to Better Understand Their Primary Contribution. Human Development, vol. 47(4), pp. 237-238.

PEW Research Center (2011). Barely Half of U.S. Adults are Married – A Record Low. Pew Research Center Report, December 14, 2011.

PEW Research Center (2012). Young, Underemployed and Optimistic. Pew Research Center Report, February 9, 2012. Downloaded from the internet on September 18, 2014. http://www.pewsocialtrends.org/2012/02/09/young-underemployed-and-optimistic/.

Price, M. T. (1992, October). A process approach to project manager training. PM Network, 17–26.

Scott, M. & Nightingale, D. (2018). The Education-Jobs "Mix-Match": How Much Opportunity is There for the College-Educated Workforce in America's Metropolitan Areas? Urban Institute. Washington, D.C.

Selingo, J. (2013). College (UN) Bound. Boston, MA.: New Harvest.

Selingo, J., Carey, K., Pennington, H., Fishman, R. & Palmer, I. The Next Generation University. New America Foundation, May 2013. Downloaded from the internet on July 19, 2013. http://higheredwatch.newamerica.net/blogposts/2013/the_next_generation_university-84378 (2013).

Simonson, M., Smaldino, S.E., Albright, M., & Zvacek, S. (2008). *Teaching and Learning at a Distance: Foundations of Distance Education.* Boston, MA: Allyn and Bacon.

Springer, M. L. (2010). Project and Program Management: A Competency-Based Approach. West Lafayette, Indiana: Purdue University Press. Pgs 40-73.

Springer, M. (2019). Project and Program Management: A Competency-Based Approach. West Lafayette, IN. 4th Ed., Purdue University Press.

Springer, M. L., & Schuver, M. T. (2018). Dwindling Graduate Student Enrollments in Distance-Based Programs: A Research-Based Exploration with Findings and Underlying Premise. *ASEE 2018 Annual Conference Proceedings.* Salt Lake City, UT.

Springer, M., Bertoline, G., & Schuver, M. (2011a). Building an Academic Center Infrastructure for Professional Development. In *Proceedings of the 2011 American Society for Engineering Education Annual Conference.* Washington, D.C.: American Society for Engineering Education.

Springer, M. L., & Schuver, M. T. (2017). In Looking at Distance Education as a Process: Activity Based Opportunities for Efficiency Gains and Cost Savings. *ASEE 2017 Annual Conference Proceedings.* Columbus, OH.

Springer, M. L., Schuver, M. T., & Dyrenfurth, M. J. (2011b). Long term alumnus performance and evaluation after graduation from a distance learning hybrid weekend master's degree program in technology. *ASEE 2011 Annual Conference Proceedings.* Vancouver, B.C.

Stinnett, W. D. (1992, May). Lone wolf teams: Reconciling the need for collaboration with the need for individual accomplishment. PM Network, 21–25.

Stokes, D. (1997). *Pasteur's quadrant: Basic science and technological innovation.* Harrisonburg, VA: The Brookings Institution.

Student Loan Hero. (2018). A Look at the Shocking Student Loan Debt Statistics for 2018. Student Loan Hero. Downloaded from https://studentloanhero.com/student-loan-debt-statistics/.

U.S. Census Bureau (2018). Older People Projected to Outnumber Children for First Time in U.S. History. United States Census Bureau, March 13, 2018. Release Number CB18-41.

Vespa, J., Armstrong, D., and Medina, L. (2018). Demographic Turning Points for the United States: Population Projections for 2020 to 2060. *Current population Reports*, P25-1144, U.S. Census Bureau, Washington, DC.

Welsh, A. (1992, May). The future of project management. PM Network, 5–6.

WICHE (2012). Knocking at the College Door: Projections of High School Graduates. Report from the Western Interstate Commission of Higher Education. Downloaded from the internet August 12, 2013. http://www.wiche.edu/knocking-8th.

Zells, L. (1992, May). Applying Japanese total quality management to software project management. PM Network, 32–35.

Chapter 13

A Sustainable Future through the 60-Year Curriculum

Amrit Ahluwalia*

The EvoLLLution: A Modern Campus Illumination, Toronto, Ontario, Canada

Abstract

The postsecondary landscape is rapidly changing which means that institutions need to change or risk obsolescence. Some of the important changes include demographics such as declining enrollments, expectations such as increased flexibility, and competition such as college credits becoming less important and credential more important. It is recommended that institutions adapt to a 60-year curriculum (60YC), recognizing that people are living longer and will need training throughout their lives (lifelong learning). Several aspects of the 60YC are discussed including learner engagement, the primacy of credentials, and learning recognition. It is argued that CEE is best positioned to help institutions adapt to the demands of the 60YC. A map is provided for higher education to adapt to the 60YC.

Keywords: 60-year curriculum

Introduction

Our postsecondary ecosystem is foundationally changing. Keeping pace with those changes—and capturing the opportunity of the disruption rather than being displaced by it—requires fundamental shifts in how colleges and universities view the market and their place within it.

This chapter will reflect on those foundational changes, introduce the 60-Year Curriculum model for those who are unfamiliar with the concept, and explore how this vision for higher education's future can establish a sustainable "next normal" for the postsecondary ecosystem. It will also highlight the role continuing and professional education divisions can play in realizing this vision for higher education's future.

The Changing Postsecondary Landscape

The postsecondary landscape is transforming for colleges and universities, posing threats for those institutions that choose to remain committed to the industry's status quo while presenting significant opportunities for those who elect to "ride the wave" of evolution. The changes are visible across a few key areas: demographics, expectations and competition.

* Corresponding Author's Email: aahluwalia@evoLLLution.com.

Demographics: These Aren't the Same Learners as Before

The demographics of the populations enrolling in higher education today are evolving, and these demographics will continue to evolve over the coming decades.

Data is woefully lacking on non-traditional student enrollments across the board (ie, registered in either non-degree or degree-bearing programming), but we know that "since 1995-96, at least 70 percent of undergraduates possessed at least one non-traditional characteristic" (U.S. Dept. of Education, 2015).

Of the approximately 73 percent of degree-seeking students who qualified as non-traditional in 2000 (the most recent available analysis of this population), 51 percent were financially independent, 48 percent attended part-time, 39 percent worked full-time, and 27 percent cared for dependents (NCES, 2002).

And for institutions dependent on traditional-age students, this population is trailing off. There had been a consistent growth rate in high school graduates since the mid-1990s, but that growth has slowed since 2012 and is expected to peak in 2025 at 3.93 million. In the decade following (2025-2037), high school graduation numbers are expected to precipitously decline year-over-year eventually reaching a 10.4 percent decline (Bransberger et al, 2020).

In short, according to Eastern Shore Community College President Jim Shaeffer, the coming enrollment cliff will create a challenging environment for higher education institutions unwilling to change.

"Kids are not graduating from high schools at the same rate as in the past," he wrote. "With the student population getting smaller, the competition for those students grows larger" (Shaeffer, 2019).

Meanwhile, adult enrollment in degree-bearing postsecondary programming is expected to continue to grow through the same period (Hussar and Bailery, 2019).

Expectations: These Learners Want Something Different

The second shift modern colleges and universities have to contend with is that the expectations of the learners they're serving are evolving.

Most students who enroll in a degree-bearing program at a college or university do so to achieve career outcomes (Strada-Gallup, 2018). However, only 35 percent of university students and 47 percent of community college students say their institutions are doing a good job connecting their education to workforce success (Strada Education, 2021). As a result, 41 percent of bachelor's degree earners are underemployed in their first job out of school (Burning Glass Technologies, 2020).

Students are also looking for a level of flexibility that modern institutions have been slow to provide. In terms of barriers to enrollment, 55 percent of Americans said work and family commitments, rigid course schedules, and transportation to campus posed major obstacles. 49 percent reported fearing a lack of success, and 48 percent cited cost as a major barrier to enrollment (Strada, July 2020).

Perhaps not surprisingly, then, 68 percent of adults considering enrolling in education in 2020 said they prefer non-degree pathways (Strada, September 16, 2020).

What's more, the modern learner is an experienced consumer. These individuals engage with major retailers (both in-store and online) on a daily basis. Their expectations for engagement and service aren't being set by their interactions with other higher education providers. Their expectations for engagement are being set by interactions with eCommerce leaders like Uber, Amazon and Netflix.

According to Belinda Elliott-Bielecki, director of marketing and communications in the College of Extended Learning at the University of New Brunswick, not recognizing the modern student's role as a customer can create significant obstacles for postsecondary providers:

> "To understand the student of today is to know they are discerning," she wrote. "They come with high expectations, and as such, universities need to treat students like customers if they want to survive. If they are not satisfied with our institution, they will go elsewhere" (Elliott-Bielecki, 2021).

This paints a troubling picture, one where college and university leaders imagine themselves to be serving a very different individual than the student who actually enrolls in a course or program.

Competition: Learners Have Numerous Options

The third shift facing modern colleges and universities is simply the reality of the market they're competing in today. Recognizing that modern learners think and act like consumers necessitates a recognition of the many options those learners have at their disposal when it comes to seeking out leaning opportunities.

As previously mentioned, the modern learner isn't as focused on earning a degree as they are on earning a credential that will allow them to achieve a specific outcome (generally a career outcome).

To that end, accreditation no longer closes the postsecondary landscape to non-institutional education providers. This is now an open market where anyone can pitch their tent and sell their wares.

This became most apparent during the recent recession created by the COVID pandemic. Historically, higher education institutions have experienced clear enrollment growth during recessions. Between 1968-1988—a period that included four recessions in the United States—colleges and universities experienced 2 percent enrollment growth for every percentage point increase in unemployment (Dellas and Sakellaris, 2003). This trend held through the Great Recession (2007-2009).

This historic multiplier suggests that higher education institutions should have seen massive enrollment growth through the COVID recession. After all, over 30 million Americans filed for unemployment in the 6 weeks following the declaration of a state of emergency (CNBC, 2020). For context, during the entire Great Recession (2007-2009), a total of 8.7 million jobs were lost.

However, higher education institutions suffered significant enrollment losses through 2020. Undergraduate enrollment fell 4.9 percent in Spring 2021, and enrollments of traditional college-age students fell 5 percent in the same period (National Student Clearninghouse Research Center, 2021).

That's not to say students responded to the recession and the stay-at-home orders by neglecting their education. Instead, learners turned to alternative education providers.

Coding bootcamps experienced a 30 percent increase in enrollment between 2019 and 2020, with an accompanying 12.6 percent increase in revenue (Career Karma, 2021).

Brian Reed (2021), associate vice provost at the University of Montana, observed that the disconnect between employer needs and institutional offerings is leading to a gap between higher education, learner expectations and labor market relevance.

"Degrees no longer enjoy primacy as the proxy for the skills needed to do a job," Reed wrote. "Since the onset of COVID 19, entry-level jobs requiring a bachelor's degree plummeted by 45 percent, suggesting employers are increasingly valuing skills and experience over academia. Thus, once institutions have a clear understanding of the labor and program demand, they need to map curricula for in-demand work skills."

The Relevance of the 60-Year Curriculum Model

Given the multifaceted shifts facing the traditional postsecondary industry, it's critical to adopt foundational changes that will support the capacity of colleges and universities to realize opportunity in this new normal—instead of being overwhelmed by the threat.

One model that provides a pathway to adapting to the industry's new normal is the 60-Year Curriculum (60YC). Shifting to (or at least toward) a 60YC structure will provide colleges and universities the necessary foundation to deliver a truly engaging and comprehensive lifelong learning infrastructure.

The Changing World of Work: Explaining the Need for a 60-Year Curriculum

The 60YC concept is borne from recognition of the fact that the working world is fundamentally transforming.

Individuals would historically graduate from a postsecondary offering, enter the world of work, and progress in a streamlined fashion toward retirement in their mid-60s. This simple three-stage model is no longer relevant.

Moore's Law and associated technology changes driving automation in the labor market mean the expectations on working professionals are constantly evolving. In a 2017 report, McKinsey (2017) found that 50 percent of paid work activities in the United States could be automated by technology that currently exists or is in development. What's more, by 2030 McKinsey expects technology to threaten the jobs of between 15-30 percent of the American workforce (McKinsey, 2017).

In the *Future of Jobs Report 2018*, published by the World Economic Forum, the reality of the near-future labor market is illustrated in even clearer terms:

> Companies expect a significant shift on the frontier between humans and machines when it comes to existing work tasks between 2018 and 2022. In 2018, an average of 71 percent of total task hours across the 12 industries covered in the report are performed by humans, compared to 29 percent by machines. By 2022 this average is expected to have shifted to 58 percent task hours performed by humans and 42 percent by machines.
>
> In 2018, in terms of total working hours, no work task was yet estimated to be predominantly performed by a machine or an algorithm. By 2022, this picture is projected to

> have somewhat changed, with machines and algorithms on average increasing their contribution to specific tasks by 57 percent. For example, by 2022, 62 percent of organization's information and data processing and information search and transmission tasks will be performed by machines compared to 46 percent today.
>
> Even those work tasks that have thus far remained overwhelmingly human—communicating and interacting (23 percent); coordinating, developing, managing and advising (20 percent); as well as reasoning and decision-making (18 percent)—will begin to be automated (30 percent, 29 percent, and 27 percent respectively).
>
> Relative to their starting point today, the expansion of machines' share of work task performance is particularly marked in the reasoning and decision-making, administering, and looking for and receiving job-related information tasks (World Economic Forum, 2018).

Increasing numbers of currently manual tasks are being automated by continuously improving technology. As these manual tasks become automated, the work expected of humans will be constantly evolving to "human-specific" tasks that cannot be automated (as outlined in Figure 1).

Compounding this reality, as outlined in Figure 1, is that life expectancy is growing. As we enter the 100-Year Life paradigm, individuals will work for longer at jobs that are constantly evolving.

Our traditional education model—where an individual effectively exclusively learns at the early stage of their lives and then only voluntarily returns to education over the course of their life—is no longer relevant.

Today and into the future, individuals' lives will be complex and multistage, with a constant learn-work cycle in effect. This is illustrated by Figure 2.

As can be seen, learning in the early stages of an individual's life focuses on building the foundational skills they will use to establish a successful career into the future. From there, individuals will enter a constant cycle of upskilling and reskilling to allow them to keep pace with required changes in their industries.

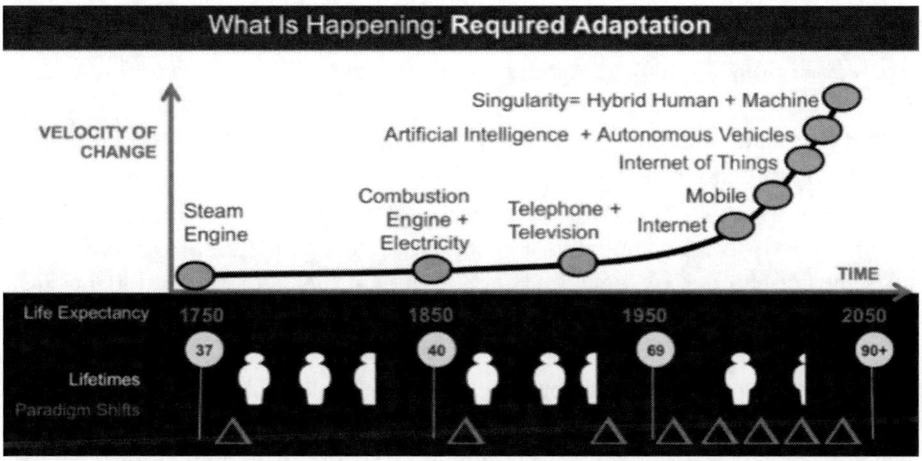

Figure 1. McGowan, H. and Shipley, C. https://medium.com/@heathermcgowan/ future-of-work-learning-to-manage-uncertainty-990601de86e8.

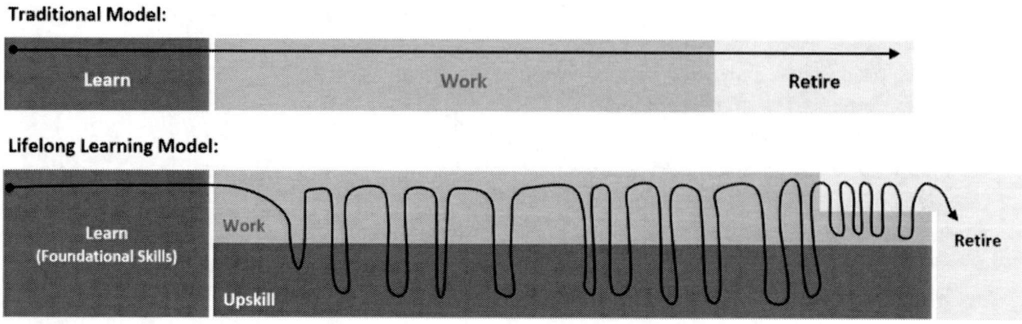

Figure 2. Pichette, J. https://evolllution.com/revenue-streams/market_opportunities/ competency-based-education-a-model-for-keeping-up-in-canada/.

The implications for higher education institutions here are significant, as outlined by Jeffrey Russell (2019), dean of continuing studies and vice provost for lifelong learning at the University of Wisconsin-Madison:

> To accommodate longer lives, we'll need to develop academic programs that stretch from childhood into old age. This will require creativity in how we deliver courses, with an emphasis on flexibility and personalization. It will also require creativity in how we provide credentials, from degrees to certificates to digital badges.
>
> In other words, we must thoroughly understand this new generation of lifelong learners. We'll have to reinvent ourselves to offer them meaning and relevance in their educational journeys, and connect with their passions over the course of their lives.
>
> Meeting these challenges will undoubtedly test leaders in higher education. They'll need a clear vision of the future—no easy task in this time of rapid change. And just as important, they'll need strategies for putting a vision into practice, and inspire conventional organizations to think in unconventional ways.

A Model for Lifelong Learning: Understanding the 60-Year Curriculum

The 60-Year Curriculum model stands as a mechanism that can help colleges and universities reorient themselves to the increased need for lifelong learning.

Right now, most colleges and universities are set up for the three-stage life. They provide access to individuals in their late-teens and early-twenties, structure programming to be full-time and semester-based, and focus on a cohort model that progresses students in lockstep toward earning an accredited credential. Once that individual graduates, they enter a workstream that they stay in over the course of their careers.

The lifelong learning environment is set up for people looking to advance beyond the preparation they receive in that traditional learning environment. As such, access is very much dependent on learners seeking out and finding pathways to upskill, reskill and advance. Lifelong learning opportunities tend to be siloed and separated from the offerings of the institutional "main campus"—offered by continuing, professional, workforce and online education units operating on the periphery of their respective institutions and departments.

This means lifelong learning infrastructures—models designed to serve anyone who falls outside the definition of a traditional postsecondary student—are generally under-resourced, misunderstood and easily overlooked in terms of their value and impact.

The 60-Year Curriculum model flips the organization of the traditional institution on its head. Instead of placing primacy on the coming-of-age experience offered by the traditional postsecondary model, it widens scope to look more honestly at the audiences that need access to the institution and the mechanisms by which the institution can serve them.

To put it plainly, the 60YC does not propose throwing the proverbial "baby" of higher education out with the "bathwater." Instead, it proposes creating entry and exit points that suit a wider array of prospective learners and meets those learners where they are in their learning lifecycle.

> Higher education was built around the concept of disciplinary education to undergraduates, master's and PhD students who are headed into a three-phase life: Childhood and education the first 18 to 25 years; then work for ages 18 to 65; then retirement for about 10 years. The model is a near-perfect fit for a growing industrial economy that needs about 20 percent college graduates.
>
> Since the emergence of the global economy, rapidly evolving knowledge work and much longer lives in the West, it is clear people will live 4-, 5-, 6-, and even 7-stage lives. Higher education needs to evolve to serve the learner from before they arrive in college through after they retire to help citizens be great professional, civic and social contributors. We tell our children they need to be lifelong learners, but so far higher education has not changed to purposefully serve their lifelong needs.
>
> The 60YC proposes that higher education providers, who happen to be best in the world at knowledge creation and dissemination through well-designed curriculum, expand that curricula concept from the current two-year AA, four-year BA, two-year master's and seven-year PhD learning models into a 60-year model inclusive of 15- to 75-year-old learners and, most likely, beyond (Lambert, 2019).

Table 1. Richards, J. and Dede, C. https://er.educause.edu/articles/2020/10/ the-60-year-curriculum-a-strategic-response-to-a-crisis

	Factory	Office	Global Network
Curriculum	Grade School	College	60 Years
Nature of Work	Manufacturing Production	Knowledge Work	Consulting Entrepreneurship
Era	Industrial	Information Processing	Digitalization
Cognition Model	Warehouse	Process	Agile Network
Learning	Information Transfer	Thinking Skills	Transferable Skills
Student Role	Clerk	Symbolic Analyst	Consultant Entrepreneur
Unit	Individual	Team	Dynamic Teams
Content	Curriculum	System	Project Deliverable
Teacher Role	Lecturer Information Resource	Facilitator Process Resource	Mentor Coach Advice Resource
Assessment Method	Test Time in Job	Authentic Degree	Performance Deliverable Portfolio
Pedagogy	Telling	Enabling	Empowering

As outlined above by Hunt Lambert, dean emeritus of the division of continuing education and extension at Harvard University, the model makes space for youth learners to access the institution through opportunities like summer camps and dual enrollment. By the same token, it allows for professional development, upskilling and reskilling opportunities for a wide array of adults who need to extend their learning to advance (or maintain!) their careers.

Rather than extending the traditional factory model of education—or the office model of education that we entered into through the 1990s and into the 21st century—the pioneers of the 60YC imagine the world to be entering a "Digitalization" era they describe as the "Global Network" (Richards and Dede, 2020). This requires some foundational changes in learner roles, institutional positioning, faculty responsibility and even programming, as outlined in Table 1.

Establishing Limits: The Boundaries of a 60-Year Curriculum

In the 21st century, learning opportunities are more accessible than they have been at any point in human history.

We're living in an era where a trip to YouTube can teach you anything from differential equations to replacing your car's rear differential. The EduPunk movement, which began to gain steam in late 2000s and helped to formalize education pathways for autodidacts, has made it clear that individuals with the desire to expand their knowledge base can do so with minimal capital outlay.

So how do we define lifelong learning in an environment where learning can happen anywhere, at any time?

According to Rovy Branon, vice provost for Continuum College at the University of Washington, there are some foundational elements to defining learning within the 60YC model:

> The term lifelong learning has been around for decades but in many circles has become synonymous with "enrichment" learning. Additionally, in an era where a 12-year-old can post a YouTube video that helps me fix a computer problem, lifelong learning can also imply all the opportunities we have for informal learning. While a formal definition of the 60-year curriculum is still emerging, some basic tenets include:
>
> - A focus on formal higher education (but not just degrees);
> - An intentionality to the learning experience;
> - And that learning needs are personal but change over a lifetime.
>
> At UW Continuum College, we have encapsulated this as a need to provide the right education for the right learner at the right time in life (Branon, June 8, 2018).

Adapting to a 60-Year Curriculum

Providing "the right education for the right learner at the right time in life" is certainly a significant shift from the traditional institutional model. Arguably, this level of personalization and flexibility is common practice in almost every other consumer-facing industry. But for higher education, the industry has maintained a rigid belief in "mother knows best" for a millennium.

To set the transformation in motion, higher education institutions—universities especially—will have to reorient themselves across three key areas: learner engagement, credential primacy and learning recognition.

Learner Engagement: Setting New Expectations and Supporting Lifelong Learning

First, and possibly most importantly, higher education institutions—and the learning-working ecosystem at large—need to reset individuals' expectations when it comes to lifelong learning.

For too many people, education is a burden—a necessary evil that helps them achieve an endpoint. According to Strada's Public Viewpoint Survey research, adults without degrees have ever-less confidence in the potential value of education. Between 2019 and 2020, 18 percent fewer adults without a degree believe education and training will be worth the cost. In the same time period, the number of adults without degrees who felt education and training would lead to a good job fell by 25 percent (Strada, September 15, 2020).

There's a tragic lack of confidence in the value of education, which could stem from the fact that many people look at learning as a one-time engagement. Of course, a single learning experience—rooted in a particular point in time and geared toward a specific labor market reality—cannot sustain an individual for a lifetime! But that's how our traditional education ecosystem has been structured.

According to Branon, this process of re-setting expectations is critical for a successful lifelong learning ecosystem:

> Universities will have to educate students that learning is not terminal and help them navigate the increasing options for learning. Our undergraduate and graduate schools have a responsibility to help students understand that career success depends on their willingness to keep their skills current and relevant. The key is to make sure that once students understand the expectation, they're pursuing the right certificate program to help themselves get ahead. The explosion of options is confusing for students, and universities need to consider how to coach students across different legitimate options. What if I want to change careers or fields, who helps me? When do I need a certificate? Which certificate do I need? Is a degree more appropriate? Universities will have to help students navigate this increasingly complex landscape beyond selecting an undergraduate major or a graduate school program (Branon, June 8, 2018).

As Branon highlights here, there's a second element to this expectation reset that needs to happen. While learners need to think differently about the education ecosystem and their place within it, institutions similarly need to rethink their role in a learner's life and reorient itself to offering a new, more personalized and robust form of customer service.

> "The rapid deployment of technologies that allow for immediate consumption of goods has flipped the script for consumers, including adult learners shopping for continuing education," wrote Heather Chakiris when she was UCLA Extension's chief student experience officer. "To my mind, "students as customers" means we don't force them through arbitrary processes that are intentionally complex" (Chakris, 2018).

Higher education is no different to any other industry in this regard. Quality customer service helps consumers build stronger relationships with service providers. In higher education, where there are a wide array of complex and confusing options available at multiple price points with multiple possible endpoints, guidance is crucial.

Part of the evolution toward offering a 60-Year Curriculum necessitates higher education institutions shifting gears to act more as a "guide on the side" than a "sage on the stage"—both inside and outside the classroom.

Higher education institutions in this model need to provide high-quality, personalized customer service that helps learners navigate numerous options to find the right learning opportunity for that individual at that point in time.

There are many benefits for higher education institutions who adopt this approach to learner engagement. While there will certainly be costs associated with implementing the infrastructure necessary to delivering this kind of robust customer experience, there will certainly be payoff in customer retention and purchasing over the long term.

In the eCommerce industry, 91 percent of consumers say they're more likely to shop with brands that provide relevant offers and recommendations. Further to that, 74 percent of consumers say "living profiles" with detailed personal preferences would be useful if they were used to curate personalized experiences, products and offers (Accenture, 2018).

The modern consumer clearly craves seamless engagement with service providers and solutions tailored to their unique needs. Higher education institutions would do well to respond to this demand—both in service of their mission and in service of the bottom line.

Credential Primacy: Making Space for Alternative Credentials, Non-Degree Credentials and Microcredentials

Secondly, the shift to a 60YC model necessitates greater recognition for the value of non-credit, alternative credentials. While generally unaccredited—and often developed without the oversight of faculty senate boards—these offerings are market-responsive and are designed to address the outcome expectations of learners unable to pay using financial aid.

This means these offerings must be relevant to specific labor market needs and deliver learners a clear and immediate return on investment. In many ways, this is a more rigorous quality standard than many accreditation processes require!

Turning back to Lambert, this shift will require creativity and openness to change on the part of traditional academia:

> Curricular integrity in the 60YC model has a different meaning than a university department. It means that in addition to the traditional, time-based learning offerings—which are likely the best in the world learning experiences for those who can immerse themselves in a full-time residential experience—the school must also offer modular, stackable and portable credentials for part-time learners who cannot control time and place of learning. This needs to be packaged for professional relevance and wrapped with highly personalized services for a relationship lasting at least 60 years (Lambert, 2019).

To that end, this shift requires colleges and universities to more honestly observe a wider Spectrum of Access that provides diversified opportunities for learning tailored for a wider audience.

The concept of the Spectrum of Access was initially presented by Oregon State University's Dave King, and it illustrated as follows (Figure 3).

As can be seen, this model requires intentional unbundling and modularization across program areas, creating stackable pathways to accessing and developing knowledge. This model also allows individuals without the financial or personal resources to enroll in a comprehensive degree offering to access the learning solution that allows them to reach their intended outcome.

At its core, this flexibility is what delivers the student-centricity inherent to the 60YC—the right learning opportunity for the right learner at the right time.

> "The benefit of certificates that stack into degrees is that shorter programs may appeal to students who need or want to work while learning," wrote Joann Kozyrev, vice president of design and development at Western Governors University. "They also create clear pathways with incremental goals, allow for more agility in curriculum design and delivery, and support a more open-loop lifelong approach to education" (Kozyrev, 2018).

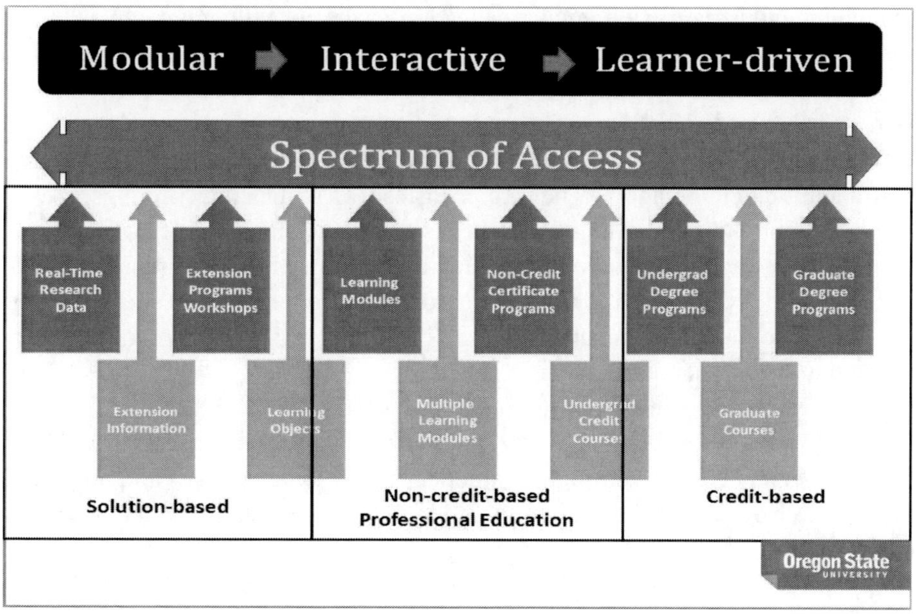

Figure 3. https://evolllution.com/opinions/honing-spectrum-learner-access/.

This approach to programming necessitates more than a culture shift. The policies, processes and technologies of the institution are designed for a traditional, faculty-centric academic model.

Shifting toward a more flexible approach to program development and student access requires the adoption of student-centric technologies designed to support non-credit program design, development, delivery and registration.

Learning Recognition: A New Role for the Institution

Finally, the shift to a 60YC model (as outlined above) necessitates a new role for the higher education institution. No longer the gatekeeper of knowledge, the modern institution evolves to serve as a learning consultant and assessment. The institution facilitates more flexible and global approaches to education and provides learners the tools to validate and credential that knowledge.

Lambert elaborates on this new reality:

> Future leaders in higher education will be the masters of information about great learning experiences that help the learners in their network get the right learning at the right time. That could be through an MBA on campus, a stackable graduate certificate, a coding bootcamp online or an intensive hybrid professional development program. Of course, each school will need to qualify each experience to meet their brand goals, but in the long run any individual school is unlikely to supply more than 20 percent of any learners' solutions from their faculty-based degree programs.
>
> The winners will have an information-based learner relationship management system business, not a teaching- and learning-based business (Lambert, 2019).

Through the industrial period, higher education institutions broadly recreated the wheel at campuses across the country. This was necessary to create opportunities to learn for a wider population of people who simply could not have otherwise accessed the knowledge the institution was able to deliver.

However, in the internet age—in an era where knowledge is widely accessible—higher education institutions must adopt a new role. Instead of replicating offerings and becoming increasingly commoditized and fungible, colleges and universities must find ways to facilitate, curate, contextualize and consult the lifelong learning journeys for diverse populations.

This is not to say the traditional undergraduate experience need disappear. Instead, it highlights the opportunity for institutions to deliver an experience that equates to something greater than the sum of their parts.

In some ways, we're already seeing this in the way colleges and universities are building on Google's IT Support Professional Certificate, offered through Coursera. While learners can enroll in the Coursera certificate program independently, some colleges and universities have built a wider offering around the program to add context and richness to the series of courses (McKenzie, 2019). It also creates access to high-quality career services and potential further learning opportunities at the institution. This represents a new role for postsecondary institutions, who are leveraging the insights and knowledge of experts and adding an additional layer of context and pedagogy to ensure learners maximize their chances at success.

Role of Continuing Education in Pioneering the 60YC

The shift toward a 60-Year Curriculum environment implies significant changes for higher education institutions in how they serve learners, how they structure programs and in how they position the institution more broadly.

Delivering on this transformation requires an array of cultural and operational changes designed to maximize student-centricity. Fortunately for the majority of colleges and universities operating today, there is a division on campus doing this work already, though they are generally under-resourced, misunderstood and siloed: the continuing, professional and workforce education (CE) division.

These units are charged with developing offerings that support labor market upskilling and reskilling. They serve the community with a wide array of offerings ranging from public health workshops to interest-based courses like wine tasting. Most importantly, they tend to be the connection point between the institution and local industry and employers—responding to labor market needs with timely offerings that bridge specifically-identified skills gaps. For some CE divisions, this even positions them to serve as consultants in structuring talent development pipelines for local employers.

They don't restrict offerings to fit the bounds of accreditation or the demands of the faculty-senate. Instead, offerings are market-responsive and their quality metrics focus on whether (and how robustly) learners achieve their goals. And as highlighted earlier, since these offerings are generally not eligible for financial aid—and therefore learners are paying out of pocket—it's critical that they deliver clear and immediate returns on learning investment.

Clearly, the CE division is best-positioned to help the institution adapt to the demands of the 60YC. And at some institutions, this vision is being realized through what's being termed a "Consolidated Administration" model (Modern Campus, nd).

The Consolidated Administration model explains the nature of collaboration between the CE division and main campus faculties. Within the structure of this model, CE pivots into an administrative service partner focused on managing programming, delivering high-quality student administrative experiences and engaging with local industry and employers. The faculty role pivots to purely developing and delivering academic offerings that address the needs of learners and employers.

This model addresses a few core issues that tend to stand in the way of scaling CE and of delivering true student-centricity.

First, it minimizes the tendency for divisions within single institutions to compete against one-another for the same corporate contracts or enrollments.

Secondly, it minimizes the tendency for institutions to have duplicated offerings delivering effectively the same learning outcomes but offered by different divisions.

These two shifts streamline engagement for both learners and external organizations looking to the institution for learning solutions. According to Kristy Davis, director for academic support resources-IT at the University of Minnesota—who have established the Consolidated Administration model for non-credit education—this creates clear benefits for stakeholders.

> "In our earlier environment, we had different units using a variety of third-party vendors to receive revenues from registration," she said. "With the single back end, supported by Academic Support Resources, we're able to achieve consistency across our non-credit offerings that today's learners—who are experienced customers—respond well to" (Davis, 2018).

Thirdly, this model allows colleges to take advantage of economies of scale in its CE offerings. In this way, institutions can streamline the management of non-degree, online, continuing and workforce education offerings that don't conform to standard administrative models into systems designed to support these complex courses and programs.

What's more, this approach improves staff efficiency and effectiveness organization-wide by taking the complexity of non-degree program administration off their plates. At Temple University, Associate Vice Provost Nicole Westrick highlighted the impact they saw immediately after adopting the Consolidated Administration model for CE:

"In our first full year with the system, our revenue was $9.6 million. This year, we're on target for over 30 percent growth and fully anticipate $12 million in revenue," Westrick said. "It makes sense because we're helping people shift their focus from administrative work to the things that really matter: program development, marketing and curriculum" (Modern Campus, nd).

CE divisions have the expertise, the insight and the customer-centric culture necessary to spearhead the shift toward a 60YC. Executing on the vision, however, requires active and conscious support from institutional executives to bring these units into the strategic center of the institution.

Conclusion

The 60YC model for higher education's future relies on a number of policy, process and culture changes taking root at colleges and universities. What's especially important to note here is that—by virtue of this being a truly student-centric enterprise—every institution's execution of the 60YC model will be unique.

Every institution is (or should be!) geared to serve different learners by leveraging their specific expertise and a unique point of view. To that end, the 60YC provides a pathway away from the commoditized and fungible higher education industry we currently operate in.

To take this concept to the next level, University of Washington's Rovy Branon proposes five key areas for strategic development that I will review here in closing:

1. Credentials: Higher education institutions must adopt new markets for formal learning achievement outside the degree, and develop more granular approaches to track skill and knowledge development.
2. The 60-Year Metacurriculum: Higher education institutions must consider the metacurricuum as "a framework for learning activities rather that a specific scope and sequence of instructional content" (Branon, November 16, 2018).
3. Learner Services: Since the 60YC lasts a lifetime, academic services and career support need also last a lifetime. Additionally, the institution must be positioned to guide students into the right offerings for their needs at specific points in time. This consultative approach requires a new approach to learner services.
4. New Academic Tech Stack: Delivering more complex program offerings and more personalized learner services requires a technology stack designed to support lifelong customer engagement.
5. Policies and Funding: It's critical to restructure financial aid models to incentivize learning over a lifetime. The current model only really incentivizes learning in the first stage of the three-stage life, which is no longer relevant.

The 60-Year Curriculum model presents an exciting and challenging future for higher education. For colleges and universities that recognize the need for this shift and pivot appropriately—leveraging their strength in CE along the way—the opportunities for growth will be great. What's more, these institutions will maximize the impact they have on their communities and local economies.

The execution of this vision depends on CE leaders being able to adequately articulate the opportunity, and senior executive leaders being willing to adapt in the face of significant threat.

References

Accenture (2018). *Making it personal: Why brands move from communication to conversation for greater personalization.* https://www.accenture.com/_acnmedia/PDF-77/Accenture-Pulse-Survey.pdf, retrieved December 21, 2021.

Branon, R. (June 8, 2018). *Addressing the increasing need for non-credit programming: The university and true lifelong education. The EvoLLLution: A Modern Campus Illumination.* https://evolllution.com/revenue-streams/market_opportunities/addressing-the-increasing-need-for-non-credit-programming-the-university-and-true-lifelong-education/, retrieved December 21, 2021.

Branon, R. (November 16, 2018). *Learning for a lifetime: A 100-year life requires a 60-year curriculum. Inside Higher Ed.* https://www.insidehighered.com/views/2018/11/16/why-longer-lives-require-relevant-accessible-curricula-throughout-long-careers, retrieved December 21, 2021.

Bransberger, P. Falkenstern, C., and Lane, P. (2020). *Knocking at the College Door: Projections of High School Graduates: December 2020. Western Interstate Commission for Higher Education*, Boulder, CO. https://knocking.wiche.edu/wp-content/uploads/sites/10/2020/12/Knocking-pdf-for-website.pdf, retrieved December 21, 2021.

Burning Glass Technologies (2020). *Bad Bets: The High Cost of Failing Programs in Higher Education.* https://www.burning-glass.com/research-project/bad-bets-high-cost-failing-programs/, retrieved December 21, 2021.

Career Karma (2021). *State of the bootcamp market 2021.* https://careerkarma.com/blog/wp-content/uploads/2021/04/Bootcamp-Market-Report-2021.pdf.

Chakiris, H. (August 31, 2018). *Creating a champagne experience on a beer budget: delivering on modern learners' customer service expectations. The EvoLLLution: A Modern Campus Illumination.* https://evolllution.com/attracting-students/customer_service/creating-a-champagne-experience-on-a-beer-budget-delivering-on-modern-learners-customer-service-expectations/, retrieved December 21, 2021.

CNBC (April 30, 2020). *US weekly jobless claims hit 3.84 million, topping 30 million over the last 6 weeks.* https://www.cnbc.com/2020/04/30/us-weekly-jobless-claims.html, retrieved December 21, 2021.

Davis, K. (January 22, 2018). *Three main benefits to centralizing non-credit program management. The EvoLLLution: A Modern Campus Illumination.* https://evolllution.com/technology/tech-tools-and-resources/three-main-benefits-to-centralizing-non-credit-program-management/, retrieved December 21, 2021.

Dellas, H. and Sakellaris, P. (2003). *On the cyclicality of schooling: theory and evidence. Oxford Economic Papers*, 55(1), 148-172. https://www.jstor.org/stable/3488876, retrieved December 21, 2021.

Elliott-Bielecki, B. (October 21, 2021). *The Benefits of Seeing Students as Customers. The EvoLLLution: A Modern Campus Illumination.* https://evolllution.com/attracting-students/customer_service/the-benefits-to-seeing-students-as-customers/, retrieved December 21, 2021.

Hussar, W., and Bailey, T. (2019). *Projection of Education Statistics to 2027. National Center for Education Statistics, U.S. Dept. of Education.* https://nces.ed.gov/pubs2019/2019001.pdf.

Kozyrev, J. (October 30, 2018). *Stackable credentials: defining their value. The EvoLLLution: A Modern Campus Illumination.* https://evolllution.com/programming/ credentials/ stackable-credentials-defining-their-value/, retrieved December 21, 2021.

Lambert, H. (March 20, 2019). *Redefining norms critical to sustained relevance in the changing postsecondary environment. The EvoLLLution: A Modern Campus Illumination.* https://evolllution.com/ revenue-streams/market_opportunities/redefining-norms-critical-to-sustained-relevance-in-the-changing-postsecondary-environment/, retrieved December 21, 2021.

McKenzie, L. (June 14, 2019). *Google's growing IT certificate.* Inside Higher Ed. https://www.insidehighered.com/digital-learning/article/2019/06/14/google-it-certificate-program-expands-more-community-colleges, retrieved December 21, 2021.

McKinsey Global Institute (2017). *Jobs lost, jobs gained: What the future of work will mean for jobs, skills, and wages. McKinsey & Company.* https://www.mckinsey.com/featured-insights/future-of-work/jobs-lost-jobs-gained-what-the-future-of-work-will-mean-for-jobs-skills-and-wages, retrieved December 21, 2021.

Modern Campus (nd-a). *Consolidated administration: The key to delivering a 60-year curriculum.* https://moderncampus.com/whitepaper/consolidated-administration-the-key-to-delivering-a-60-year-curriculum.html, retrieved December 21, 2021.

Modern Campus (nd-b). *Non-Credit at Temple University.* https://moderncampus.com/case-study/non-credit-at-temple-university.html, retrieved December 21, 2021.

National Center for Education Statistics (NCES; 2002). *Nontraditional Undergraduates: Findings from the Condition of Education 2002. U.S. Dept. of Education.* https://nces.ed.gov/pubs2002/ 2002012.pdf, retrieved December 21, 2021.

National Student Clearinghouse Research Center (Spring 2021). *Overview: Spring 2021 Enrollment Estimates.* https://nscresearchcenter.org/wp-content/uploads/ CTEE_Report_Spring_2021.pdf, retrieved December 21, 2021.

Reed, B. (October 11, 2021). *Re-imagining higher education through a career-readiness lens. The EvoLLLution: A Modern Campus Illumination.* https://evolllution.com/revenue-streams/workforce_development/re-imagining-higher-education-through-a-career-readiness-lens/, retrieved December 21, 2021.

Richards, J. and Dede, C. (October 26, 2020). *The 60-year curriculum: A strategic response to a crisis. Educause.* https://er.educause.edu/articles/2020/10/the-60-year-curriculum-a-strategic-response-to-a-crisis, retrieved December 21, 2021.

Russell, J. (January 28, 2019). *Lifelong learning for the 100-year life. The EvoLLLution: A modern campus illumination.* https://evolllution.com/revenue-streams/extending_lifelong_learning/ lifelong-learning-for-the-100-year-life/, retrieved December 21, 2021.

Shaeffer, J. (April 5, 2019). *Stepping Back from the Cliff: Facing New Realities of Changing Student Demographics. The EvoLLLution: A Modern Campus Illumination*, April 5, 2019. https://evolllution.com/managing-institution/higher_ed_business/stepping-back-from-the-cliff-facing-new-realities-of-changing-student-demographics/, retrieved December 21, 2021.

Strada (July 15, 2020). *Public Viewpoint: Enrolling in Education: Motivations, Barriers, and Expectations*. https://cci.stradaeducation.org/pv-release-july-15-2020/, retrieved December 21, 2021.

Strada (September 15, 2020). *Confidence in the value of education has fallen among adults without degrees seriously considering additional education.* https://stradaeducation.org/press-release/confidence-in-the-value-of-education-has-fallen-among-adults-without-degrees-seriously-considering-additional-education/, retrieved December 21, 2021.

Strada (September 16, 2020). *Public Viewpoint: Interested But Not Enrolled: Understanding and Serving Aspiring Adult Learners.* https://cci.stradaeducation.org/pv-release-september-16-2020/, retrieved December 21, 2021.

Strada Education (January 27, 2021). *Public Viewpoint: Covid-19 Work and Education Research.* https://cci.stradaeducation.org/wp-content/uploads/sites/2/2021/01/Report-January-27-2021.pdf, retrieved December 21, 2021.

Strada-Gallup (2018). *Why Higher Ed? Top Reasons U.S. Consumers Choose Their Educational Pathways. Gallup.* https://news.gallup.com/reports/226457/why-higher-ed.aspx, retrieved December 21, 2021.

U.S. Dept. of Education (2015). Demographic and Enrollment Characteristics of Nontraditional Undergraduates: 2011-2012. https://nces.ed.gov/pubs2015/ 2015025. pdf, retrieved December 21, 2021.

World Economic Forum (2018). *The Future of Jobs Report* 2018. https://www3.weforum.org/docs/WEF_Future_of_Jobs_2018.pdf, retrieved December 21, 2021.

Chapter 14

Sustainability Education and Research in Action (SERINA): An IACEE Perspective for Sustainable Career Pathways in Engineering

Errol la Grange[1,*], Anthony Smith[2] and Ragna Ann Berge[3]

[1]Cahoot Learning, Melbourne, Australia

[2]IACEE, Melbourne, Australia

[3]Norwegian University of Science and Technology, Trondheim, Norway

Abstract

This chapter sets out to present the aims and outcomes of the journey of the *International Association for Continuing Engineering Education (IACEE)* in building and continuously expanding on a framework of hope that highlights inspirational examples of active practice of sustainability initiatives from around the world. This is done through a key initiative of the organisation, *Sustainability Engineering Education and Research in Action (SERINA)*, which connects the sustainability initiative examples to the United Nations 17 Sustainable Development Goals (SDGs) and Fundamental Human Needs (FHNs) as proposed by Max-Neef.

Keywords: SERINA, IACEE

Introduction

In 2020 and 2021, the world was turned upside down by a global pandemic. Looking back at this period in ten, twenty, thirty years, will this period prove to have been a catastrophe for humanity, or the catalyst for necessary transformation?

The passing of time affords the luxury, wisdom and relative certainty of hindsight. But what about for now and the path ahead where we do not yet know what we don't know or could reasonably predict based on what we do know?

The answer to that depends on one's perspective. Most likely both outcomes will form part of the reality to varying degrees, in addition to other foreseen and unforeseen consequences to a lesser or greater extent.

Of the many possible reflections of the events of 2020 and 2021, a standout may be that just about anything and everything is possible. This relates to the work of nature, the nature of work, human behaviour, and the ensuing consequences both positive and negative. There has undoubtedly been a rewrite of the playbook of life in that regard.

[*] Corresponding Author's Email: errol@cahootlearning.com.

Perhaps the most impactful takeout that has become much more clearly evident, is that people (as individuals and collectively) can and do make a real difference at micro and macro levels, driven by the power of their beliefs, nature of their intent, and extent of commitment of their actions in alignment with their beliefs and intent.

The IACEE [1] has long held the view (over the more than thirty years since its inception), that engineers from all corners of our planet, learning, working, and collaborating together are a significant force for positive change in solving structural challenges that impede the development of societies and their wellbeing.

In terms of sustainability of people and the planet the IACEE recognises the scale and magnitude of the structural challenges that lie ahead.

In 2015, the organisation embarked on a journey to engage with its members from around the world to find a means whereby it could make a meaningful and positive contribution to tackling the grand challenges of the 21st Century. This led to the Porto Declaration of the IACEE in 2016 in Porto, Portugal and the subsequent creation and official launch of SERINA in Monterrey Mexico in 2018 [2].

The IACEE

The IACEE [1] serves both individuals and organizations interested in the roles of and needs for professional development of engineers worldwide. These activities enable the work performed by professionals to be of the highest quality worldwide. It also enables the members to link their engineering stories and work worldwide to create a sustainable future for all humankind and our world.

The Porto Declaration

During the IACEE World Conference of 2016 in Porto, Portugal, the members developed and signed up to the *Porto Declaration*

Porto Declaration, May 20, 2016

"Whereas the International Association for Continuing Engineering Education (IACEE) was founded in 1989 to foster a global network of organizations promoting lifelong engineering education.

Whereas the IACEE recognises the scale and complexity of the gap between existing solutions and the needs facing our planet and that the IACEE is uniquely placed to act on this opportunity.

Whereas the IACEE seeks to pivot the organisation to connect individuals, universities, industry, government and NGO organizations to meet the grand challenges facing humanity.

Now therefore in keeping with its dedication to leading lifelong learning, the IACEE will develop global initiatives to address those 21st century challenges threatening the survival of humankind through collaboration, design, creative thinking and engineering.

We the undersigned do hereby declare this at the IACEE 2016 Global Conference in Porto Portugal and pledge our commitment in actioning this call to service."

Global Sustainability Initiatives and the SERINA Journey

As a means to activate the Porto Declaration, the IACEE created the portfolio of Global Sustainability Initiatives as a key strategic direction for the organisation.

In response to this, a research project was proposed and initiated with the task of identifying what was already being achieved in relation to sustainability within the membership's body. The initial aim was to research and create a list of sustainability practices within the IACEE membership bodies so as:

> "… to prioritise the low hanging fruit of influential organisations where we have our best connections and who would be most likely to be working in the sustainability space. Sustainability matters in the 21st Century can be broad as we have very many global challenges to consider" [3]

The above proposal was put into action and explored by researching each of the IACEE 2017 membership's institutions websites. The IACEE Executive at its quarterly meeting in Beijing, in late 2017, made the decision as a benchmark evaluation tool to use the United Nations 17 Sustainable Development Goals (SDGs) [4]. The concept of Fundamental Human Needs (FHNs) [5, 6] was added as a second based tool for evaluating the meeting of sustainable practices across the IACEE membership.

This then led to the development of the SERINA research project [7]. This initiative was unveiled to members at IACEE's 16th World Conference in Monterrey, Mexico in May 2018 and formed a strategic pathway to this next World Conference hosted by the Norwegian University of Science and Technology, Trondheim, Norway in May 2021.

The resulting research provides evidence of significant activity in engineering sustainability initiatives taking place around the world, many of which involved the active practice of sustainable engineering within the IACEE membership. Below is a brief description of the research steps taken:

1. Accessing the membership websites and use of their website search feature to look for and list sustainable initiatives.
2. Categorising the initiatives against the United Nations 17 sustainable development goals and Manfred Max Neef's Fundamental Human Need criteria. Utilising the criteria of the UN Sustainable Development Goals classifications and Fundamental Human Needs in the case of SERINA.
3. Looking at a sample of organisations and initiatives beyond that of the immediate membership base for possible best practices.
4. Cherry picking some of the most relevant sustainability links to act as examples and to make these available via an IACEE website portal, being a website on which the IACEE would have an explanation of what this project is all about, being connected to the Porto Declaration, meeting certain UN Goals and Fundamental Human Needs.

5. Taking SERINA live for the first time in Facebook [7] in July 2018 and then also as a stand-alone IACEE website [8] and finally on LinkedIn [9].

SERINA posts are created weekly from internet searches using primarily "Sustainable Engineering Education," "Sustainable Engineering Research" or "Sustainable Engineering Active Practice," "Circular Economy" amongst others. There was also some reliance on using members suggestions.

Once a relevant post is found, it is analysed against the following three sets of criteria:

1. Under which of the main heading or headings can the post be classified?
   - Sustainable Engineering Education
   - Sustainable Engineering Research
   - Sustainable Engineering Active Practice
   - The Circular Economy [10]
2. Which UN SD Goals then are being met?
   - Goal 1 *End Poverty*
   - Goal 2 *Zero Hunger*
   - Goal 3 *Good Health and Well-Being*
   - Goal 4 *Quality Education*
   - Goal 5 *Gender Equality*
   - Goal 6 *Clean Water and Sanitation*
   - Goal 7 *Affordable and Clean Energy*
   - Goal 8 *Decent Work and Economic Growth*
   - Goal 9 *Industry, Innovation, and Infrastructure*
   - Goal 10 *Reduce Inequality*
   - Goal 11 *Sustainable Cities and Communities*
   - Goal 12 *Responsible Consumption and Production*
   - Goal 13 *Climate Action*
   - Goal 14 *Life Below Water*
   - Goal 15 *Life on Land*
   - Goal 16 *Peace, Justice and Strong Institutions*
   - Goal 17 *Partnerships for the Goals*
3. Which Fundamental Human Needs are being met?
   - Subsistence
   - Protection
   - Understanding and Meaning
   - Participation
   - Identity
   - Creation

This research approach has provided a positive narrative of *Sustainability Education, Research and Action* initiatives, which shifts the focus away from a doom and gloom framing of the challenges to one of positive active practice in meeting the UN Goals and humankind's Fundamental Human Needs.

The intention of IACEE with SERINA is to be solution focused and to highlight, encourage, support and grow the incredible work that is being done at all levels by people and organisations throughout the world to empower and inspire others to do the same.

A Snapshot of SERINA

As of October 2021 IACEE has made 425 SERINA research posts, which has reached 9,552 interested parties, resulting in 2,009 engagements.

Below is a list in order of the most popular SDG engagements:

Goal 13	*Climate Action*	197 posts
Goal 11	*Sustainable Cities and Communities*	176 posts
Goal 9	*Industry, Innovation, and Infrastructure*	149 posts
Goal 7	*Affordable and Clean Energy*	122 posts
Goal 12	*Responsible Consumption and Production*	109 posts
Goal 17	*Partnerships for the Goals*	85 posts
Goal 4	*Quality Education*	90 posts
Goal 17	*Partnerships for the Goals*	85 posts
Goal 6	*Clean Water and Sanitation*	80 posts
Goal 14	*Life Below Water*	58 posts
Goal 15	*Life on Land*	55 posts
Goal 8	*Decent Work and Economic Growth*	54 posts

The post which reached the most people (489) and engagements (64) was headed:

"Readying Todays Higher Ed Students to tackle the Worlds Grand Challenges"

"An ABET Issue Brief Fall 2018 SUMMARY: Advancing sustainable solutions to the world's most pressing environmental challenges will take more than technical skills. Students of higher education, especially those in "STEM/STEAM" fields, need diverse and global perspectives to understand—and address—long standing problems in new ways" [11]

Science, Technology, Engineering and Mathematics (STEM) [12] or at a deeper level Science and Technology as interpreted through Engineering and the Arts and based in Mathematical elements (STEAM) [13] plays a crucial part in thinking in a sustainable way particularly for young learners in both High School and University. Both of these approaches provide a hands-on activity embedded methodology by applying science knowledge to real-world applications through technology, engineering, art and math. The SERINA knowledge base contains many examples of these approaches to education initiatives.

Sharing Some of the Most Read and Shared Stories, to Engage and Enlighten, as Examples of the Enormous Engagement from Around the World [7]

The importance of the engineer, SDG2 Zero Hunger: An eye-opening introduction to engineering's role in the SDGs.

THE United Nations has 17 Sustainable Development Goals to eradicate extreme poverty and hunger, promote economic growth and prosperity, improve health and education and protect the planet.

Engineering has been identified as one of the professions that will be critical to achieving these goals. Turns out, engineers may just be the superheroes who save the world!

http://www.bbc.com/.../western-australia-sustainable...

Goal 13 Climate Action. 197 posts

- *What is the circular economy and how do you start to explain it?*

"Four of the Ellen MacArthur Foundation's experts will talk you through four different ways to start the conversation about the circular economy as a fundamentally different way of designing, making and operating, while sharing plenty of examples of it in action along the way.

Do you have your own way of introducing the concept? Come prepared to interact here in SERINA by sharing below!"

- Sustainable Engineering: "Active Practice" "Circular Economy"
- UN SD Goals: 11 "Sustainable Cities and Communities," 12 Responsible Consumption and Production," 13 "Climate Action"
- Fundamental Human Need for "Protection," "Understanding and Meaning," "Participation" and "Creation"
  https://www.facebook.com/SERINA.iacee/posts/4391185884264128
  https://www.youtube.com/watch?v=X20aR9ykpyE
  https://www.facebook.com/SERINA.iacee/posts/1992957747420299

- *What an awesome African initiative!*

South Africa's Nelson Mandela University's Sustainability Research Unit: "The Sustainability Research Unit is home to a group of critical thinkers dedicated to promoting the sustainability and resilience of social-ecological systems. We focus on the crossing point between ecosystems and society, on understanding the processes that influence human behaviour and therefore the sustainability of society, water, earth, plant and animal resources".

- Sustainable Engineering: "Education," "Research," "Active Practice" "Circular Economy"

- UN SD Goals: 4 "Quality Education," 5 "Gender Equality," 13 "Climate Action," 14" Life Below Water," 15 "Life on Land," and 17 "Partnerships for the Goals"; https://sru.mandela.ac.za/About-us

Sustainable Engineering Practice: COURSE LEVEL Undergraduate

Course aim: to introduce students to the engineering profession through course and project work and interaction with industry. All are designed to help students to develop some of the essential personal and professional skills needed for sustainable professional practice.

Course content: Lectures and tutorials will introduce students to a variety of areas in sustainable and professional engineering, including the role of the professional engineer, graduate attributes of engineering professionals, career management skills, critical analysis and reflective practice, principles of sustainable engineering, professional ethics, intercultural issues including Indigenous protocols, basic project management principles, effective teamwork, team roles, evaluation of team performance, engineering report writing, effective presentations, locating and using information and referencing."

- Sustainable Engineering: "Education"
- UN SD Goals: 4 "Quality Education," 5 "Gender Equality," 9 "Industry, Innovation, and Infrastructure," 11 "Sustainable Cities and Communities," 13 "Climate Action".
- Fundamental Human Need for "Protection," "Understanding and Meaning," "Participation" and "Creation".
  https://www.facebook.com/SERINA.iacee/posts/3468509893198403
  https://study.unisa.edu.au/courses/101752.

Goal 11 Sustainable Cities and Communities. 176 posts

- *Batteries of the future set to be cheaper and better*

"Australian researchers say they've opened the path to a new generation of batteries that could allow an electric vehicle to drive from Melbourne to Sydney on a single charge... Electric vehicles (EVs), mobile phones and other consumer electronics mostly use lithium-ion batteries that were commercialised in the 1990s.

They're made using toxic and exotic materials such as cobalt, nickel and manganese that are in increasingly short supply around the world."

- Sustainable Engineering: "Research"
- UN SD Goals: 7 "Affordable and Clean Energy," 9 "Industry, Innovation, and Infrastructure," 11 "Sustainable Cities and Communities," 12 Responsible Consumption and Production," 13 "Climate Action"
- Fundamental Human Need for "Protection," "Participation," and "Creation"
  https://www.facebook.com/SERINA.iacee/posts/4291281777587873
  https://www.abc.net.au/.../batteries-lithium.../100457492

- *Untapped Potential being tapped!*

TRA is among a handful of companies finding solutions to the textile waste problem. Others include Sydney-based Blocktexx, which believed it was one of only three companies globally to find a way to chemically break clothing down into its "building blocks".

A man, slightly frowning, in a black suit standing in front of a black brick wall. Adrian Jones said he was inspired to find another way to deal with unwanted textiles after a career in the fashion industry. "Bed sheets, or work shirts, or football uniforms … through a chemical process we separate those back into polyester, and cellulose, which can both be reused as high value, raw materials back into other industries,"- Adrian Jones, co-founder of Blocktexx, said. The company has patented its technology and will begin constructing a facility to handle industrial quantities of textiles near Brisbane, later this year.

Mr Jones said more than 30,000 tonnes had already been committed, and the new plant will be able to process 3,800 tonnes a year initially, and up to 10,000 tonnes a year long-term."

""We've already sold the output, two or three times over, because people want to use it locally, into injection moulding, into building materials," he said.""

- Sustainable Engineering: "Active Practice" "Circular Economy"
- UN SD Goals: 9 "Industry, Innovation, and Infrastructure," 11 "Sustainable Cities and Communities," 12 Responsible Consumption and Production," 13 "Climate Action,"
- Fundamental Human Need for "Protection," "Understanding and Meaning," "Participation" and "Creation"
  https://www.facebook.com/SERINA.iacee/posts/3634894569893267
  https://www.abc.net.au/news/2021-01-27/bathurst-regional-council-textile-recycling-clothing-waste/13087864

- *What is the circular economy and how do you start to explain it?*

Four of the Ellen MacArthur Foundation's experts will talk you through four different ways to start the conversation about the circular economy as a fundamentally different way of designing, making and operating, while sharing plenty of examples of it in action along the way. Do you have your own way of introducing the concept? Come prepared to interact and share below!

- Sustainable Engineering: "Active Practice" "Circular Economy"
- UN SD Goals: 11 "Sustainable Cities and Communities," 12 Responsible Consumption and Production," 13 "Climate Action"
- Fundamental Human Need for "Protection," "Understanding and Meaning," "Participation" and "Creation"
  https://www.facebook.com/SERINA.iacee/posts/4391185884264128
  https://www.youtube.com/watch?v=X20aR9ykpyE

Goal 9 Industry, Innovation, and Infrastructure. 149 posts

- *An awesome project!!!!*

"A tidal-powered turbine, which its makers say is the most powerful in the world, has started to generate electricity via the grid in Orkney. The Orbital O2 has the capacity to meet the annual electricity demand of 2,000 homes for the next 15 years."

- Sustainable Engineering: "Active Practice"
- UN SD Goals: 7 "Affordable and Clean Energy," 9 "Industry, Innovation, and Infrastructure," 13 "Climate Action"
- Fundamental Human Need for "Protection," "Participation," and "Creation"
  https://www.facebook.com/SERINA.iacee/posts/4153476714701714
  https://www.bbc.com/.../uk-scotland-north-east-orkney...

- *Self-Driving Electric Ferries in Norway*

"In the next 10 years, the way we travel day-to-day will change dramatically with transport innovations that make mobility easier and greener. Autonomous electric ferries like Zeabuz's will allow people living in cities with plentiful waterways to commute via driverless boats. Click through to see the innovations that will transform the commutes of the future."

- Sustainable Engineering: "Research," "Active Practice"
- UN SD Goals: 7 "Affordable and Clean Energy," 9 "Industry, Innovation, and Infrastructure," 11 "Sustainable Cities and Communities," 12 Responsible Consumption and Production," 13 "Climate Action"
- Fundamental Human Need for "Protection," "Participation" and "Creation"
  https://www.facebook.com/SERINA.iacee/posts/3449695768413149
  https://edition.cnn.com/.../norway-self.../index.html

- *London Is "Rewilding"*

"The British capital is resurrecting its long lost natural habitats to foster biodiversity...."

"London's Wild West End project is a prime example. Launched in 2015, the scheme set out to turn the city's fragmented natural landscape into a place where native species could thrive, weaving pockets of greenery across rooftops, at street level and along the sides of buildings. The goal was simple: to create pathways of natural habitat along which wildlife can travel and flourish unfettered by human activity."

- Sustainable Engineering: "Active Practice"
- UN SD Goals: 3 "Good Health and Well-Being, 9 "Industry, Innovation, and Infrastructure," 11 "Sustainable Cities and Communities," 13 "Climate Action".
- Fundamental Human Need for "Protection," "Understanding and Meaning," "Participation" and "Creation"
  https://www.facebook.com/SERINA.iacee/posts/3575067462542645

https://reasonstobecheerful.world/london-is-rewilding.../

Goal 7 Affordable and Clean Energy. 122 posts

- *Solar panels and water canals could form a real power couple in California*

"California has around 4,000 miles of canals that shuttle clean water throughout the state. New research shows that these canals can do way more than bringing California's residents with drinking water—paired with solar panels, these canals might also be a way to both generate solar power and save water.

This new study presents an analysis from researchers at the University of California Merced and University of California Santa Cruz that quantifies the economic feasibility of building a "solar canal" system in the state."

- Sustainable Engineering: "Active Practice" "Circular Economy"
- UN SD Goals: 7 "Affordable and Clean Energy," 9 "Industry, Innovation, and Infrastructure," 13 "Climate Action"
- Fundamental Human Need for "Protection," "Participation" and "Creation"
https://www.facebook.com/SERINA.iacee/posts/4303971219652262
https://www.popsci.com/story/environment/solar-panels-cover-water-canals

- *Tasmania poised to play a key role in country's renewable future*

"Tasmania's wind and hydro power assets will play a significant role in helping the rest of the country reach net zero greenhouse gas emissions." A young family has seen their yearly power bills shrink by thousands of dollars after installing a solar power system.

- Sustainable Engineering: "Active Practice" "Circular Economy"
- UN SD Goals: 7 "Affordable and Clean Energy," 9 "Industry, Innovation, and Infrastructure," 11 "Sustainable Cities and Communities," 13 "Climate Action" and 17 "Partnerships for the Goals"
- Fundamental Human Need for "Protection," "Participation," and "Creation"
https://www.facebook.com/SERINA.iacee/posts/4374443829271667
- Tasmania poised to play a key role in country's renewable future
- A solar panel in space is collecting energy that could one day be beamed to anywhere on Earth

"Scientists working for the Pentagon have successfully tested a solar panel the size of a pizza box in space, designed as a prototype for a future system to send electricity from space back to any point on Earth. The panel -- known as a Photovoltaic Radiofrequency Antenna Module (PRAM) -- was first launched in May 2020, attached to the Pentagon's X-37B unmanned drone, to harness light from the sun to convert to electricity. The drone is looping Earth every 90 minutes..."

- Sustainable Engineering: "Research"
- UN SD Goals: 7 "Affordable and Clean Energy," 9 "Industry, Innovation, and Infrastructure," 11 "Sustainable Cities and Communities," 13 "Climate Action" and 17 "Partnerships for the Goals"
- Fundamental Human Need for "Protection," "Participation" and "Creation"
  https://www.facebook.com/SERINA.iacee/posts/3715351091847614
  https://edition.cnn.com/2021/02/23/americas/space-solar-energy-pentagon-science-scn-intl/index.html

Goal 12 Responsible Consumption and Production. 109 posts

- *Straws are the Solution?*

Wheat vs Others: Wheat-stem drinking straws offer an excellent, environmentally sustainable and economically feasible alternative to disposable plastic straws.

WHAT? A WHEAT STEM?

... these straws can revolutionize how we consume beverages. Sometimes to go ahead it is worth to look back. Wheat stems were the drinking straws that we used on a daily basis before the nasty invention of plastic ones... Sustainable, Biodegradable, Compostable. WHEAT is the best alternative for drinking straws. There are no other straws on the market that is so life-friendly."

- Sustainable Engineering: "Active Practice"
- UN SD Goals: 6 "Clean Water and Sanitation," 12 Responsible Consumption and Production," 13 "Climate Action," 14" Life Below Water," 15 "Life on Land"
- Fundamental Human Need for "Protection," "Participation and "Creation"
  https://www.facebook.com/SERINA.iacee/posts/2271295682919836
  https://www.whatawheat.com

- *Why does toothpaste come in a box?*

In Island 90% av the toothpaste comes without a box.
https://www.facebook.com/SERINA.iacee/posts/2080650165317723

- Sustainable Engineering: "Active Practice" "Circular Economy
- UN SD Goal 11 "Sustainable Cities and Communities," "12 "Responsible Consumption and Production"
- Fundamental Human Need for "Protection".

Principles of Sustainable Engineering

"Engineering in context, engineering with a conscience, engineering for a finite planet and the indefinite future"-Benoit Cushman-Roisin

"Sustainable engineering should be based on principles that support sustainable development, as defined in the upper sections of this lesson. Engineering forms an interface between the design (i.e., the idea how to provide a sustainable solution to a technical problem) and implementation and production. In case of technology development, engineering phase is linked to almost every level of technology readiness spectrum. Sustainable engineering principles should be contemplated and applied early to ensure that technology development and scale-up follow the environmentally benign route. It will be hard to turn back to redo and redesign things from later stages! In that sense, the sustainable engineering principles should be taken into account in decision making for both research and industrial projects as well as in policy making and decisions regarding funding of technological research."

- Sustainable Engineering: "Education," "Research," "Active Practice" "Circular Economy"
- UN SD Goals: 1 "End Poverty," 2 "Zero Hunger," 3 "Good Health and Well-Being, 4 "Quality Education," 5 "Gender Equality," 6 "Clean Water and Sanitation," 7 "Affordable and Clean Energy," 8 "Decent Work and Economic Growth," 9 "Industry, Innovation, and Infrastructure," 10 "Reduce Inequality," 11 "Sustainable Cities and Communities," 12 Responsible Consumption and Production," 13 "Climate Action," 14" Life Below Water," 15 "Life on Land," 16 "Peace, Justice and Strong Institutions" and 17 "Partnerships for the Goals"
- Fundamental Human Need for "Subsistence," "Protection," "Understanding and Meaning," "Participation," "Identity" and "Creation"
  https://www.facebook.com/SERINA.iacee/posts/3665413946841329
  https://www.e-education.psu.edu/eme807/node/688

For the rest of the goals we share the main theme and you can visit the websites for more information;

SDG 1 End poverty

- Engineering for Sustainable Human Development: A Guide to Successful Small-Scale Community Projects. Book
  https://www.facebook.com/SERINA.iacee/posts/4187734861275899
- Engineering for a Sustainable Human Development: A Guide to Successful Small-scale Community Projects (webpage)
  https://ascelibrary.org/doi/abs/10.1061/9780784413531?fbclid=IwAR2w5pA5XBqvYpFTfIDdRTawMwo5B8_5bC7LA8O7XIu4V7yCCGDah9eB8B4
- Give-power, non-governmental organization. Active practice
- Kenya installs the first solar power plan to transform ocean water into drinking water.
  http://SERINA.iacee.org/what-an-awesome-active-practice-project/

https://www.designboom.com/technology/givepower-kenya-solar-power-plant-water-farm-ocean-water-11-27-2019/?fbclid=IwAR18Ho6WGZT0lUFFluPlDGpsR9knOAZEUcVhzqkvzKMIhMNkTSjD8BBbmUY

- How a local community is reducing Co2 emissions, as well as contributing to meeting the other UN goals
- Generating electricity by utilizing clean solar energy and reduce greenhouse gas emissions.
  https://www.goldstandard.org/projects/solar-power-project-rajasthan
  https://www.facebook.com/SERINA.iacee/posts/1900846816631393

SDG 2 Zero Hunger

- An eye-opening introduction to engineering's role in the SDGs: Three stories.

https://www.facebook.com/SERINA.iacee/posts/2378692338846836
News article on the stories

- Sowing the seeds of sustainability. News article

https://www.facebook.com/SERINA.iacee/posts/2175583125824426
News article on the story

SDG 3 Good Health and Well Being

- Inspiring Woman Invents Refugee Tents That Collect Rainwater and Store Solar Energy
  https://www.facebook.com/SERINA.iacee/posts/2523585047690897
  https://egyptianstreets.com/2018/12/27/female-architect-invents-refugee-tents-that-collect-rainwater-and-store-solar-energy/
- Economic Empowerment for women
  http://www.za.undp.org/.../limpopo-women-preserve.../
- Calling all engineers! It's time to redesign the old air conditioner! Newsarticle
  https://www.facebook.com/SERINA.iacee/posts/2645632782152789
  https://edition.cnn.com/style/article/global-cooling-prize-india/index.html

SDG 4 Quality Educaton

- Free, open educational resources from the world's leading experts on sustainable development
  https://www.facebook.com/SERINA.iacee/posts/1926780034038071
  https://sdgacademy.org
- Underrepresented Population in STEM Careers: Why Is Progress So Slow. Two speeches
  https://www.facebook.com/SERINA.iacee/posts/2763632383686161

https://mediaspace.gatech.edu/.../Underreprese.../1_5k9jouay

- Catholic schools on the road to solving 'STEM problem. Project
https://www.facebook.com/SERINA.iacee/posts/4295920757123975
News article

SDG 5 Gender Equality

- Some awesome points in encouraging a young woman to think about becoming an engineer. Paper (?)
https://www.facebook.com/SERINA.iacee/posts/2168217183227687
https://alltogether.swe.org/2020/07/5-steps-raising-female-engineers/
- Contributing to a more sustainable future for our oceans. Awesome research project.
https://www.facebook.com/SERINA.iacee/posts/2392274184155318
https://roaring40skayaking.com.au/blog/reef-life-survey...

SDG 6 Clean Water and Sanitation

- Simple nets can turn fog into 9,500 gallons of water in just one day. Video
https://www.facebook.com/SERINA.iacee/posts/1971489936233747
https://www.facebook.com/attn/videos/1403404646457560/
- Stopping Australian sewers from going to waste. Active Practice
https://www.facebook.com/SERINA.iacee/posts/3430846523631407
https://www.createdigital.org.au/home-grown-innovation...
- Smart Cities. Active Practice
https://www.facebook.com/SERINA.iacee/posts/2770177063031693
https://www.youtube.com/watch?v=bKmfGI5Ogik

SDG 8 Decent Work and Economic Growth

- How sustainability impacts business success in engineering and construction. Active Practice
https://www.facebook.com/SERINA.iacee/posts/3329070603809000
https://createdigital.org.au/free-download-whitepaper-embracing-the-power-of-sustainability/
- What do you think of the need for a Circular Economy? Blog Article (?)
https://www.facebook.com/SERINA.iacee/posts/2406206276095442
https://thisspaceshipearth.org/.../accounting-for.../
- Renewable Energy Investment in Australia. Paper
https://www.facebook.com/SERINA.iacee/posts/2935737926475605
https://www.rba.gov.au/.../renewable-energy-investment-in...

SDG 10 Reduce Inequalities

- How is engineering practice addressing SDG 10?
https://www.facebook.com/SERINA.iacee/posts/4391078390941544
https://www.raeng.org.uk/global/sustainable-development-goals/sdg10-reduced-inequalities
https://www.youtube.com/watch?v=xzMS00FqJzk&list=TLPQMTYxMDIwMjGStQr9M1VXOg&index=3

SDG 14 Life below Water

- Scientific research to protect our Earth
https://www.facebook.com/SERINA.iacee/posts/3105140959535300
https://www.uwa.edu.au/.../restoring-and-maintaining...
- Universities 'essential' to climate action. UN secretary General, António Guterres
https://www.facebook.com/SERINA.iacee/posts/3504716366244422
https://www.universityworldnews.com/post.php...
- How do you help to rid waterways of plastic? Personal Initiative, boy diving for plastic
https://www.facebook.com/SERINA.iacee/posts/2540868055962596
https://www.bbc.com/news/av/world-asia-50376439

SDG 15 Life on Land

- Engineers need to be part of a team that is thinking and acting on Climate Change!! Newsletter article
https://www.facebook.com/SERINA.iacee/posts/2227826080600130
Newsletter article
- Sustainability in Engineering. Video
https://www.facebook.com/SERINA.iacee/posts/2185168691532536
https://www.youtube.com/watch?v=c2eNrFK5M8I
- Farmers on South Australia's Eyre Peninsula adapting to climate change with water-saving research and renewable technology. News Article
https://www.facebook.com/SERINA.iacee/posts/4187775751271810
https://www.abc.net.au/.../climate-change-eyre.../100330400

SDG 16 Peace, Justice, and Strong Institutions

- Achieving the SDGs in cities and regions. OECD Guide: Indicator Framework.
https://www.facebook.com/SERINA.iacee/posts/2331732590209478
- Website OECD
- What an AWESOME movie!!!!! Dream Big: Engineering our World.
https://www.facebook.com/SERINA.iacee/posts/2371397346243002

https://dreambigfilm.com/

SDG 17 Partnerships for the Goals

- We are shaping A DIVERSE ENGINEERING WORKFORCE!
  https://www.facebook.com/SERINA.iacee/posts/4397026407013409
  https://www.powerofengineering.org/
- Representatives from leading world universities convened at Rutgers University. Article
  https://www.facebook.com/SERINA.iacee/posts/2440672012648868
  Article
- Sustainable Development Goals and civil engineering. Paper
  https://www.facebook.com/SERINA.iacee/posts/2593202387395829
  https://www.designingbuildings.co.uk/.../Sustainable...

Discussion

The UN 2030 Goals for Sustainable Development delivery presents a plan of action, and engineers are surely at the forefront of this endeavour. Engineers are challenged to deliver on the UN Goals and the meeting of Fundamental Human Needs, using their scientific knowledge and experience to turn innovative ideas into sustainability projects for the benefit of all [14, 16].

It can be seen that *STEM/STEAM*, *Engineering Education (EE)* and *Continuing Engineering Education (CEE)* [15] platforms all play a crucial role in the handling of the scale and complexity of the gap between existing and past engineering practices and the meeting of sustainable needs facing our planet and humanity. Engineers of all kinds are uniquely placed to act and share knowledge in this regard as evidenced by SERINA's knowledge base. Ultimately, the engineering profession itself needs to be reshaped to address today's pressing issues and to promote a sense of global responsibility towards achieving the necessary innovations. This is what the IACEE is attempting to achieve via its Porto Declaration and SERINA's knowledge base [15, 16].

SERINA's implementation and exploration has found strong evidence of substantial sustainable engineering initiatives taking place around the world. These involving the implementation of sustainable active practice and research both feeding back into engineering education as linked to the UN's 17 Sustainable Goals and the meeting of Fundamental Human Needs [16].

The ongoing systematic approach in applying SERINA's methodology and exposition facilitates the knowledge search of and for website users and promotes interaction within these active initiatives.

The real benefit of the SERINA platform is its provision of examples of practical solutions to sustainable engineering problems and initiatives by highlighting successful active case studies and examples. Thereby, encouraging stakeholders and supporting engineers in their practices as well as organizations looking for possible answers, tools, techniques and possible partners, so as to grow the work done by peoples facing the threats of global warming and the socioeconomic challenges of our day. The SERINA website, its Facebook page and LinkedIn repository reach out in a practical and accessible way to all levels of people and organisations throughout the world thereby empowering and inspiring hope and encouraging others to do the same [14].

Continuing Engineering Education (CEE) knowledge sharing, research and active practice has developed and will continue to develop global initiatives to address many of the twenty-first century challenges threatening the survival of humankind and our planet. This requires a *Lifelong Learning* approach emphasising sustainable thinking practices through collaboration, design and creative thinking within engineering initiatives.

SERINA's implementation has and can be used to motivate collaboration within the engineering community and influence a majority of stakeholders so as to engage with a framework of global sustainable development and human need satisfaction via its varied websites.

Engineering Education (EE) and CEE can influence in the short and long term the involvement of the engineering community and its related sectors to a global commitment in implementing this call to service.

This change and improvement can be mostly achieved via education and training of the engineering community around the world, as engineers and related stakeholders have a major influence in the world's development. It is crucial that within a global and international arena that engineers engage in sustainable measures to ensure a future for the world [15].

SERINA and the Porto Declaration can act as beacon and motivation for all and especially for active engineers and for future engineers. The contribution of these initiatives arises from examples of related activities, the role of online learning in CEE sustainability courses, some guidelines for online sustainable courses and the provision of training and education for a sustainable world.

Engineers in many countries have compulsory periodic CEE and *Continuing Professional Development (CPD)* to keep their status as engineers. Professional organisations, government agencies, companies, non-governmental organizations and policy makers could transform the competences of active engineers through CPD and CEE to include those related with sustainability. SERINA can play an active role in supporting this goal [15].

A possible incentive to engage active engineers in CPD and in CEE on topics of sustainability can be the recognition of qualifications in sustainability practices. The title of sustainable engineer maybe awarded to those willing to be trained. The title could bring prestige to the engineer, higher employability and possible increase in salaries. CPD and CEE can provide formal learning in these areas, organized and structured through formal evaluation and assessment in traditional education systems or non-formal learning embedded in planned activities containing important learning sustainable elements. The overarching aim of enhancing sustainability competences is to have in place a system to document and to validate the competences that will be easily recognized by companies, professional organizations and society. Again, SERINA and the IACEE can play a part in supporting this endeavour.

Another possible enhancement is to have mandatory sustainability education and training for engineering programs at all levels. A challenge is that many engineering programs from academia do not comply fully with the demands of the professional qualification frameworks and many graduates do not have the necessary sustainability competences in terms of knowledge, skills and attitudes.

There is a need to develop ways of thinking sustainability as an engineer. SERINA's research brings out a number of ways to scale best practices in education programs. The need to:

- be purposeful in the integration of global issues knowledge and being aware of the differing cultural factors across and within nations and communities
- create opportunities and methodologies for learners and practitioners of sustainable practices to connect the traditional, the unexpected and the exceptional in going about daily practice
- keep learners at the centre when shaping educational experiences
- allow each educational experience to bleed into practical professional experiences, and vice versa
- prioritize immersive experiences that reinforce diverse thinking, but don't underestimate the need to have a diverse classroom from the start, and
- bring about continuous sustainable thinking when planning and implementing a project.

Conclusion

On-going educating of engineers for the implementation of the Sustainable Development Goals requires not only new competencies. It also requires the inclusion of creative learning and thinking, complex problem-solving opportunities, as well as interdisciplinary and international cooperation and a code of sustainable engineering ethics. It demands a change in engineering education itself!

On-going educating of engineers necessitates a shift from today's academic, technical, knowledge-focused path to a much broader interdisciplinary approach to learning, as seen in *STEAM* education approaches. It requires movement away from a teacher-centric focus to one that is more student-centred and problem-based. It will require building a structured approach, with its related quality assurance and accreditation, in order so as to promote lifelong learning and professional development for a real world.

SERINA has shown that it is possible to support the above approaches in gathering examples and case studies valuable for preparation of future graduates and active engineers in the area of sustainability practice. SERINA is open to replication by other stakeholders so as to become databases of examples and of case studies, of training programs and of initiatives that can bring confidence for those concerned with the future. Bringing hope and optimism whilst developing an active mind set in respect of both the UN Sustainable Development Goals and the Fundamental Human Needs enacting thoughtful implementation in planning and executing future projects.

Adapting to change is one of the most difficult and complex human challenges to navigate successfully. Yet no person or community has any immunity to change. The dawn of the third decade of the 21st Century has swiftly heralded unimaginable shifts and challenges across the globe without exception. So too, it has revealed signs of tremendous accomplishments, of ingenuity and of adaptability, buoyed by resilient human spirit and by desire to succeed against immense adversity. Against this backdrop, the decade ahead for engineers could hardly be more significant, demanding, challenging, and exciting for the best of engineers and of academics as problem solvers.

As engineering students, professionals, educators, and administrators, one can look to the enormity of 2030 deadline targets of the UN Sustainable Development Goals with a renewed sense of purpose, urgency, and optimism, knowing that the world has demonstrated its ability to adapt at a scale of unprecedented solving measures at the blink of an eye. SERINA carries many examples (and will continue to do so) of what can be achieved.

References

Circular Economy - https://ellenmacarthurfoundation.org/topics/circular-economy-introduction/overview

E-mail: July 2017- IACEE, Vice President - Global Initiatives.

https://www.abet.org/wp-content/uploads/2018/11/ABET_Sustainable-Engineering_Issue-Brief.pdf.

http://www.iacee.org.

http://www.iacee.org/docs/PORTO_DECLARATION1.pdf

Max-Neef, M., The Human Scale Development: Conception, Application and Further Reflection, Apex Press, New York, ISBN: N 0-945257-35-X, 1991, http://www.wtf.tw/ref/max-neef.pdf.

SERinA at LinkedIn - https://www.linkedin.com/groups/13680684/

SERinA on Facebook - https://www.facebook.com/serina.iacee.

SERinA website, Sustainability, *Education and Research in Action*, 2018, http://serina.iacee.org/, accessed 27 September 2021.

Smith A. : 'The nature of IACEE's SERINA's content, relative to the Association's Porto Declaration' (2021), IACEE2021 *Proceedings 17th World Conference on Continuing Engineering Education*, NTNU Trondheim.

Smith, A. E., The Role of Social Capital in Negotiating Socio-Economic Needs, Research Master thesis, University of Tasmania, 2001, https://eprints.utas.edu.au/21606/, accessed 27/09/2021.

Soeiro, A., Smith, A. and La Grange, E.: 'Continuing Engineering Education and Sustainability: IACEE Contribution with SERINA and Porto Declaration,' TEEM'18: *Proceedings of the Sixth International Conference on Technological Ecosystems for Enhancing Multiculturality*, (2018), 515–520, doi: 10.1145/3284179.3284266.

Soeiro A. and Smith A. : 'Nature of SERINA's Project related with 2016 Porto Declaration and linked to United Nations SDGs' - 'The role of education for Civil Engineers in the implementation of the SDGs' (2021), *1st Joint Conference of EUCEET and AECEF,* Thessaloniki, Greece

STEAM - https://www.acsa.edu.au/pages/images/STEAM%20into%20STEM%20Linking%20to%20the%20Australian%20Curriculum.pdf

STEM - https://www.industry.gov.au/policies-and-initiatives/science-technology-engineering-and-mathematics-stem.

UNITED NATIONS, Sustainable Development Goals, 2015, https://sdgs.un.org/goals, accessed 27 September 2021.

Chapter 15

Future Research Areas

Charles E. Baukal, Jr.*

Koch Engineered Solutions Institute^{SM}, John Zink Hamworthy Combustion, Tulsa, Oklahoma, USA

Abstract

This brief chapter recommends some potential areas of future research on the subject of continuing engineering education. This is a fruiful area for research as relatively little has been done to date, especially compared to the amount of research done on undergraduate engineering education. This chapter is not intended to be comprehensive nor exhaustive but rather representative, with specific focus on the topics discussed in this book. The chapter is divided into four research areas: working engineers' learning preferences, curriculum, virtual and augmented reality, and multimedia. Some conclusions and recommendations are presented.

Keywords: continuing engineering education, future research

Introduction

A considerable amount of research has been done on the subject of undergraduate engineering education (UEE), yet much still remains to be done (Baukal, 2020). There continue to be calls for improving UEE. The U.S. National Academy of Engineering (2004) established a Committee on Engineering Education to answer the question "What will or should engineering be like in 2020?" The Phase 2 report (National Academy of Engineering, 2005) from that committee titled Educating the Engineer of 2020 calls for the reinvention of engineering education. An important recommendation of that study was the importance of academic management such as engineering deans endorsing engineering education. This includes study into how undergraduate engineering students learn. This included how to better serve students with different learning styles and how to determine pedagogical approaches that excite them. The report notes that while progress has been made in determining best practices for engineering education, much remains to be done. The *Journal of Engineering Education* (Anonymous 2006) recommended five research areas for engineering education: engineering epistemologies, engineering learning mechanisms, engineering learning systems, engineering diversity and inclusiveness, and engineering assessment. Duderstad (2008) recommended (p. v) "a systematic, research-based approach to innovation and continuous improvement of

* Corresponding Author's Email: charles.baukal@kes.global.

engineering education." The U.S. National Academy of Engineering (2008) identified 14 grand challenges in engineering. One of those challenges is to advance personalized learning that recognizes individual preferences and aptitudes to help motivate learners to become more self-directed. While that challenge was targeted at the development of learning software by computer engineers, it applies to all types of learning and learners, including working engineers.

Johri and Olds (2014) have edited a substantial book on engineering education research which is divided into six parts and contains 36 chapters. While it is targeted at UEE, many of the principles are applicable to continuing engineering education (CEE). Male and Baillie (2014) argue that teaching practices should be research-based with proven improvements in student learning. Atman, Eris, McDonnell, Cardella, and Borgford-Parnell (2014) recommend that research into engineering design specifically should be based on solid research.

Fortenberry. Sullivan, Jordan, and Knight (2007, p. 1175) argue the ultimate aims of engineering education research should include:

- the creation of education programs that attract more, and more diverse, students to the study of engineering,
- retain more of the students who are enrolled,
- deepen students' understanding of engineering concepts,
- broaden students' appreciation of engineering's role in meeting the needs of a global society, and
- better prepare students for further study or professional practice.

Research has shown that working engineers have much different learning strategy and verbal-visual preferences than the general population (Baukal, 2017; Chapter 3). They prefer more problem solving and more visual materials. The research-based learning preferences of working engineers suggests that CEE should be designed differently than traditional education which is often highly verbal and less visual with little problem solving.

While considerable research has been done on UEE, relatively little research has been done on CEE so much remains to be done (Nørgaard, 2019). There are many reasons for the dearth of research on CEE. Since faculty do the vast majority of research related to education, they often focus on research of personal interest and value. Engineering educators are typically motivated to improve their instructional materials and methods. Since relatively few of them are involved in CEE, there is naturally less attention to that subject. Another reason is that the primary business of many CEE providers (see Chapter 11) is not education. For example, industry subject matter experts (SMEs) who teach internal training courses are primarily valued for their technical knowledge as it is applied to the business of making a profit, not to training their colleagues. For those for-profit organizations who do conduct CEE research, they are less likely to share results externally for research they have funded and for which they believe may provide a competitive advantage. It may be easier for faculty to have their students participate in engineering education research than for CEE instructors to have working engineers participate in education research. Further, some might argue that there is little difference between UEE and CEE so that separate research on CEE is duplicative. It is the opinion of this author that while there are clearly similarities, there are enough differences between UEE and CEE that each should be studied separately.

This chapter focuses on potential future CEE research topics. It is not intended to be exhaustive or even comprehensive. It is clearly influenced by the author's experiences and biases. It is intended to present some topics that in the opinion of the author could benefit from more disciplined research study.

Working Engineers' Learning Preferences

One of the major goals of discipline-based education research (DBER) related to engineering education is to "understand how people learn the concepts, practices, and ways of thinking of science and engineering" (Singer, Nielsen, and Schweingruber, 2012). A National Academy of Engineering report (2005) argues the importance of pursuing student-centered education and finding out how students learn as well as what they learn. As noted in Chapter 4 of this book, working engineers have statistically significantly different learning preferences compared to the general population. Their learning strategy preference has a much higher proportion of problem solvers and their verbal-visual preference is much more highly visual. However, the data to date is very limited. There is not enough yet to draw generalizable conclusions.

The research reported in Chapter 3 merely scratches the surface for the learner characteristics of working engineers. There are not enough data in many cases to conduct valid statistical analyses. For example, six age categories were reported but in half of them, there were seven or fewer participants. More characteristics, such as specific multimedia preferences, should be measured in future studies. Nearly all of the participants in the study worked in industry. Much more data should be collected for: engineers >55 years old, more data on various engineering disciplines, African Americans (none participated in the study), and Hispanic/Latinos. A wide-open area of research relates to working engineers in other countries. While there were some in the present study, there were nowhere near enough to draw any conclusions for countries besides the U.S. There may be important cultural differences that should be investigated to determine if they need to be incorporated into instructional materials and methods.

Another important variable that warrants further investigation is the industry where engineers work. In the reported study, the participants worked in the oil and gas industry. It is possible that engineers working in other industries such as automotive, aerospace, and academia may have different preferences compared to those found in the reported study.

Much more research needs to be done to better understand the learner characteristics of working engineers, both as a professional group and as individuals; the content nature and complexity; and how to use media options to maximize the effectiveness and efficiency of their continuing education needs and requirements.

Curriculum

Research should be done to help organizations develop more effective curricula for CEE. This includes the who, the what, the when, and the how. The who refers to who delivers the content (see Chapter 11). Many options are available including employers, vendors, professional societies, trade organizations, government, and consultants. Research could help organizations better determine the appropriate mix of providers to maximize learning and the return on their investment.

What should be covered in an individual course is important to ensure that learners learn what they need to know but no more than that as superfluous materials cost valuable time and money. It is tempting for subject matter experts to share everything they know on a given topic. In many cases that would be over the head of most learners and not necessary as the goal in most training courses is not to produce more SMEs, but rather to move the participants from novices to at least capable in a given area of technology. Research can help instructors better focus training content to find a good balance between enough information but not too much.

The *when* for curricula is the timing of the delivery of CEE. Covid-19 has shown that a lot more can be handled at a distance than previously believed with many workers working from home for months at a time. Related to that is distance training where the delivery technology and software is now good enough to realistically teach courses over the Internet. For virtual courses (see Chapter9), it is recommended that complicated content be spread out over multiple sessions, rather than the traditional model of having face-to-face classes for eight hours or more each day to minimize travel and living expenses. Research should be done to find out what is the proper balance of time to cover various types of content. For example, is two hours in the morning and two hours in the afternoon a good balance for training using Zoom or TEAMS?

The *how* refers to the delivery method. What is the most effective method for each type of training? In particular, while much research has been done on in-person classes, much less has been done on virtual classes using current technologies. This type of research could help determine the proper balance for lecture vs. participant interaction. A current challenge of online learning is that it is easy for participants to be working on other things while attending a course. That is harder to do in in-person classes as the instructor can see all of the students sitting in front of them in a classroom.

Another aspect of curriculum research is how to customize training and tailor it to individual learners (Nørgaard, 2019). While this is obviously a preferred mode of learning, there is also a cost so customized training needs to be both efficient and effective so that it is also provides a sufficient return on investment. Research into this topic can help determine what types of training might be customized and how to best accomplish that.

Virtual and Augmented Reality

Two important areas of emerging engineering education research are spatial ability and the use of representations, although relatively little research in these areas has been done to date (Singer, Nielsen, and Schweingruber, 2012). Kober (2015, p. 4) wrote, "Animations, interactive computer simulations, virtual models, and other technology-based representations are widely used in the practice of science and engineering and are becoming increasingly popular in undergraduate education." The research discussed in Chapter 3 shows that engineers are highly visual and in Chapter 5 shows that engineers prefer images over text.

Some relatively newer visual technologies that can be used for educational purposes are virtual reality (VR) and augmented reality (AR) as discussed in Chapter 7. While VR and AR have been used for some time in gaming, they have only been used relatively recently in educational contexts. At the time of this writing there is relatively little research on the efficacy of using these technologies for learning purposes (e.g., Ausburn, Ausburn, Martens, Baukal,

Agnew, and Dionne, 2019). The technologies are changing rapidly and the costs have come down dramatically, making them much more affordable to use for educational purposes. However, the software is still somewhat expensive and can be time intensive.

A sister company to the author's organization developed some VR simulations of a fired heater in one of their refineries which they use to train operators. If the operators fail to respond correctly in one of the scenarios, it is possible for the heater to explode. The simulator is similar in concept to the simulators used in training airline pilots who encounter dangerous situations virtually so they will be prepared to handle them if they should ever occur in reality. VR is particularly powerful for simulating dangerous situations which are difficult or impossible to demonstrate in reality because of the possibility of equipment damage and personnel injuries. The sister company ran parallel operator training classes: one was in a traditional classroom and the other was using the VR simulations. Besides the fact that the operators liked the VR training better, they also scored higher in their qualification tests in the VR classes. Published research studies are needed to show the power of using VR and AR for training engineers. Until software costs come down significantly, research can help show what types of training are better done using VR and AR compared to traditional lecture-style classes. For example, simulating conditions that are difficult to demonstrate in reality and have serious consequences if mistakes are made would seem to be good candidates for VR and AR simulations.

Multimedia

One area of theoretical interest is collecting more data on concrete vs. abstract multimedia as a function of prior knowledge. In the study reported in Chapter 6, working engineers with less prior knowledge slightly preferred more abstract multimedia over more concrete multimedia. This is not what current realism theory (more concrete or realistic materials should be used with novice or low prior knowledge learners and more abstract materials should be used with expert or high prior knowledge learners) would suggest. However, as previously discussed there may be some confounding variables that could explain this apparent discrepancy. Much more data is required before any definitive conclusions can be made such as discarding the realism theory.

In the data reported in Chapter 5, the technology considered was combustion equipment used in the oil and gas industry. Many more types of technologies should be investigated to determine if there are any differences in learner preferences based on technology that could impact instructional materials and methods.

No qualitative studies on CEE were found at the time of this writing. Qualitative research has the potential to provide much richer information than quantitative research that could be used to improve CEE. As is sometimes the case, completely new areas of research could be identified.

Clark (1989) listed five types of technical training content: procedures, concepts, factual information, processes, and principles. The study reported in Chapter 5 investigated only one particular type of subject matter – factual information. Future studies could investigate multimedia preferences for the other four types of technical content.

Manjit Sidhu (2010) argues there is a need for developing multimedia software for engineering. Hardware and software costs have come down significantly but it still requires specialized training and knowledge to develop VR and AR simulations. It is expected that virtual and augmented reality software will become easier and more ubiquitous in the future.

For now, it may be necessary to hire vendors or programmers to develop sophisticated simulations. This takes time and money which means that for the near future organizations will likely have to pick and choose what is important enough to simulate.

Conclusions and Recommendations

Duderstadt (2008) argues, "the goal of American engineering schools and industry training programs should be to focus more on quality, producing engineers capable of adding exceptional value through innovation, entrepreneurial skills, and global competence." Ferrando-Bataller, Antonio-Daviu, Ripoll-Soler, and Turro (2010, p. 3) wrote, "Continuing Engineering Education (CEE) is necessary to maintain the skills of our engineers and is fundamental to build new competences around their current skills. Still nowadays many CEE activities are based on traditional learning methods, like face-to-face seminars, conferences or short training courses." It is ironic that engineers regularly work with cutting-edge technologies, yet they are often subjected to antiquated teaching methods. Research is required to determine the best ways to train working engineers.

The challenge of engineering education research is translating research results into effective practice in the classroom (Singer, Nielsen, and Schweingruber, 2012). National surveys have shown that engineering faculty are among the least likely to employ student-centered or collaborative instructional techniques.

References

Anonymous (2006). The Research Agenda for the New Discipline of Engineering Education. *Journal of Engineering Education 95*(4), 259-261.

Atman, C., Ozgur Eris, Janet McDonnell, Monica Cardella, and Jim Borgford-Parnell, "Engineering Design Education," in Aditya Johri and Barbara Olds (eds.), *Cambridge Handbook of Engineering Education Research*, New York: Cambridge University Press, 2014.

Ausburn, L., Ausburn, F., Martens, J., Baukal, C., Agnew, I. and Dionne, R. (2019). User Characteristics, Trait vs. State Immersion, and Presence in a First-Person Virtual World, *Journal for Virtual Worlds Research 12*(3), December.

Baukal, C. (2020). Future Research Areas. In Baukal, C. (ed.), *Mechanical Engineering Education Handbook.* New York: Nova.

Baukal, C. E. (2017), *Learner Preferences for Continuing Engineering Education.* Mauritius: Lambert.

Clark, R. C. (1989). *Developing technical training: A structured approach for the development of classroom and computer-based instructional materials. Reading,* MA: Addison-Wesley.

Duderstadt, J. J. (2008). Engineering for a changing world: A roadmap to the future of engineering practice, research, and education. Ann Arbor, *Michigan: The Millennium Project*, The University of Michigan. Retrieved from: http://milproj.dc.umich.edu/.

Ferrando-Bataller, M., Antonio-Daviu, E., Ripoll-Soler, C., and Turro, C. (2010). Continuing Engineering Education and Life-Long Learning, *Proceedins of the Fourth European Conference on Antennas and Propagation*, April 12-16, 2010, Barcelona.

Fortenberry, N., Jacquelyn Sullivan, Peter Jordan, and Daniel Knight, "Engineering Education Research Aids Instruction," *Science317*, 1175-1176, August 31, 2007.

Johri, A. and Barbara Olds (eds.), *Cambridge Handbook of Engineering Education Research*, New York: Cambridge University Press, 2014.

Kober, L., *Reaching Students: What Research Says About Effective Instruction in Undergraduate Science and Engineering,* Washington, DC: National Academies Press, 2015.

Male, S. and Caroline Baillie, "Research-Guided Teaching Practices," in Aditya Johri and Barbara Olds (eds.), *Cambridge Handbook of Engineering Education Research*, New York: Cambridge University Press, 2014.

National Academy of Engineering, *Educating the Engineer of 2020: Adapting Engineering Education to the New Century,* Washington, DC: National Academies Press, 2005.

National Academy of Engineering, *Grand Challenges for Engineering,* Washington, DC: National Academies Press, 2008.

National Academy of Engineering, *The Engineer of 2020: Visions of Engineering in the New Century*, Washington, DC: National Academies Press, 2004.

Nørgaard, B. (2019). Implications of facilitated work-based learning implemented as an approach to continuing engineering education. *European Journal of Engineering Education 44*(4), 629-642.

Sidhu, M., Technology-Assisted Problem Solving for Engineering Education: Interactive Multimedia Applications, *Hershey: Engineering Science Reference,* 2010.

Singer, S., Natalie Nielsen, and Heidi Schweingruber (eds.), *Discipline-Based Education Research: Understanding and Improving Learning in Undergraduate Science and Engineering,* Washington, DC: National Academies Press, 2012.

About the Editor

Charles E. Baukal, Jr. is the Director of the Koch Engineered Solutions Institute which is part of John Zink Hamworthy Combustion (Tulsa, Oklahoma) where he has been since 1998. He previously worked for 13 years at Air Products and Chemicals, Inc. (Allentown, Pennsylvania). He has been working since 1978 in the general areas of industrial combustion, pollution emissions, heat transfer, and engineering education. He is currently an adjunct engineering instructor for John Brown University (Tulsa, OK), Oral Roberts University (Tulsa, OK), and the University of Tulsa (Tulsa, OK). He has been an adjunct since 1984. He has a Ph.D. in Mechanical Engineering from the University of Pennsylvania and an Ed.D. in Applied Educational Studies from Oklahoma State University. He is a licensed Professional Engineer in the State of Pennsylvania and a Qualified Environmental Professional (QEP). He has over 250 publications/presentations including authoring/editing 17 books in the fields of industrial combustion and engineering education. He is an inventor on 11 U.S. patents. He is a member of the Combustion Institute, the American Society for Mechanical Engineers, the American Society for Engineering Education and the Air & Waste Management Association. He is a member of many honor societies and serves on many advisory boards.

About the Authors

Amrit Ahluwalia is the Editor-in-Chief of The EvoLLLution and Senior Director of Strategic Insights at Modern Campus. He was part of the team that founded and launched The EvoLLLution—which is the leading source of insights and opinions on non-traditional higher education—in 2011. He regularly speaks on topics related to the changing higher education environment at conferences across Canada and the United States, advises college and university leaders to help frame the strategic visions for their institutions, and has authored numerous book chapters. Ahluwalia earned his BA (Honors) in Political Studies from Queen's University and his MA in International Politics from McMaster University. He lives in London, Ontario.

Lynna Ausburn, Ph.D. is Professor Emerita from Oklahoma State University (OSU), Stillwater, Oklahoma, USA. She holds a PhD in instructional technology from the University of Oklahoma. Dr. Ausburn retired as Program Coordinator for Workforce and Adult Education after 14 years of teaching and extensive research and publication. While at OSU, she produced more than 50 publications and more than 100 professional presentations and was awarded the Oklahoma State Board of Regents Distinguished Research Award, the Phoenix Award for Outstanding Member of the Graduate Faculty, and a Governor's Commendation. Prior to going to OSU, Dr. Ausburn completed 14 years in technical workforce development in Australia and 10 years in curriculum development at the Oklahoma State Department of CareerTech. Her career also featured working in the area of workforce training and development with major international funding organizations such as World Bank, USAID, United Nations, and several government agencies in 19 countries. Her research focused on learning styles, technology development and virtual reality applications for training.

Ragna Ann Berge, Head of Section, Centre for Continuing Education and Professional Development, Division of Education and Quality, Norwegian University of Science and Technology, Trondheim, Norway. Ragna Ann Berge has been Head of Section at the Centre for Continuing Education and Professional Development since 2011 and has a long career at the Norwegian University of Science and Technology. From 2007 to 2011, she was project manager for the successful conference NOKIOS, an annual Norwegian conference for eGovernment. At the university, Berge serves as a member of the Project Management Program steering committee and "KOMPiS", a program for continuing education/professional development for teachers K-12. She is part of the management team at the Division of Education and Quality at NTNU. Berge has over 20 years of experience working with development and delivery of Continuing Education in collaboration with the faculties at the university. The Centre for Continuing and Professional Development Berge heads annually administers over 350 ECTS courses at bachelor and masters level with 8-9000 participants. The center also administers 40 to 50 conferences every year, the last two years delivered mostly as hybrid or digital conferences, a huge transition for both the centre and the conferences. In 2021 one of these conferences was the IACEE World Conference. Because of the pandemic the conference was postponed one year and it became the first ever IACEE conference to be delivered online. Berge received her bachelor's degree in Environmental Studies at Telemark University College and her master's degree in Human Geography at the Norwegian University of Science and Technology in Trondheim. Berge has been a member of IACEE since 2014 and currently serves on the IACEE Executive Committee both as Vice President (VP) of Member Engagement and Communication, as the World Conference Chair and previously as the VP for Processes and Systems.

Wes Bussman, Ph.D. is a senior research and development engineer for the John Zink Co. LLC. He received his Ph.D. in mechanical engineering from the University of Tulsa. Dr. Bussman has 27 years of basic scientific research work, industrial technology research and development, and combustion design engineering. He holds ten patents and has authored several published articles and conference papers and has been a contributing author to several combustion related books. He has taught engineering courses at several universities and is a member of Kappa Mu Epsilon Mathematical Society and Sigma Xi Research Society.

John Carpinelli, Ph.D. is a Professor of Electrical and Computer Engineering at the New Jersey Institute of Technology. He received his Bachelor of Engineering degree in Electrical Engineering from Stevens Institute of Technology, and his Master of Engineering degree in Electrical Engineering and Ph.D in Computer and Systems Engineering from Rensselaer Polytechnic Institute. He has approximately 100 publications, including his textbook *Computer Systems Organization and Architecture*. He served as Executive Director of NJIT's Center for Pre-College Programs from 2012 to 2014, and most recently served as Associate Director of NJIT's Institute for Teaching Excellence. His research focuses on engineering education, for which he has received numerous grants from the National Science Foundation and other sources. He has been designated a Master Teacher at NJIT.

Myra N. Fanning is a Koch Engineered Solutions Institute Training Leader, which is part of John Zink Hamworthy Combustion (Tulsa, Oklahoma). She was employed with Koch Industries (Wichita KS) 1998-2000 and transferred to John Zink Hamworthy Combustion in June 2000. She began working for Unified School District #259 (Wichita, KS) during summers in college and full-time starting 1982. She is a graduate from Wichita State University (Wichita, KS) with a BA in speech communications. She has been a Power Member of the Association for Talent Development (ATD) for many years. She has served in several board positions of the Northeast Oklahoma Chapter (Tulsa OK), including Chapter President. She has received several training certificates through ATD and a training generalist certificate through Langevin Learning Services. "Having fun while learning," is her motto.

Regina M. Godfrey is a Training Leader for the John Zink Institute which is part of John Zink Hamworthy Combustion (Tulsa, Oklahoma) since 2019. She previously worked for 7 years at Noria Corporation (Tulsa, Oklahoma). Her primary areas of focus have been marketing communications, manufacturing event management, and logistics for engineering education courses. She has a Bachelor of Science degree in Journalism/Public Relations from Oklahoma State University, as well as a minor in Marketing from the same university. She has served in leadership roles with the Public Relations Society of America (Tulsa, Oklahoma Chapter) and American Marketing Association (Tulsa, Oklahoma Chapter). She has served on committees for Women in Reliability and Asset Management and is currently a committee lead for the North American Process Technology Alliance.

Hiro Iino is the Manager of the Professional Development Program in Engineering-LAS Online Learning at Iowa State University. The program offers non-credit short courses to working engineers who are seeking just-in-time courses, continuing education units (CEU), and professional development hours (PDH). He has over twenty-two years of experience working with faculty and industry experts to develop online educational programs. Throughout his career, he held various roles, including instructional design, learning management system administration, developing and delivering online degree programs. He also held the open-source learning management administrator position for campus-wide English Composition Programs to several thousand students annually. His interests are Online Education, User Experience, and Digital Accessibility. Hiro received a B.S. in Journalism & Mass Communications and Psychology and an M.S. in Human-Computer Interaction from Iowa State University.

Paul G. Jewell has dedicated his professional career to supporting education through the relevant and innovative application of instructional technology, service to professional organizations, and commitment to mentoring students. Paul's career began in the San Diego Unified School District. He used video as an instructional tool to create programs and enhance the curriculum of high school teachers in the district. Some of these programs were used in a series of national and international interactive satellite teleconferences, which led to Paul's induction in 1988 as a Fellow of the Christa McAuliffe Institute for Educational Pioneering. Paul was later integral in the creation of a distance learning network in Point Barrow, Alaska. He used live, interactive video classrooms to connect nine remote Arctic village schools across the 88,000 square-mile North Slope Borough School District. In 1993, Paul began work with Iowa State University's College of Engineering Distance Education (EDE), first as an instructional technologist, then later as the Director of Industry Relations. In 2010, EDE became Engineering-LAS Online Learning (ELO), which currently serves about 10,000 student enrollments annually. Over the years, Paul led the way on campus with innovative technology in classroom services for online learning and in the creation of Collaborative Technology Services, which offers broadcast-quality video production and streaming services across campus and beyond. For over 17 years, Paul was also active in the American Society for Engineering Education, Continuing and Professional Development Division, including serving as chair of the division. After 28 years, Paul retired in 2021, leaving an award-winning legacy of high-quality, professional leadership and service to Iowa State University.

Darshana P. Juvale is an Instructional Design Supervisor at Engineering-LAS Online Learning (ELO), Iowa State University. She has been working with ELO for over thirteen years bringing applied experience in designing, developing, and facilitating faulty to design online and blended courses. She collaborated with the faculty to help determine appropriate pedagogical approaches to asynchronous online courses following best practices and instructional design principles. Her professional interests include: instructional design, faculty development and blended/flipped learning approach. Darshana holds a Master's degree in Education with an emphasis on Curriculum and Instructional Technology.

Howard Kimmel, Ph.D. is professor-emeritus of chemical engineering at New Jersey Institute of Technology (NJIT). He was a member of the NJIT faculty for 46 years. Dr. Kimmel earned his Ph.D. in Chemistry from the City University of New York. He has 145 publications as well as chapters in 6 books and co-authored 3 books including "A History and Legacy of STEM Education at NJIT". Dr. Kimmel collaborated with colleagues providing numerous workshops for faculty on instructional practices required for ABET accreditation. He was Executive Director of NJIT's Center for Pre-College Programs from its creation in 1979 until his retirement in 2012. As Executive Director, he had a primary focus on teacher professional development practices, curriculum and assessment. As a result in 2000, he served as co-PI for a $2.5 million grant "High-Tech Workforce Excellence Grant" from the New Jersey Commission on Higher Education. In addition, he had been the recipient of numerous federal grants from the National Science Foundation (NSF), U.S. Dept. of Energy, and U.S. Department of Education. Dr. Kimmel has received national recognition for his efforts and accomplishments, including receipt of The Presidential Awards for Excellence in Science, Mathematics and Engineering Mentoring (PAESMEM). PAESMEM is a Presidential award established by the White House. It recognizes individuals for their mentoring of persons from underrepresented racial and ethnic groups, women, persons with disabilities. On the national level, he has also received recognition from the American Society for Engineering Education (ASEE), including the CENTENNIAL MEDALION for "Significant Lasting Impact on Engineering Education" in 1993.

Errol la Grange is the co-founder and executive chairman of Cahoot Learning www.cahootlearning.com. Established in 2008, Cahoot Learning remains at the forefront of relationship based, cohort driven online continuing education for professionals. In his role Errol has developed key relationships with leading organizations in both the private and public sectors around the world, which has helped enable Cahoot Learning to deliver 12,000 online learning events to 2 million professionals in 134 countries, and across many industries. Errol is a passionate advocate for what he calls professional education 4.0. This is about connecting heart, mind, work, and purpose through learning, the aim of which is to live a joyful and fulfilling life. As a Vice President of the International Association for Continuing Engineering Education, IACEE, from 2016 to 2021, Errol was the lead of Global Sustainability Initiatives for the organization. SERINA, Sustainable Engineering Education and Research in Action, is one of those initiatives that commenced under his watch. A qualified lawyer and mediator, Errol is registered to practice in 3 countries. His perspective on matters legal is that it is better to mediate than litigate a dispute, and better still to learn how not to require either from a personal, organizational or nation state perspective. This, in part, is what lead Errol into the world of professional and workforce learning. Errol is a recreational pilot who loves taking photos and videos of nature, be that from above, or on a hike down below.

Brian D. Laughlin, Ph.D. is a Technical Fellow and Systems Engineer at Boeing Defense Systems, working on Advanced Visualization and Point of Use technologies in the area of Mobile, Wearables and Internet of Things (IoT), for Space and Launch Systems, serving both the Commercial and Military businesses at the Enterprise level. Dr. Laughlin has over 32 years of experience in the Aerospace industry. He began his Aerospace career at Beechcraft in 1990, on the Starship, T1A Jayhawk Trainers and King Air 350s as a Composites and Sheetmetal Mechanic. He joined Boeing in late 1991, where he has filled a plethora of diverse rolls, providing him a broad range and depth of expertise, including: Certified Sheet Metal and Composites Mechanic, Process Engineering Analyst (liaison planner, engineer, and tool coordinator) Computing Process Analyst, Software Developer, Certified Lean coach, IT Architect, and Course Developer/ Instructor/ consultant on many topics including: Human Centered Design, Systems Thinking, Lean, Innovation, Leadership/ Coaching/ Mentoring, and Internet and Mobile Computing technologies, and most recently, Systems Engineer, to name a few. Dr. Laughlin also served as a Graduate Teaching and Graduate Researching Assistant, for the National Institute for Aviation Research (NIAR) in the Software Engineering lab where he helped pioneer web technologies in the early 90s. Dr. Laughlin has over 125 invention disclosures and patents (both US and international) across a multitude of technology domains (AR/VR, Supply Chain Systems, Optical Inspection, Machine Learning, Security, etc.) that have resulted in a plethora of awards for his innovative work. His awards include: 1 Technical Replication, 3 Special Invention Awards, 5 Meritorious Invention Awards, and an Ed Well's Initiative Award. His expertise and diverse background, coupled with his passion for problem solving and teaching have made him a sought after speaker, author, course creator, teacher, industrial/academic advisory board member, and mentor across educational (MIT, Virginia Tech, Carnegie Mellon, etc) and industrial sectors (military and commercial). Laughlin holds a Bachelor of Arts in Psychology, a Master of Arts in Experimental Psychology, and a Ph.D. in Human Factors Psychology, all from Wichita State University.

Dane Laughlin is an Innovation Engineer at INVISTA, a Koch Industries subsidiary in Wichita, Kansas, focused on integrating emergent technology across INVISTA's global operations space for process improvement. Acting as an internal specialist and project manager, Dane has worked on projects in environments ranging from a corporate cubicle to the manufacturing floor. Utilizing his biomedical engineering background, his core focus is to streamline the interface between people and technology to improve execution accuracy and increase efficiency. In addition to being a technologist, Dane is also an Inventor with four patents in the visualization of deep brain stimulation. Dane's core competency is in AR/VR implementation, wearables, IOT, and data analytics. Using these tools, he is looking to change the way that employees interact with their work as well as their training.

Rema Nilakanta, Ph.D. is the Manager of Instructional Design and Delivery at Engineering-LAS Online Learning (ELO) at Iowa State University, Ames, Iowa. She leads a dynamic team of professional and student instructional designers to develop, support, and research online and blended courses at the Colleges of Engineering and the Liberal Arts and Sciences. She also oversees the design, development and support of training programs for working professionals in engineering. Rema has a Ph.D. in Education with emphasis in Curriculum and Instructional Technology from Iowa State University. Before she joined ELO, Rema managed an outreach initiative for young adults at the Institute for Transportation (InTrans), Iowa State University. It involved overseeing the development, production, and distribution of Go!, an e-magazine designed for young adults on the exciting career possibilities in transportation. Rema has also worked as a research associate in a 3-year NSF-funded research project on high school girls exploring engineering as a career option. She has taught, researched, and published on the implementation of technology in teaching and learning, with an emphasis on online learning. In addition to her university experiences, Rema has a Master's in German from Rice University, Houston, TX. She also owned a publishing company that designed and typeset textbooks for academic publishers.

Ronald H. Rockland, Ph.D. is a professor emeritus in Engineering Technology from New Jersey Institute of Technology (NJIT). He received his Ph.D. in bioengineering and electrical engineering from New York University and received an M.B.A. in marketing from the University of St. Thomas. After almost 25 years of industrial experience in research, engineering, marketing and sales management and general management with several high technology corporations, he joined New Jersey Institute of Technology (NJIT) in 1995 as an Assistant Professor and spent 23 years there. Prior to teaching at NJIT, he was an adjunct professor in marketing for eight years at two community colleges in New Jersey. Dr. Rockland was the chair and professor of the Department of Engineering Technology, with a joint appointment in the Department of Biomedical Engineering. Prior to that he served as Associate Dean, Undergraduate Studies for the Newark College of Engineering of NJIT, and was responsible for recruitment and retention. His research in industry was in the area of pacemakers and defibrillation, and his research at Medtronic Inc led to five patents. At NJIT, he was a principal investigator for a three-year, $1 million NSF grant entitled Medibotics: The merging of medicine, robotics and IT, was a co-principal investigator for a $2.5 million grant on pre-engineering workforce enhancement from the New Jersey Commission on Higher Education, as well as a principal investigator for a Whitaker Foundation grant. His research at NJIT was in biological signal processing, related to cardiovascular signals, and in enhancing STEM education through use of engineering principles and technologies. He has written over 70 articles in journal and conference proceedings, in both the educational and biomedical fields. Dr. Rockland was very active in the Engineering Technology community, including a term as Chair for the Engineering Technology Division (ETD) of the American Society of Engineering Technology (ASEE), where he is also a Fellow of the ASEE. He has received numerous university and national awards for teaching and service, including the Berger Award of ASEE, and was named a Master Teacher at NJIT. He was one of the founding faculty for the Institute of Teaching Excellence (ITE) and was responsible for coordinating and creating numerous workshops for new faculty on teaching.

Tony Smith, M.Ed.(Research), B.Ed.(Adult and Workplace Education (Hons)) is the Lead Researcher for the International Association for Continuing Engineering Education's (IACEE) SERINA Research Project and Council Member of the IACEE. Tony has over many years gained extensive practical, research and theoretical experience in a number of fields such as leadership research, organisation management, education, social services, community development and production management. High School mathematics and science teaching; lecturing in adult education and researching with the University of Tasmania's Centre for Research and Learning in Regional Australia, organisational learning and leadership, community development and education programming, systems and processes; knowledge management research with Hydro Tasmania; social services management delivery and systems with Anglicare Tasmania from an individual to a collective social psychology level. Tony is actively involved in Tasmania with the University of Tasmania and in the wider Tasmanian primary and secondary education community concerning education and active practice mitigating against Climate Change.

Gale Tenen Spak, Ph.D. established Build Their Future, LLC in 2019 to provide in-person and online talent and workforce in-person and STEM and durable skill development training and education consultancy appropriate for “cradle to gray” generations based on what she has learned over four decades in higher education. Most recently, she was Associate Vice President of Continuing and Distance Education at New Jersey Institute of Technology, Newark New Jersey. Her oversight of that unit focused on providing working learners with professional STEM development programming which often was funded externally, involved digital instruction, and established collaborations with academe, industry and government. These are topics about which she presents and writes broadly, including one which won an ASEE Best Paper award. Earlier in her career, Dr. Spak was Dean of the School of Professional and Continuing Education at New York Institute of Technology, Old Westbury, New York, and, during America’s first energy crisis, served as the Director of the Center for Energy Policy and Research and authored “how to” reports which were sent to every American Governor and honored by President Jimmy Carter. She earned her Doctor of Philosophy in Political Psychology and Master of Arts from Yale University, and her Bachelor of Arts, magna cum laude, Phi Beta Kappa, in Political Science from Brooklyn College of City University of New York.

Mitchell L. Springer, PhD., PMP, SPHR, SHRM-SP, has extensive theoretical, higher education-, and defense industry-based practical experience spanning four disciplines: software engineering, systems engineering, program management and human resources. He possesses a significant strength in pattern recognition, analyzing and improving organizational systems, with expertise in program/project management, organizational restructuring, change management, negotiation, and mediation processes. Dr. Springer is internationally recognized. He has contributed to scholarship more than 400 books, articles, presentations, editorials, and reviews on software development methodologies, management, organizational restructuring, change management, program/project management, and diversity leadership. Dr. Springer is the recipient of numerous awards and recognitions, including local, regional, and national recognitions for leadership in diversity, equity, and inclusion, with over 100 attendant contributions to scholarship. Dr. Springer received his Bachelor of Science in Computer Science from Purdue University, his MBA and Doctorate in Adult and Community Education with a Cognate in Executive Development from Ball State University. He is certified as a Project Management Professional (PMP), Senior Professional in Human Resources (SPHR & SHRM-SCP), in Alternate Dispute Resolution (ADR), and, in civil and domestic mediation. Dr. Springer is a State of Indiana Registered domestic mediator and member of the Indiana Association of Mediators.

Ann D. Thompson, Ph.D., University Professor at Iowa State University, has devoted her career to designing, studying, and implementing effective uses of technology in education. She is the Founding Director of the Center for Technology in Learning and Teaching at Iowa State University and has been involved in preservice teacher preparation and educational technology research and development for more than 30 years. In June 2011, Technology and Learning magazine named her one of its "10 most influential people in ed tech for 2011." She served as

Chair of the Department of Curriculum and Instruction for seven years and as chair was responsible for overseeing the education of 1,000 undergraduate majors, 500 secondary education students, and 200 graduate students. The department received the AACTE Award for Best Practice in Technology Integration in 2000. She is currently pursuing initiatives in technology pedagogical content knowledge, technology professional development, and technology in teacher education. She developed the Technology Mentoring Program in the College of Education at Iowa State University that is recognized as a national model. Ann is a past president of the Society for Information Technology and Teacher Education (SITE), co-editor of *The Journal of Digital Learning in Teacher Education*, and has led grant supported projects totaling more than $5 million. She was elected to the Iowa Academy of Education in 2001. She is the author of more than 70 refereed journal articles, 3 books, and numerous book chapters. Before pursuing her PhD, Ann taught mathematics and English in the San Mateo High School District and the Conejo Valley Unified School District.

Index